Lecture Notes in Electrical Engineering

Volume 72

Qi Luo (Ed.)

Advances in Wireless Networks and Information Systems

Springer

Qi Luo
School of Electrical Engineering
Wuhan Institute of Technology
Wuhan 430070
China
E-mail: witluo@ieee.org

ISBN 978-3-642-14349-6	e-ISBN 978-3-642-14350-2

DOI 10.1007/978-3-642-14350-2

Library of Congress Control Number: 2010933250

© 2010 Springer-Verlag Berlin Heidelberg

This work is subject to copyright. All rights are reserved, whether the whole or part of the material is concerned, specifically the rights of translation, reprinting, reuse of illustrations, recitation, broadcasting, reproduction on microfilm or in any other way, and storage in data banks. Duplication of this publication or parts thereof is permitted only under the provisions of the German Copyright Law of September 9, 1965, in its current version, and permission for use must always be obtained from Springer. Violations are liable to prosecution under the German Copyright Law.

The use of general descriptive names, registered names, trademarks, etc. in this publication does not imply, even in the absence of a specific statement, that such names are exempt from the relevant protective laws and regulations and therefore free for general use.

Typeset & Coverdesign: Scientific Publishing Services Pvt. Ltd., Chennai, India.

Printed on acid-free paper

9 8 7 6 5 4 3 2 1

springer.com

Preface

The purpose of WNIS 2009, the 2009 International Conference on Wireless Networks and Information Systems, is to bring together researchers, engineers and practitioners interested on information systems and applications in the context of wireless networks and mobile technologies.

Information systems and information technology are pervasive in the whole communications field, which is quite vast, encompassing a large number of research topics and applications: from practical issues to the more abstract theoretical aspects of communication; from low level protocols to high-level networking and applications; from wireless networking technologies to mobile information systems; many other topics are included in the scope of WNIS 2009.

The WNIS 2009 will be held in Shanghai, China, in December 2009. We cordially invite you to attend the 2009 International Conference on Wireless Networks and Information Systems. We are soliciting papers that present recent results, as well as more speculative presentations that discuss research challenges, define new applications, and propose methodologies for evaluating and the road map for achieving the vision of wireless networks and mobile technologies.

The WNIS 2009 is co-sponsored by the Institute of Electrical and Electronics Engineers, the IEEE Shanghai Section, the Intelligent Information Technology Application Research Association, Hong Kong and Wuhan Institute of Technology, China. The purpose of the WNIS 2009 is to bring together researchers and practitioners from academia, industry, and government to exchange their research ideas and results and to discuss the state of the art in the areas of the symposium. In addition, the participants of the conference will have a chance to hear from renowned keynote speakers.

This volume contains revised and extended research articles written by prominent researchers participating in WNIS 2009 conference. Topics covered include Wireless Information Networks, Wireless Networking Technologies, Mobile Software and Services, intelligent computing, network management, power engineering, control engineering, Signal and Image Processing, Machine Learning, Control Systems and Applications, The book will offer the states of arts of tremendous advances in Wireless Networks and Information Systems and also serve as an excellent reference work for researchers and graduate students working on Wireless Networks and Information Systems.

Qi Luo

Contents

Web Services Discovery in Metric Space through Similarity Search 1
Ming-hui Wu, Fan-wei Zhu, Jing Ying

An Effective Measure of Semantic Similarity 9
Songmei Cai, Zhao Lu, Junzhong Gu

Improved Run-Length Coding for Gray Level Images Using Gouraud Shading Method 19
Wei Huang, Shuai Chen, Gengsheng Zheng

Factors Affecting Consumer Intentions to Acceptance Banking Services in China 27
Sun Quan

Comparative Study of Workload Control Methods for Autonomic Database System 35
Yan Qiang, Juan-juan Zhao, Jun-jie Chen

Consensus of Mobile Agent Systems Based on Wireless Information Networks and Pinning Control 43
Hongyong Yang, Lan Lu, Xiao Li

Dynamic Evolution in System Modeling of Knowledge-Intensive Business Services' Organizational Inertia 53
Min Liu

Research on Java Imaging Technology and Its Programming Framework 61
M.A. Weifeng, Mao Keji

The Performance of SFN in Multi-Media System 69
Zhang Naiqian, Jin Libiao

A Domain-Oriented Goal Elaborate Method *Yonghua Li, Yingjie Wu*	77
A Hybrid Genetic Routing Algorithm in Wireless Sensor Networks *Lejiang Guo, Bengwen Wang, Qian Tang*	87
The Application of Fast Multipole-BEM for 3-D Elastic Contact Problem with Friction *Gui Hai-lian, Huang Qing-xue*	93
Simulation Method Research of Ground Target IR Scene Based on Aerospace Information *Chen Shan, Sun Ji-yin*	101
Research on Role-Based Agent Collaboration in WSN *Lei Yan, Xinying Wang, Dongyang Zhang*	111
Spectral Matting Based on Color Information of Matting Components *Jia-zhuo Wang, Cui-hua Li*	119
The Correlation of Conditional Time Series of Sunspot Series *Wenguo Li, Haikun Zhou, Hong Zhang*	131
Hurst Exponent Estimation Based on Moving Average Method *Nianpeng Wang, Yanheng Li, Hong Zhang*	137
Comprehensive Evaluation of Regional Independent Innovation Ability: Based on Chinese Enterprise Level *Zishuo Feng, Shukuan Zhao*	143
Research of Chord Model Based on Grouping by Property *Hongjun Wei, Jun Yan, Xiaoxia Li*	153
Research of Some Autopilot Controller Based on Neural Network PID *Jinxian Yang, Bingfeng Li, Hui Tao*	161
Dependent Failure Reliability Assessment of Electronic System *Wenxue Qian, Xiaowei Yin, Liyang Xie*	171
Research of RFID Authentication Protocol Based on Hash Function *Li Heng, Gao Fei, Xue Yanming, Feng Shuo*	177

Energy Efficient Message Routing in a Small World
Wireless Sensor Network 183
Suvendi Chinnappen-Rimer, Gerhard P. Hancke

Based on the Integration of Information Technology and
Industrialization to Propel and Accelerate Industrialization
of HeBei Province 193
Aifen Sun, Jinyu Wei

Pivoting Algorithm for Mean-CVaR Portfolio Optimization
Model 201
Yanwu Liu, Zhongzhen Zhang

The Ontology-Cored Emotional Semantic Search Model 209
Juan-juan Zhao, Jun-jie Chen, Yan Qiang

Multi-agent Task Allocation Method Based on Auction 217
Xue-li Tao, Yan-bin Zheng

Error Bound for the Generalized Complementarity
Problem in Engineering and Economic Equilibrium
Modeling 227
Hongchun Sun

Research of Tag Anti-collision Technology in RFID
System 235
Zhitao Guo, Jinli Yuan, Junhua Gu, Zhikai Liu

An Algorithm Based on Ad-Hoc Energy Conservation in
Wireless Sensor Network 245
Jianguo Liu, Huojin Wan, Hailin Hu

The Design of the Hardware and the Data Processing
Technology of Interbus 253
Jianguo Liu, Huojin Wan, Hailin Hu

Broadband MMIC Power Amplifier for X Band
Applications 259
*Jiang Xia, Zhao Zhengping, Zhang Zhiguo, Luo Xinjiang,
Yang Ruixia, Feng Zhihong*

The Selection of Dry Port Location with the Method of
Fuzzy-ANP 265
Jinyu Wei, Aifen Sun, Jing Zhuang

Growing and Declining of Aged Population Asymmetric
Real Estate Price Reactions: The Proof of China 275
Jinqiu Xu

Mobile Learning Application Research Based on iPhone 281
Zong Hu, Dongming Huang

Research of Improved Frame-Slot ALOHA Anti-collision Algorithm 289
Shuo Feng, Fei Gao, Yanming Xue, Heng Li

Research on Transmission of Power Telecontrol Information Based on IEC 61850/OPC 295
Changming Zhang, Yan Wang, Zheng Li

E-Commerce Leading Development Trend of New Sports Marketing 303
Zhongbin Yin, Binli Wang, Lina Wang

A Modified Robust Image Hashing Using Fractional Fourier Transform for Image Retrieval 309
Delong Cui, Jinglong Zuo, Ming Xiao

Short-Circuit Current Calculation of Distribution Network Based on the VDNAP 317
Yungao Gu, Yuexiao Han, Jian Li, Chenghua Shi

People's Event Features Extraction 327
Wen Zhou, Ping Yi, Bofeng Zhang, Jianfeng Fu, Ying Zhu

A Server-Based Secure Bootstrap Architecture 333
Qiuyan Zhang, Chao Chen, Shuren Liao, Yiqi Dai

Design and Implementation of the Integration Platform for Telecom Services Based on SOA 343
Xiaoxiao Wei, Xiangchi Yang, Pengfei Li

An Online Collaborative Learning Mode in Management Information System Experimental Teaching 351
Hanyang Luo

Statistical Properties Analysis of Internet Traffic Dispersion Networks 359
Cai Jun, Yu Shun-Zheng

Design and Implementation of Mobile Learning System Based on Mobile Phone 365
Qianzhu Shi

Extract Backbones of Large-Scale Networks Using Data Field Theory 371
Zhang Shuqing, Li Deyi, Han Yanni, Xing Ru

An Efficient Collaborative Recommendation Algorithm
Based on Item Clustering................................... 381
Songjie Gong

An Intelligent Solution for Open Vehicle Routing Problem
in Grain Logistics.. 389
Hongyi Ge, Tong Zhen, Yuying Jiang, Yi Che

E-Commerce Comparison-Shopping Model of Neural
Network Based on Ant Colony Optimization................. 397
Kang Shao, Ye Cheng

PPC Model Based on ACO................................... 405
Li Yancang, Hou Zhenguo

Image Algorithm for Watermarking Relational Databases
Based on Chaos 411
Zaihui Cao, Jianhua Sun, Zhongyan Hu

The Building of College Management Information System
Based on Team Collaboration............................... 419
Yu-you Dong, Shuang Chen

The Study on Decision Rules in Incomplete Information
Management System Based on Rough Sets 425
Xiu-ju Liu

Collaboration CAD Design Based Virtual Reality Modeling
Language in Heterogeneity Assembly Environment 435
Jian Yingxia, Huang Nan

The Study on the Sharing of Data Sources in CAD
Environment Based on XML 443
Jian Yingxia

Semantic Mapping Approach for Logistics Services
Integration in 4PL... 451
Qifeng Wang

The Study on Distributed Database Security Strategy 459
Yan bing

An Improved Symmetric Key Encryption Algorithm for
Digital Signature 467
Xiuyan Sun

Application of Soft Test Method to Intelligent Service
Robots 475
Wang Hongxing

The Research of the Role Information of the Marketing
Channel Management Personnel of the Domestic Leisure
Garment Brands... 483
Zhang Junying

The Study on Data Mining to CRM Based on Rough Set.... 491
Zhang Wei-bo

The Data Mining Method Based on Rough Sets in
Economic Practice... 499
Luo Shengmin

The Study on Adaptive Routing Protocol in Mobile Adhoc
Network Based on Rough Set................................ 507
Pan Shaoming, Cai Qizhong, Han Junfeng

The Evaluation Model of Network Security Based on Fuzzy
Rough Sets 517
Yaolong Qi, Haining An

A Novel Image Fusion Method Based on Particle Swarm
Optimization 527
Haining An, Yaolong Qi, Ziyu Cheng

Relational Database Semantic Access Based on Ontology.... 537
Shufeng Zhou

Dynamic Analysis of Transmission Shaft of Width
Equipment for Square Billet 547
Xianzhang Feng, Hui Zhao

Mobile Software Testing Based on Simulation Keyboard..... 555
Hua Ji

VLSI Prototype for Mpeg-4 Part 2 Using AIC................ 563
Kausalya Gopal, Kanimozhi Ilambarathi, Riaz Ahmed Liyakath

Author Index... 569

Web Services Discovery in Metric Space through Similarity Search

Ming-hui Wu[1,2], Fan-wei Zhu[2], and Jing Ying[1,2]

[1] Department of Computer and Engineering, Zhejiang University City College,
Hangzhou, 310015, China
[2] College of Computer Science and Technology, Zhejiang University
Hangzhou, 310027, China
mhwu@zucc.edu.cn

Abstract. Most current semantic web services discovery approaches focus on the matchmaking of services in a specific description language such as OWL-S, and WSML. However, in practical applications, effective services discovery is expected to have the ability to deal with all heterogeneous and distributed web services. This paper proposes a novel semantic web service discovery method using the metric space approach to resolve this problem. In the method, all heterogeneous web services are modeled as similar metric objects regardless of concrete description languages, and thereby the discovery problem can be treated as similarity search in metric space with a uniform criterion. In the matchmaking process, both the functional semantics and non-functional semantics of the web services are integrated as selection conditions for similarity query. And two types of similarity queries: range query and an improved nearest neighbor query are combined to produce a sorted result set so that the method can be better applied to practical situation.

Keywords: Semantic web service, metric space, similarity search, pkNN, range query.

1 Introduction

The web has evolved from solely a repository of pages to a collection of complex and heterogeneous services that distributed over the Internet [4]. The increasing number of Web services and the widespread distribution makes it difficult for the user to find a interested service and therefore has led to much interest in the area of service discovery [1, 2, 3] for both research and commercial application.

Discovery of web services is the first and crucial step of service-oriented computing. When you want to use a web service, composite it or deploy it, you first have to locate it correctly. And the performance of the service-oriented computing depends on the quality of web services discovery. In order to find web services efficiently, the semantic description of web service such as OWL-S, WSML has introduced and replaced the keyword-based discovery mechanism supported by UDDI [16] and most early service search engines like GOOGLE and Baidu.

As an attempt to resolve the heterogeneity at the level of web service specifications and to enable automated discovery and composition of web services, the Semantic Web Services (SWS) have becoming a major research focus in web service discovery[5]. However, most current semantic web services (SWS) discovery approaches focus on the matchmaking of services in a specific description language such as DAML-S [6], and WSML. While in practical applications, effective SWS discovery is expected to have the ability to deal with all web services specified in different languages. In this paper, we propose a novel semantic web service discovery method using the metric space approach to meet the need. Our method can be viewed as a meta-model on the existing web services description language, in which, all heterogeneous web services are modeled as similar metric objects regardless of concrete description languages, and thereby the discovery problem can be treated as similarity search in metric space with a uniform criterion. In the matchmaking process, both the functional semantics and non-functional semantics of the web services are integrated as selection conditions for similarity query. And two types of similarity queries are combined to produce a set of sorted results.

The remainder is organized as follows. Section 2 describes the SWS discovery problem and Section 3 presents the metric space model for SWS and introduces the matchmaking approach. Section 4 presents a prototype based on our method and related works are reviewed in Section 5. Finally, Section 6 concludes the research and provides directions for future work.

2 Problem Statement

Existing approaches have showed good performance in discovering SWS specified in a specific description framework, such as [7] for OWL-S and [8] for WSML. Different with these approaches, our paper focuses on building a meta-model on top of existing description frameworks to support heterogeneous SWS discovery. Due to space limitation, we only refer to two prominent SWS description frameworks: OWL-S and WSML while other frameworks like SWSL and the DIANE will be discussed in our future work.

Both OWL-S and WSML have adopted ontologies for adding semantics to web services descriptions which includes functional properties and non-functional properties. However, they are different in the structure for representing web services semantics. OWL-S consists of three sub-ontologies[17]: the Profile, the Process Model, and the Grounding to describe the web services while in WSML the description is structured in terms of service capability, imported ontologies, and the interface. Table 1 gives a detailed comparison between OWL-S and WSML.

From the comparison, we can see that, no matter what elements are used for representing SWS, the concerns of discovery are similar, that is, the functional semantics of service signature and the non-functional semantics composed of Quality of Service (QoS) and context policies. Specifically speaking, the functional parameters: hasInput, hasOutput, precondition and effect in OWL-S and the capability elements: precondition, postcondition, assumption and effect in WSML concern on the same aspects for service discovery.

Table 1. Comparison between OWL-S and WSML

	OWL-S	WSML
Components	Service profile Service process model Service grounding	Service capability Ontologies Service interface
Elements for Discovery	Service profile	Service capability
Functional parameters	hasInput hasOuput Precondition Effect	Preconditon Postcondition Assumption Effect
Semantics described	Functional Non-functional semantics	Functional Non-functional semantics
Representation framework	Structured formal	Structured formal
Support reasoning	Yes	Yes
Typical matchmaker	OWLS-MX [7]	WSMO-MX [8]

Therefore, to support SWS discovery on heterogeneous web services, we propose to build a meta-model based on the fundamental factors in both OWL-S and WSML so that the different structures in description framework can be ignored and thereby the heterogeneous SWS can be matched in service discovery.

3 Discover SWS in Metric Space

Since the functionality of SWS can be characterized by the factors mentioned in section 2, we can model both the requester and the advertised web services as objects with the capability features as dimensions in metric space. Thereby the discovery problem turns to similarity search problem in the metric space [9] composed by a collection of web service objects and a request object.

In the following, we define the basic concepts in respect with metric space and similarity search and introduce the algorithm for matchmaking and ranking respectively.

3.1 Basic Definitions

In our approach, the discovery of SWS is processed in metric space, therefore all heterogeneous web services are modeled as similar metric objects regardless of concrete description languages and the matchmaking of request service and advertised services is modeled as distance measuring between metric objects accordingly.

Definition 1 (Metric Space). A metric space M = <D,d> is defined for a domain of objects D and a distance function d. And for SWS discovery:

- D is a collection of web services
- d is the semantic similarity between requester and web services.

Definition 2 (Semantic Web Service). A semantic web service modeled in metric space is a triple SWS=<FS, NFS, FW> such that:

- FS is the functional semantics
- NFS is the non-functional semantics
- FW is the description framework

Definition 3 (Functional semantics). The functional semantics of SWS is defined as a quadruple FS=<EX, RE, PR, OU> such that:

- EX describes what a web service expects for enabling it to provide its service
- RE describes what a web service returns in response to its input
- PR describes conditions over the world state to met before service execution
- OU describes the state of the world after the execution of the service

Definition 4 (Non-functional semantics). The non-functional semantics of SWS is defined as a triple NFS = <Q, SP, CP> such that:

- Q is a set of QoS parameters offered by the service
- SP is a set of service specific parameters
- CP is context policy

Definition 5 (Distance measure). Distance measure is a way of quantifying the closeness of SWS in metric space. In our approach, we define the distance measure as semantic similarity based on functional semantics of SWS that:

$$d_M(SWS_R, SWS_A) = \sqrt{\sum_{i=1}^{n} w_i (SWS_{Ri} - SWS_{Ai})^2} \tag{1}$$

where SWS_R is the request object and SWS_A is the advertised service object. W is a weight matrix reflecting user's preference on the n dimensions of request object respectively.

3.2 Matchmaking Algorithm

Based on the basic definitions in Section 3.1, we have build up a uniform metric space model for heterogeneous SWS despite of the concrete description languages. And as we discussed before, the matchmaking between request and advertised SWS can be treated as similarity query problem, that is, searching for close objects to the given request object in the metric space. More specifically, the selection condition of geographic closeness is the semantic similarity of SWS descriptions.

In order to improve the recall and precision rate, we combined two elementary types of similarity query to consider both functional semantics and non-functional semantics in SWS discovery.

1) **Recall and precision.** Recall and precision are the two standard measures for evaluating the performance of a discovery approach. The recall of a matchmaking algorithm is defined as:

$$Recall = RtR / TRt \qquad (2)$$

where *RtR* stands for the number of relevant web services retrieved, and *TRt* represents the total number of relevant web services.

And the precision is defined as:

$$Precision = RtR / TR \qquad (3)$$

where *RtR* has the same meaning in Recall and *TR* stands for the total number of retrieved web services.

2) **Nearest neighbor query on functional semantics.** The basic version of Nearest Neighbor query finds the closest object to the given query object, that is the nearest neighbor of query object [9]. However, in practical application, a generalized *k*NN query is often used for looking for the *k* nearest neighbors. A formal definition of *k*NN(q) in metric space M is:

$$kNN(q) = \{R \subseteq M, | R |= k \wedge \forall x \in R, y \in M - R : d(q,x) \leq d(q,y)\} \qquad (4)$$

Though *k*NN(q) query had successfully used in many applications especially the geographic applications, considering our case for SWS discovery, the traditional *k*NN(q) has bad performance in recall and precision rate for the reason that the if the collection to be searched consists of fewer than k objects, the query will return the whole collection. Therefore, in this case, irrelevant results may be returned due to limited advertised services so that the precision rate will be low.

---

p*k*NN Search Algorithm

Input: request object *r*, number of neighbors *k*, total number of SWS *t*.
Output: response set *RS* of cardinality *k/t*.

Fill *RS* with *k/t* objects randomly selected from metric space
Calculate the distance between *r* and each object $o_j \in RS$
Sort objects in *RS* with decreasing distance with *r*
Designate the maximum distance as *td*
While $RS \neq \Phi$
   Extract the first object o_n from *RS*
   Foreach object o_j **do**
     If $d(r,o_j) \leq td$ **then**
       Update *RS, td* by inserting *oj* and removing on from *RS*
   Enddo
   Sort objects in *RS* with decreasing distance with *r*
Enddo

Fig. 1. Search algorithm for p*k*NN

To solve this problem and improve the traditional kNN query algorithm by searching for the top k% nearest neighbor, pkNN for short, such that the number of retrieved services is determined by the parameter k specified by user and the total number of the services in the metric space.

By taking the total number of advertised services into consideration, our algorithm can improve the precision of SWS discovery defined in Section 3.2. Example 1 gives a comparison between kNN query and our pkNN query.

Example 1: Suppose the total number of SWS in the collection is 30, and there are 3 services relevant with user's request and the parameter k is set as 10. If we use the kNN query, then the Precision=3/10 which is relatively low. On contrary, if our pkNN query is applied, Precision=3/3 which reflects a huge improvement compared with kNN.

Our pkNN query algorithm is presented in figure 1.

3) Range query on non-functional semantics. Range query is specified by a query object q, with some query radius r as the distance constraint [9]. The query retrieves all objects found within distance r of q in metric space M. The formal definition of range query is:

$$R(q,r) = \{o \in M, d(o,q) \leq r\} \tag{5}$$

In SWS discovery, the distance constraint of range query is defined on non-functional semantics. Specifically, we have three types of query radius according to Definition 4: Q, SP, and CP. Therefore, the range query in our algorithm applies these selection conditions in turn and the final results should meet all the distance constraints.

The extended range query is formally defined as:

$$R(q, r_Q, r_{SP}, r_{CP}) = \{o \in M, d(o,q) \leq r_Q \wedge d(o,q) \leq r_{SP} \wedge d(o,q) \leq r_{CP}\} \tag{6}$$

In our SWS discovery, range query is executed on the response set returned by pkNN query. In other words, we look for the services satisfying the non-functional semantics constraints only in the collection of the services with a desired similarity in functional semantics. The algorithm for executing the extended range query listed in figure 2.

Extended Range Query Algorithm

Input: response set *RS* of cardinality *k/t*
(output of *pk*NN Search Algorithm)
Output: final response set *FRS*.

$FRS = \Phi$
While $RS \neq \Phi$ **do**
  Foreach object $o_j \in RS$ **do**
    If $d(r, o_j) \leq r_Q$ and $d(r, o_j) \leq r_{SP}$ and $d(r, o_j) \leq r_{CP}$ **then**
      Add o_j to *FRS*
  Enddo
Enddo

Fig. 2. Search algorithm for extended range query

4 Related Work

Currently, several techniques have been proposed to deal with semantic service discovery based on formal description of Web services. These techniques can be divided into three categories according to the services characteristics taken into consideration for matchmaking: IO matchmaker, IOPE matchmaker and Role-based matchmaker.

IO matchmakers, like LARKS [10], the OWL-S/UDDI [11], the RACER [12], the MaMaS (MatchMaker-Service) [13], the HotBlu [14] and the OWLS-MX [7], mainly take the input parameters, output parameters and the categories of the service descriptions for comparing. While IOPE matchmaker not only compare input and output parameters between request and advertised services, but also evaluate the similarity of pre-conditions and effects between two services, typically represented by PCEM [15]. And role-based matchmakers exploit common organizational concepts such as roles and interaction types to improve the efficiency of matchmaking, such as ROWLS [15] which is built around the matching between two roles in the taxonomy and the similarity between two services depends on the level of match and the distance in the taxonomy.

Alternatively, existing SWS discovery approaches also are differentiated by their description frameworks such as OWLS-MX [7] for discovery web services described in OWL-S and WSMO-MX [8] for web services modeled in WSML.

5 Conclusions and Future Work

In this work we proposed a metric space approach for semantic web services discovery. Our approach is based on three strategies: a) exacting common semantic features of heterogeneous web services ignoring their description frameworks; b) modeling both quest and advertise web services as uniform objects in metric space; c) combining the functional properties and non-functional properties as similarity measure for services matchmaking.

The contributions of this research are manifold: a) transforming service discovery to similarity search problem in metric space; b) extending the kNN search algorithm to kpNN search algorithm for a better precision in service discovery; c) adjusting weights of selection conditions based on user preference for flexible discovery and ranking.

We also presented a general framework for semantic web services discovery, in which domain ontologies are used for semantic similarity calculation and the pkNN Search Algorithm and Extended Range Query Algorithm are integrated in the matchmaking module to return a more accurate, user-adjustable response set.

Acknowledgements. This work is partly supported by the National High-Tech Research Plan, China (Grant No.2007AA01Z187) and the National Natural Science Foundation, China (Grant No.60805042).

References

1. Benatallah, B., Hacid, M., Leger, A., Rey, C., Toumani, F.: On Automating Web Service Discovery. VLDB Journal, 84–96 (2005)
2. Nayak, R.: Facilitating and Improving the Use of Web Services with Data Mining, Research and Trends in Data Mining Technologies and Applications, Taniar (2007)
3. Wang, H., Huang, J.Z., Qu, Y., Xie, J.: Web services: Problems and Future Directions. Journal of Web Semantics, 309–320 (2004)
4. Pereira Filho, J.G.: Web service architecture: Semantics and context-awareness issues in web service platforms (2003)
5. Hess, A., Kushmerick, N.: Learning to attach semantic metadata to web services. In: Fensel, D., Sycara, K., Mylopoulos, J. (eds.) ISWC 2003. LNCS, vol. 2870, pp. 258–273. Springer, Heidelberg (2003)
6. Ankolekar, A., Burstein, M., Hobbs, J.R., Lassila, O., Martin, D., McDermott, D., McIlraith, S.A., Narayanan, S., Paolucci, M., Payne, T.R., Sycara, K.: DAML-S. Coalition. Daml-s: Web service description for the semantic web. In: Horrocks, I., Hendler, J. (eds.) ISWC 2002. LNCS, vol. 2342, p. 348. Springer, Heidelberg (2002)
7. Klusch, M., Fries, B., Sycara, K.: Automated Semantic Web Service Discovery with OWLS-MX. In: Proc. 5th Intl. Conference on Autonomous Agents and Multi-Agent Systems (AAMAS). ACM Press, Japan (2006)
8. Kaufer, F., Klusch, M.: Hybrid Semantic Web Service Matching with WSMO-MX. In: Proc. 4th IEEE European Conference on Web Services (ECOWS). IEEE CS Press, Switzerland (2006)
9. Zezula, P., Amato, G., Dohnal, V., Batko, M.: Similarity search-The metric space approach. Springer Press, New York (2006)
10. Sycara, K., Widoff, S., Klusch, M., Lu, J.: LARKS: Dynamic Matchmaking Among Heterogeneous Software Agents in Cyberspace. Journal of Autonomous Agents and Multiagent Systems (2002)
11. Paolucci, M., Kawamura, T., Payne, T., Sycara, K.: Semantic matching of Web Services capabilities. In: Horrocks, I., Hendler, J. (eds.) ISWC 2002. LNCS, vol. 2342, pp. 333–347. Springer, Heidelberg (2002)
12. Li, L., Horrocks, I.: A software framework for matchmaking based on semantic web technology. In: Proc. of the twelfth international conference on World Wide Web, pp. 331–339. ACM Press, New York (2003)
13. Martin, D., Burstein, M., Hobbs, J., Lassila, O., McDermott, D., McIlraith, S., Narayanan, S., Paolucci, M., Parsia, B., Payne, T., Sirin, E., Srinivasan, N., Sycara, K.: OWL-S 1.1 Release
14. http://www.daml.org/services/owls/1.1/overview/ (2004)
15. Constantinescu, I., Faltings, B.: Efficient matchmaking and directory services. In: Proc. of IEEE/WIC International Conference on Web Intelligence (2003)
16. Fernandez, A., Vasirani, M., Caceres, C., Ossowski, S.: A Role-Based Support Mechanism for Service Description and Discovery. In: Huang, J., Kowalczyk, R., Maamar, Z., Martin, D., Müller, I., Stoutenburg, S., Sycara, K. (eds.) SOCASE 2007. LNCS, vol. 4504, pp. 132–146. Springer, Heidelberg (2007)
17. W3C Web Services Description Language, http://www.w3.org/TR/WSDL/

An Effective Measure of Semantic Similarity

Songmei Cai, Zhao Lu, and Junzhong Gu

Department of Computer Science and Technology,
East China Normal University,
200241, Shanghai, China
smcai@ica.stc.sh.cn, zlu@cs.ecnu.edu.cn, jzgu@cs.ecnu.edu.cn

Abstract. Measuring semantic similarity between two concepts is an important problem in web mining, targeted advertisement and domains that need semantic content matching. Nevertheless, developing a computational method capable of generating satisfactory results close to what humans would perceive is still a difficult task somewhat owed to the subjective nature of similarity. This paper presents an effective measure of semantic similarity between two concepts. It relies on hierarchical structure of WordNet 3.0, and considers not only semantic distance but also depth sum and depth difference between two concepts. The correlation value of the proposed semantic similarity measure compared with the human ratings reported by Miller and Charles for the dataset of 28 pairs of noun is higher than some other reported semantic similarity measures for the same dataset.

Keywords: Semantic Similarity; Semantic Relatedness; WordNet.

1 Introduction

Human beings have an innate ability to determine whether two concepts are related or not. For example, most would agree that the automotive senses of car and bus are related while car and tree are not. However, assigning a value that quantifies the degree to which two concepts are related proves to be more difficult [1].

Semantic similarity, semantic relatedness, and semantic distance are sometimes confusing in research literatures. They however, are not identical [2]. Semantic relatedness measures usually include various kinds of relationships, such as hypernym, hyponym, subsumption, synonym, antonym, holonym, and meronymy. Semantic similarity is a special case of semantic relatedness which only considering the synonym relationships and the subsumption relationships. For example, the two words (apple and vitamin) are more closely related in a functional context than apple and pear, but apple and pear are more similar with respect to a subsumption context. Semantic distance is the inverse of semantic relatedness. The more two words are semantically related, the more semantically close they are.

WordNet [3][4] is a lexical database for English . It is created and maintained by the Cognitive Science Laboratory of Princeton University directed by Professor

George A. Miller. Differing from other traditional lexicons, WordNet groups words into sets of synonyms called synsets, provides short, general definitions, and records various semantic relations between these synonym sets. Its taxonomies usually represent the lexical knowledge implicit in languages by means of hierarchical structures which reflect concepts of words and their relationships.

WordNet connects concepts or senses, but most words have more than one sense. In Yang and Powers' evaluation [5], word similarity can be determined by the best conceptual similarity value among all concept pairs. Suppose the word w_1 corresponds with m concepts, and the word w_2 corresponds with n concepts, then $m \times n$ semantic similarity should be considered when calculate the similarity between words w_1 and w_2. It can be defined as follows:

$$sim(w_1, w_2) = \max_{(i,j)}[sim(c_{1i}, c_{2j})] \quad (1)$$

Where c_{1i} ($1 \leq i \leq m$) and c_{2j} ($1 \leq j \leq n$) are the senses of words w_1 and w_2 respectively [6].

The rest of this paper is organized as follows. Section 2 describes some related works. Our proposal of semantic similarity measure between concepts appears in Section 3. Section 4 presents some experiments to evaluate the proposed semantic similarity measure. Finally, discussion and further work are discussed in Section 5.

2 Related Work

In this section, two kinds of traditional semantic similarity measure approaches are discussed. Until now, several measures to compute similarities between concepts have been proposed. According to various parameters used in measure approaches, they can be classified into three main categories, distance-based measure, information-based measure, and hybrid measure which combines above two kinds measures [7].

2.1 Distance-Based Measures

The basic idea of distance-based measure is to select the shortest path among all possible paths between two concepts. This measure assumes that the shorter the distance, the more similar the concepts are.

In 1989, Rada et al. [8] uses the minimum length of path connecting two concepts containing words as a metric for measuring the semantic similarity of words. Their work forms the basis of distance-based similarity methods.

In 1994, Wu & Palmer measure [9] (shortly WP measure) calculates semantic similarity through considering depths of two synsets, along with the depth of the least common subsume (LCS) of two concepts in WordNet taxonomy, expressed by:

$$sim_{WP}(c_1,c_2) = \frac{2 \times depth(lcs(c_1,c_2))}{depth(c_1) + depth(c_2)} \qquad (2)$$

In 1998, Leacock & Chodorow measure [10] (shortly LC measure) calculates semantic similarity through considering the length of the shortest path that connects two concepts and the maximum depth of the WordNet taxonomy, expressed by:

$$sim_{LC}(c_1,c_2) = -\log \frac{len(c_1,c_2)}{2 \times \max_{c \in WordNet} depth(c)} \qquad (3)$$

Recently Yang and Powers [5] proposed to augment the focus on is-a relationships of previous approaches to the use of additional information, specifically, equivalence, and part–whole relationships. They introduced a bi–directional depth limit search, which obtained the best results.

In summary, the distance-based measure obviously requires a lot of information on detailed structure of lexical database. Therefore it is difficult to apply or directly manipulate it on a generic lexical database, which originally is not designed for similarity computation [11].

2.2 Information-Based Measures

Information-based methods were introduced to take advantages of usage of external corpora avoiding unreliability of path distances and taxonomy.

In 1995, Resnik measure [12][13][14] (shortly R measure) brought together lexical database and corpora firstly, and calculates similarity through considering information content (IC) of LCS between two concepts. If LCS between two concepts is identical, the similarity is same. That's the deficiency of the measure.

In 1997, in view of the deficiency of R measure, Jiang & Conrath measure [15] (shortly JC measure) calculates similarity by considering not only the IC of the LCS of two concepts, but also every concept's IC, expressed by:

$$dist_{JC}(c_1,c_2) = IC(c_1) + IC(c_2) - 2 \times IC(lcs(c_1,c_2)) \qquad (4)$$

Where semantic similarity is represented by semantic distance. Since they are inverse relationship, so the bigger the distance between two concepts, the less similar the two concepts are.

In 1998, similar to JC measure, Lin measure [16] (shortly L measure) uses the same elements as Jiang & Conrath, but in a different fashion:

$$sim_L(c_1,c_2) = \frac{2 \times IC(lcs(c_1,c_2))}{IC(c_1) + IC(c_2)} \qquad (5)$$

The result is the ratio of the information shared in common to the total amount of information possessed by two concepts.

Recently Banerjee and Pedersen introduced a new method based on dictionary definitions. It relies on the number of overlapping words in the long description (gloss) of each concept. Later, Patwardhan and Pedersen in [17] reported better results with the use of gloss information, modeling the semantic similarity by means of second order cooccurrence vectors in combination with structural information from WordNet.

To conclude, unlike the distance measures, the information theoretic measures require less structural information of the lexical database. Since the information theoretic measures are generic and flexible, they have been used in many applications with different types of lexical database. However, when it is applied on hierarchical structures, it does not differentiate the similarity of concepts as long as their minimum upper bounds are the same.

2.3 Hybrid Measures

T. Hong-Minh and D. Simth [11] proposed a hybrid measure for measuring semantic similarity which is derived from the information-based measure by adding depth factor and link strength factor. Zhou [6] also proposed a hybrid measure which combines the path length and IC value. Although the hybrid measure proposed combines both advantages of distance-based measure and information-based measure, their accuracy is not very close to what humans would perceive.

3 New Measure of Semantic Similarity

This section presents an effective semantic similarity measure between concepts. It considers factors which affect accuracy of semantic similarity measure as many as possible. Firstly, we will give some related definitions in our semantic similarity measure.

3.1 Related Definitions

A lexical database model is mainly composed of a group of concept aggregation and a group of semantic relation aggregation, which can be represented by a concise DAG [18]. The graph is concise because only *is-a* and *part-whole* relationships are mined from WordNet. Figure 1 is a DAG fragment of WordNet 3.0 [19].

Definition 1 Path: A path between concepts c_i and c_j in an ontology graph is represented as follows: $\{c_i, p_{i,i+1}, c_{i+1}, \cdots, c_k, p_{k,k+1}, c_{k+1}, \cdots, c_{j-1}, p_{j-1,j}, c_j\}$, where k=i+1,i+2,$\cdots$,j-2 , c_k are all nodes representing concepts, and $p_{k,k+1}$ are all edges appearing in the path of two concepts .

Definition 2 Path Length: The path length is the sum of all edges in a path between concepts c_i and c_j, and it can be defined as follows:

$$pl(c_i, c_j) = \sum_{k=i}^{j-1} p_{k,k+1} .$$

Definition 3 Shortest Path: The shortest path is defined as the path with minimum path length among all paths between concepts c_i and c_j.

Definition 4 Distance: The distance between a concept c_i and a concept c_j ($i \neq j$) is the number of all edges appearing in the shortest path between the concept c_i and the concept c_j, marked as $len(c_i, c_j)$. Such as in Figure 1, there is $len(bicycle\#1, car\#2) = 2$.

Definition 5 Concept Depth: In an ontology graph, suppose the root concept depth is 1, there is $depth(root) = 1$, then a not-root concept c depth is defined as follows: $depth(c) = depth(parent(c)) + 1$. Where $parent(c)$ is the deepest father concept of the concept c. Such as in Figure 1, there is $depth(engine) = 5$.

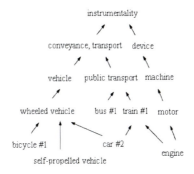

Fig. 1. A DAG fragment of WordNet 3.0

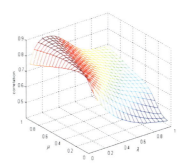

Fig. 2. Correlation between proposed semantic similarity (7) and human judgments versus λ and μ

3.2 The Proposed Measure

It is intuitive that the semantic similarity between two concepts grows higher if the depth of the least common superconcept (LCS) between them increases in the lexical hierarchy [20], and semantic similarity is proportional to the ratio of the corresponding part of common and different features. That can be defined as follows:

$$sim(c_1, c_2) = \frac{f(\lambda \times 2d)}{f(\lambda \times 2d) + f(\mu \times len)} \quad (6)$$

Where d is the depth of LCS in the hierarchical taxonomy; len is the shortest path length between concepts c_1 and c_2; $\lambda \in (0, 1]$ and $\mu \in (0, 1]$ are adjustable parameters; f is the transfer function for d and len. Let $f(x) = e^x - 1$, then the expression is changed by as follows:

$$sim(c_1, c_2) = \frac{e^{\lambda \times 2d} - 1}{e^{\lambda \times 2d} + e^{\mu \times len} - 2} \quad (7)$$

Besides that, in the hierarchical taxonomy, From top to down, the concept classification is from abstract to detail, such as the second layer is the first layer's refinement, and the third layer is the second layer's refinement, and so forth. When distance between two concepts is same, semantic similarity will become bigger with the depth sum becoming bigger, and become smaller with the depth difference becoming bigger [19]. That can be defined as follows:

$$sim(c_1, c_2) = 1 - \alpha^{\frac{\beta \times (l_1 + l_2)}{(len + \beta) \times \max(|l_1 - l_2|, 1)}} \quad (8)$$

Where len is defined as same as (7); $l_1 = depth(c_1)$; $l_2 = depth(c_2)$; $\alpha \in (0,1)$ and $\beta \in (0,1]$ are adjustable parameters.

Therefore we should take into account not only the shortest path, but also the depth in the hierarchical taxonomy. The new measure is defined as follows:

$$sim(c_1, c_2) = k(\frac{e^{\lambda \times d} - 1}{e^{\lambda \times d} + e^{\mu \times len} - 2}) + (1-k)(1 - \alpha^{\frac{\beta \times (l_1 + l_2)}{(len + \beta) \times \max(|l_1 - l_2|, 1)}}) \quad (9)$$

Where $k \in [0,1]$ is a changeable factor so as to adjust the weight of the two items of the equation.

In order to select the most suitable parameters $\lambda, \mu, \alpha, \beta$, firstly, only use (7) as the measure of semantic similarity between concepts. Compute different correlation coefficients between human judgments and proposed measure corresponding to different λ, μ values, the ultimate λ, μ value corresponding to the maximum correlation coefficient. When λ, μ take different values, the results are presented in Figure 2, which is a three-dimensional grid using matlab software to draw.

Figure 2 show that the correlation coefficient obtains the maximum at $\lambda = 0.25$, $\mu = 1$. In the same way, only use (8) as the measure of semantic similarity between concepts, and the correlation coefficient reaches the maximum at $\alpha = 0.2$, $\beta = 0.1$. Thus the new measure equation will be as follows:

$$sim(c_1, c_2) = k(\frac{e^{0.25 \times d} - 1}{e^{0.25 \times d} + e^{len} - 2}) + (1-k)(1 - 0.2^{\frac{0.1 \times (l_1 + l_2)}{(len + 0.1) \times \max(|l_1 - l_2|, 1)}}) \tag{10}$$

In the following part, we will use this equation to evaluate the performance of our proposed measure.

4 Evaluation

In order to evaluate the performance of our proposed semantic similarity measure, we will compare several traditional semantic similarity measures listed in Section 2 with the proposed measure. The experiments focused on 28 pairs of nouns (shortly D) carried out by Miller and Charles. We set up an experiment to compute the similarity of dataset D, and examine the correlation between human judgments and proposed measure. In order to make fair comparisons, we decide to use an independent software package that would calculate similarity values using previously established strategies while allowing the use of WordNet 3.0. The freely available package using here is that of Siddharth Patwardhan and Ted Pederson [21], which implements semantic similarity measures described by Wu and Palmer, Leacock and Chodorow, Resnik, Lin, etc. When k take different values, the experiment results are show in Figure 3.

Figure 3 show that the correlation coefficient reach the maximum value 0.9133 at $k = 0.8$. When compared with other previous measures, see Table 1 [19].

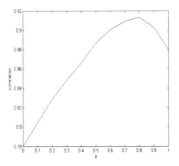

Fig. 3. The relationship between k and correlation

Table 1. Correlation coefficient between human judgments and some classical measures

Measures	Correlation coefficient
Wu and Palmer	0.768
Leacock and Chodorow	0.840
Yang and Powers	0.921
Resnik	0.825
Lin	0.853
T. Hong-Minh and D. smith	0.880
Zili Zhou	0.880
proposed measure	0.9133

Form Table 1, it is clear that the result obtained by the proposed measure outperforms previous results, with the exception of Yang and Powers. Note that Yang and Powers is only based on the distance of concepts, and this measure need 7 parameters to be fine tuned. Anyway, the proposed measure in this paper has certain advantages than previous measures.

5 Conclusion and Further Work

This paper proposed an effective measure to calculate semantic similarity between concepts, which considers not only semantic distance but also depth sum and depth difference of two concepts. The experimental results show that the correlation with average human judgments on the standard 28 word-pair dataset is 0.9133, which outperforms some traditional semantic similarity measures.

The main task for future work is to further improve the proposed measure with some other factors which affect accuracy of semantic similarity measure. We will attempt to evaluate the proposed measure in some practical applications.

Acknowledgments. This paper is partially supported by the NNSFC (No. 90718004 and No. 60703004) and the STCSM project (No.075107006). The authors wish to thank other members of ECNU-ICA for their useful suggestions.

References

1. Banerjee, S., Pedersen, T.: Extended gloss overlaps as a measure of semantic relatedness. In: Proceedings of the Eighteenth International Joint Conference on Artificial Intelligence, Acapulco, pp. 805–810 (2003)
2. Kolb, P.: Experiments on the difference between semantic similarity and relatedness. In: NODALIDA 2009 Conference Proceedings, pp. 81–88 (2009)
3. WordNet (2009), http://wordnet.princeton.edu/
4. Fellbaum, C.: WordNet: An electronic lexical database. MIT Press, Cambridge (1998)

5. Yang, D., Powers, D.M.W.: Measuring semantic similarity in the taxonomy of WordNet. In: Proceedings of the 28th Australasian Computer Science Conference, Australia, pp. 315–322 (February 2005)
6. Zhou, Z., Wang, Y., Gu, J.: New Model of Semantic Similarity Measuring in Wordnet. In: Proceedings of the 2008 3rd International Conference on Intelligent System and Knowledge Engineering, vol. 1, pp. 256–261 (November 2008)
7. Varelas, G., Voutsakis, E., Raftopoulou, P.: Semantic similarity methods in wordNet and their application to information retrieval on the web. In: Proceedings of the 7th annual ACM international workshop on Web information and data management, Bremen, Germany, pp. 10–16 (2005)
8. Rada, R., Mili, H., Bichnell, E., Blettner, M.: Development and application of a metric on semantic nets. IEEE Trans. Systems, Man, and Cybernetics 9(1), 17–30 (1989)
9. Zhibiao, W., Palmer, M.: Verb semantics and lexical selection. In: Proceedings of the 32nd Annual Meeting of the Association for Computational Linguistics, Las Cruces, NM, pp. 133–138 (1994)
10. Leacock, C., Chodorow, M.: Combining local context and WordNet similarity for word sense identification. In: Christiane Fellbaum, pp. 265–283 (1998)
11. Hong-Minh, T., Simth, D.: Word similarity in WordNet. In: Proceedings of the Third International Conference on High Performance Scientific Computing, Hanoi, Vietnam, pp. 1–10 (March 2006)
12. Resnik, P.: Disambiguating noun groupings with respect to WordNet senses. In: Third Workshop on Very Large Corpora, Cambridge. Association for Computational Linguistics, pp. 55–68 (1995)
13. Resnik, P.: Using information content to evaluate semantic similarity in a taxonomy. In: Proceedings of the 14th International Joint Conference on Artificial Intelligence, Montreal, Canada, pp. 448–453 (1995)
14. Marton, Y., Mohammad, S., Resnik, P.: Estimating semantic distance using soft semantic constraints in knowledge-source–corpus hybrid models. In: Proceedings of the 2009 Conference on Empirical Methods in Natural Language Processing, Singapore, pp. 775–783 (August 2009)
15. Jiang, J.J., Conrath, D.W.: Semantic similarity based on corpus statistics and lexical taxonomy. In: Proceedings of International Conference on Research in Computational Linguistics, TaiWan, pp. 19–33 (1997)
16. Lin, D.: An Information Theoretic definition of similarity. In: Proceedings of 15th international Conf. On machine learning, pp. 296–304. Morgan Kaufmann Publishers Inc., San Francisco (1998)
17. Patwardhan, S., Pedersen, T.: Using WordNet-based Context Vectors to Estimate the Semantic Relatedness of Concepts. In: Proceedings of the EACL Workshop on Making Sense of Sense: Bringing Computational Linguistics and Psycholinguistics Together, Trento, Italy, pp. 1–8 (April 2006)
18. Qin, P., Lu, Z., Yan, Y., Wu, F.: A New Measure of Word Semantic Similarity based on WordNet Hierarchy and DAG Theory. In: WISM 2009 (2009)
19. Cai, S., Lu, Z.: An Improved Semantic Similarity Measure for Word Pairs. In: IC4E 2010, Sanya, China (January 2010) (in press)
20. Liu, X.-Y., Zhou, Y.-M., Zheng, R.-S.: Measure semantic similarity in WordNet. In: Proceedings of the Sixth International Conference on Machine Learning and Cybernetics, Hong Kong, August 2007, pp. 19–22 (2007)
21. Pedersen, T., Michelizzi, J.: CGI (2009), http://marimba.d.umn.edu/cgi-bin/similarity/similarity

Improved Run-Length Coding for Gray Level Images Using Gouraud Shading Method

Wei Huang[1,2], Shuai Chen[2], and Gengsheng Zheng[2]

[1] Hubei Province Key Laboratory of Intelligent Robot,
Wuhan Institute of Technology, Wuhan 430073, China
[2] School of Computer Science and Engineering,
Wuhan Institute of Technology, Wuhan 430073, China
huangw2046@gmail.com, chsh_349@163.com,
zhenggengsheng@sina.com

Abstract. While improving compression ratio is a persistent interest in image representation, the encoding and decoding time is critical in some applications such as real-time image retrieval and communication. Based on the Run-Length Coding (RLC) and the Gouraud shading method, an Improved Run-Length Coding (IRLC) for gray level images is presented in this paper. The theoretical analysis and the experimental studies illustrate that the IRLC is superior to the S-Tree Compression (STC) in respect of the encoding and decoding time, without the compromises of the bit rates and the image quality.

Keywords: Run-length coding, Gouraud shading, Image representation, Image processing.

1 Introduction

Image representation has been an active research interest in image processing. Most of the efforts have been focused on improving the compression performance [1-3], and have produced some famous lossless compression algorithms such as the JPEG-2000 algorithm [4]. Although the compression ratio is always critical to image representation, however, in some applications such as real-time image retrieval and communication, the encoding and decoding time is important as well.

Distasi et al. [5] presented a storage-saving image compression method referred to as the B-Tree Triangular Coding (BTTC), which takes $O(n\log n)$ time of encoding and $O(n)$ time of decoding. Although the BTTC's bit rates are higher than the JPEG's by a factor of about 2, the encoding and decoding time of the former is much less than that of the latter. Based on the Gourand shading method and a modified S-Tree data structure, Chung et al. [6] presented the S-Tree Compression (STC) method, which partitioned an image into some homogenous blocks based on the binary tree decomposition principle and then represented the image by using the modified S-Tree. The STC further reduced the encoding time, compared with the BTTC, without the compromises of the compression ratio and the image quality.

Based on the Run-Length Coding (RLC) [7] and the Gouraud shading method, this paper presents an Improved Run-Length Coding (IRLC) for gray-level images. Like [5-6], the encoding and decoding of the IRLC take respectively $O(n\log n)$ time and $O(n)$ time. However, the IRLC needs one multiplication, one division, and four additions to estimate the gray level of a pixel within a run length, which are less than three multiplications, two divisions, and ten additions required by the STC to estimate the gray level of a pixel within a rectangular block. Since both the methods require a large number of the gray level estimations, it is not surprising that the IRLC is superior to the STC in respect of the encoding time.

The remainder of this paper is organized as follows: Section 2 briefly introduces the S-Tree Compression (STC). Section 3 presents the Improved Run-Length Coding (IRLC) for gray-level images. Next, the experimental studies are performed in section 4. Finally, section 5 concludes this paper.

2 S-Tree Compression

The S-Tree Compression (STC) [6] repeatedly partitioned an image into two equal sub-images, (from here on out, we refer to these sub-images as 'blocks'.) alternately in the horizontal and vertical directions, until the block is homogenous. And then, the STC used a modified S-Tree to encode the obtained blocks.

A block is considered to be 'homogenous' if the estimated gray level of any pixel in this block is an appropriate (according to a predefined tolerance) approximation of the real gray level of this pixel. Suppose that the coordinates of the most-top-left pixel, the most-top-right pixel, the most-bottom-left pixel, and the most-bottom-right pixels in a block B are (x_1, y_1), (x_2, y_2), (x_3, y_3), and (x_4, y_4), respectively, and the real gray levels of these pixels are g_1, g_2, g_3, and g_4, respectively. The estimated gray level g_{est} of a pixel (x, y) in B is

$$g_{est}(x, y) = g_5 + \frac{g_6 - g_5}{x_2 - x_1}(x - x_1) \tag{1}$$

where

$$g_5 = g_1 + \frac{g_3 - g_1}{y_3 - y_1}(y - y_1) \tag{2}$$

and

$$g_6 = g_2 + \frac{g_4 - g_2}{y_4 - y_2}(y - y_2) \; . \tag{3}$$

The block B is called 'homogenous' if every pixel (x, y) in B satisfies the formula given below:

$$\left| g_{est}(x, y) - g(x, y) \right| < \varepsilon \tag{4}$$

where $g(x, y)$ is the real gray level of the pixel (x, y) and ε is the predefined tolerance.

IRLC for Gray Level Images Using Gouraud Shading Method 21

Noting that $(y-y_2)/(y_4-y_2) = (y-y_1)/(y_3-y_1)$, Chung et al. [6] concluded that their method required three multiplications, two divisions, and ten additions to estimate the gray level of a pixel.

Once partitioning an image into some homogenous blocks, we obtain a binary tree whose leaf nodes all are the homogenous blocks. The STC uses two linear lists, i.e. a linear tree table and a color table, to encode the obtained binary tree. The STC scans the binary tree in the Breath-First-Search manner, and inserts a '0' ('1') into the rear of the linear tree table if an inner (leaf) node is encountered. Meanwhile, if a leaf node is encountered, a 4-tuple (g_1, g_2, g_3, g_4) is inserted into the rear of the color table where g_1, g_2, g_3, and g_4 are the gray levels of the most-top-left, the most-top-right, the most-bottom-left, and the most-bottom-right pixels, respectively. The number H_{STC} of the bits required to encode an image of bit depth m is

$$H_{STC} = N_{lt} + 4mN_{ct} \qquad (5)$$

where N_{lt} is the number of the elements in the linear tree table (i.e. the number of the nodes in the binary tree) and N_{ct} is the number of the elements in the color table (i.e. the number of the leaf nodes in the binary tree).

3 Improved Run-Length Coding

This section presents the Improved Run-Length Coding (IRLC) for gray level images. Due to the simple geometrical decomposition, the STC [6] has reduced the encoding time by a factor of approximate 2, compared with the BTTC. However, since the STC calculated the increments in two dimensions when estimating the gray level of a pixel, (please refer to (2) and (3).) its encoding process is still time-consuming. Moreover, the STC considered an image as a two-dimension array, which implies that the entire image must be loaded into RAM before the encoding process can be started. In some smart devices such as the capsule endoscopy [8], loading the entire image into RAM is difficult if not impossible.

In the IRLC, an image is considered as a data stream and the concept of the 'blocks' is replaced by the concept of the 'run length'. The 'run length' is an alias of a sequence of pixels and it is represented by a 3-tuple (l, g_1, g_2) where l is the number of the pixels in the sequence, g_1 and g_2 are respectively the gray levels of the first pixel and last pixel in this sequence. The estimated gray level g_{est} of the xth pixel in a run length (l, g_1, g_2) is

$$g_{est}(x) = g_1 + \frac{g_2 - g_1}{l-1}(x-1) \qquad (6)$$

where $x = 1, 2, ..., l$. A un length (l, g_1, g_2) is called 'homogenous' if the predefined tolerance ε, the estimated gray level g_{est} and the real gray level g of the xth pixel in this run length satisfy the inequality as follow:

$$|g_{est}(x) - g(x)| < \varepsilon . \qquad (7)$$

From (6), we can conclude that, in order to estimate the gray level of a pixel in a run length, the IRLC requires one multiplication, one division, and four additions, which are less than three multiplications, two divisions, and ten additions required by the STC [6]. Since both the STC and the IRLC require a large number of the gray level estimations, the IRLC is superior to the STC in respect of the encoding time due to the lesser number of the arithmetic operators.

When encoding an image, the IRLC reads *MAX* pixels from the image stream (*MAX* is predefined in the encoding algorithm.) and determines whether or not these *MAX* pixels belong to a homogenous run length. If so, a 3-tuple (*MAX*, g_1, g_2) is recoded, or else these *MAX* pixels are divided repeatedly into two halves until all halves are homogenous. Then, the IRLC reads the next *MAX* pixels and repeats the above process. At last, a sequence of the homogenous run lengths are obtained, which can be considered as a representation of the original image. The encoding algorithm of the IRLC is described as follows:

IRLC Encoding Algorithm

Input: Tolerance ε, maximum length *MAX*, and a data stream ($a_1a_2a_3...a_n$) where a_i (i = 1, 2, ..., n) is the real gray level of the ith pixel.
Output: The sequence *I* of the homogenous run lengths.
Step 1: Initialize *I* as a empty sequence and let index *j* equal to 1.
Step 2: Let the sequence $S = (a_ja_{j+1}...a_{j+MAX-1})$ and increase *j* by *MAX*.
Step 3: Determine whether or not *S* is homogenous by using (6) and (7). If *S* isn't homogenous, go to step 5.
Step 4: Insert 3-tuple (*MAX*, a_j, $a_{j+MAX-1}$) into the rear of *I*. go to step 8.
Step 5: Divide *S* into two halves $S_1 = (b_1b_2...b_m)$ and $S_2 = (c_1c_2...c_m)$ such that $S = (S_1S_2)$.
Step 6: Determine whether or not S_1 is homogenous. If so, insert (m, b_1, b_m) into the rear of *I*, or else let *S* equal to S_1 and recursively execute from step 5 to step 7.
Step 7: Determine whether or not S_2 is homogenous. If so, insert (m, c_1, c_m) into the rear of *I*, or else let *S* equal to S_2 and recursively execute from step 5 to step 7.
Step 8: If *j* is less than or equal to *n*, go to step 2.
Step 9: Exit with *I*.

The decoding process is simpler than the encoding process. In order to decoding an image represented by the IRLC, we scan the sequence of the run lengths and assigned the estimated gray level to every pixel.

The decoding algorithm of the IRLC is described as follows:
IRLC Decoding Algorithm:
Input: a sequence $I = (r_1r_2r_3...r_m)$ of the homogenous run lengths.
Output: the data stream $A = (a_1a_2a_3...a_n)$ where a_i (i = 1, 2, ..., n) is the estimated gray level of the ith pixel.
Step 1: Initialize the stream *A* as an empty stream.
Step 2: Let index *j* equal to 1.
Step 3: Let 3-tuple (l, g_1, g_2) equal to the *j*th run length r_j in *I*.
Step 4: Let increment *increment* equal to $(g_2 - g_1)/(l-1)$.
Step 5: Let index *k* equal to 0.

Step 6: Estimate the gray level $g_{est} = g_1 + increment \times k$ and insert g_{est} into the rear of A.
Step 7: If k is less than $l-1$, increase k by 1 and go to step 6.
Step 8: If j is less than m, increase j by 1 and go to step 3.
Step 9: Exit with A.

The length l of a homogenous run length varies in a large dynamic rang. However, in practice, the distribution of the lengths is far from uniform. In order to improve the storage efficiency, the IRLC encodes the length of the run lengths by using Huffman code. As a result, the IRLC representation of an image is a Huffman dictionary followed by a sequential list where every element is a homogenous run length. The storage structure of the IRLC is illustrated in Fig. 1.

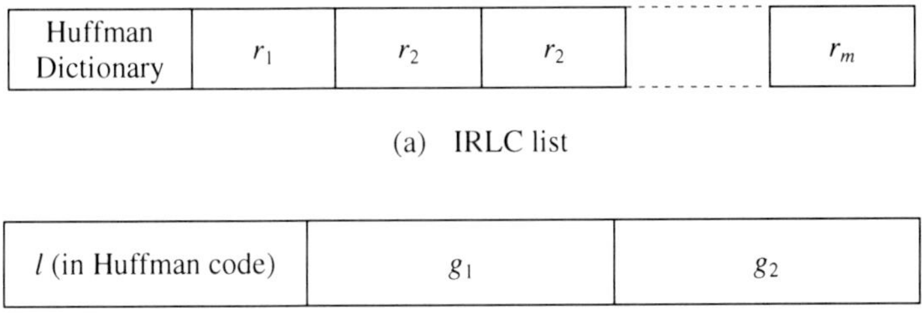

Huffman Dictionary	r_1	r_2	r_2		r_m

(a) IRLC list

l (in Huffman code)	g_1	g_2

(b) A element in the IRLC list

Fig. 1. The storage structure of the Improved Run Length Coding (IRLC)

4 Experimental Studies

The presented algorithm has been tested in a notebook PC with Intel Core 2 Duo 1.6GHz CPU and 2G RAM. Both the IRLC encoding algorithm and the STC encoding algorithm are programmed by using Visual C++ 7.0. The four 256×256 images used in the experiments are illustrated in Fig. 2.

(a) Lena   (b) Peppers   (c) F-16   (d) Sailboat

Fig. 2. Four 256×256 images used in the experiments

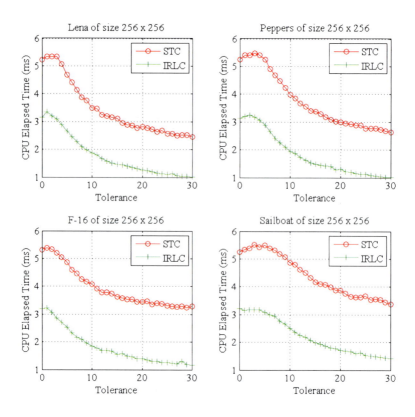

Fig. 3. Comparison of the CPU elapsed time

Fig. 3 gives the comparison of the CPU elapsed time required by the encoding algorithm of the STC and the encoding algorithm of the IRLC. In Fig. 3, we have plotted the CPU elapsed time of encoding the test images with the varied tolerances from 0 to 30. It is obvious that the CPU elapsed time used by the encoding algorithm of the IRLC is reduced to about 1/2, compared with that used by the encoding algorithm of the STC.

Fig. 4 illustrates the comparison of the Peak Signal Noise Ratio (PSNR) with the varied bit rates. The PSNR is defined as follow:

$$PSNR = 10\log_{10} \frac{255^2 \times 256^2}{\sum_{x=1}^{256}\sum_{y=1}^{256}[g(x,y) - g_{est}(x,y)]^2} . \qquad (8)$$

From Fig. 4, we can see that the PSNRs of the IRLR are slightly higher than those of the STC for all four test images, especially for the image 'Sailboat'. Fig. 5 gives the images reconstructed from the IRLC (with the bit rates 2.25) and the STC (with the bit rates 2.27). We can conclude from Fig. 5 that there is no obvious difference between the IRLC and the STC in respect of the reconstructed quality.

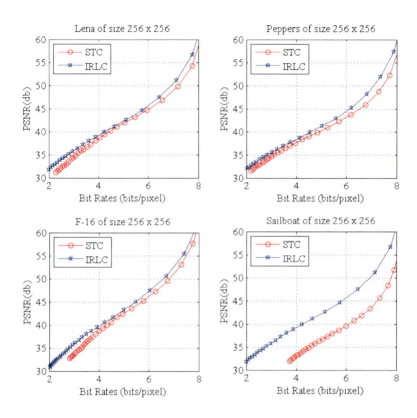

Fig. 4. Comparison of the PSNR with varied bit rates

(a) Reconstructed from STC (2.27)   (b) Reconstructed from IRLC (2.25)

Fig. 5. Comparison of the reconstructed qualities for the close bit rates

From the experimental results given above, we can safely conclude that the IRLC is superior to the STC in respect of the encoding time without the compromises of the bit rates and the image quality.

5 Conclusion

On the basis of the Run length Coding and the Gouraud shading method, this paper has presented a novel method, referred to as the Improved Run Length Coding (IRLC), to compress gray level images. In order to estimate the gray level of a pixel, the IRLC requires one multiplication, one division, and four additions, which is less than three multiplications, two divisions, and ten additions required by the STC [6]. The theoretic analysis and the experimental studies have shown that the IRLC is superior to the STC in respect of the encoding time without the compromises of the bit rates and the image quality.

Acknowledgment. We thank the support of the Science Research Foundation of Wuhan Institute of Technology under the granted no. 12096021 and the Research Foundation of Education Bureau of Hubei Province, China under the granted no. B20081506.

References

1. Nathanael, J.B., Mahmoud, R.E.: Grayscale True Two-dimensional Dictionary-Based Image Compression. Journal of Visual Communication and Image Representatio 18, 35–44 (2007)
2. Reichel, J., Menegas, G., Naddenau, M.J., Kunt, M.: Integer Wavelet Transform for Embedded Lossy to Lossless Image Compression. IEEE Transactions on Image Processing 10, 383–392 (2001)
3. Martin, M.B., Bell, A.E.: New Image Compression Techniques Using Multiwavelets and Multiwavelet Packets. IEEE Transactions on Image Processing 10, 500–510 (2001)
4. Taubman, D.S., Marcellin, M.W.: JPEG2000: Image Compression Fundamentals, Standards, and Practice. Kluwer, Boston (2002)
5. Distasi, R., Nappi, M., Vitulano, S.: Image Compression by B-tree Triangular Coding. IEEE Transactions on Communication 45, 1095–1100 (1997)
6. Chung, K.L., Wu, J.G.: Improved Image Compression Using S-Tree and Shading Approach. IEEE Transactions on Communications 48, 748–751 (2000)
7. Golomb, S.W.: Run-Length Encodings. IEEE Transactions on Information Theory IT12, 399–401 (1966)
8. Seibel, E.J., Carroll, R.E., Dominitz, J.A., Johnston, R.S., Melville, C.D., Lee, C.M., Seitz, S.M., Kimmey, M.B.: Tethered Capsule Endoscopy, A Low-Cost and High-Performance Alternative Technology for the Screening of Esophageal Cancer and Barrett's Esophagus. IEEE Transactions on Biomedical Engineering 55, 1032–1042 (2008)

Factors Affecting Consumer Intentions to Acceptance Banking Services in China

Sun Quan

Business Department, Suzhou Vocational University,
Suzhou, China
sunxjtu@163.com

Abstract. This study compared the factors influencing consumer acceptance banking services in China. Five factors were identified: service quality, customer satisfaction, trust, commitment, behavioural intentions. Structural equation modelling (SEM) was computed to determine the relationships among these factors. The main results are as follows: there is 1) a positive relationship between service quality and customer satisfaction; 2) a positive relationship between customer satisfaction and behavioral intentions; 3) a positive relationship between customer satisfaction and trust; 4) a positive relationship between trust and commitment; 5) a positive relationship between commitment and behavioral intentions. The findings aim to enhance service quality and performance in banking.

Keywords: Banking Services, Service Quality, Customer Satisfaction, Trust, Commitment.

1 Introduction

Globalization and deregulations have increased competition in the marketplace, as nowadays it has become much easier for companies to cross borders and compete internationally. Technological advancement, sophisticated and swiftly changing customers' expectations and the resulting high market competitiveness are giving prominence to the issue of the quality of customer service in the services industry sector, leaving practitioners in the sector with no choice but to properly address the issue in order to be still competitive in the marketplace. Services industry has expanded rapidly, has come to play an increasing role in the world economy, has occupied a great majority of the markets in many countries and has contributed a total of 66.3 percent of world gross domestic product. Also, the service industry has sump based the manufacturing industry and agriculture and become the main stream of economic in China. China has stepped into the stage of "Service Economy".

Parasuraman et al. posited that delivering superior service quality to customers is essential for success and survival in the competitive market environment [1]. Additionally, provision of high quality service helps meet several requirements such as customer satisfaction and its consequent loyalty and market share, attracting new customers through word-of-mouth advertising, improving productivity; financial performance and profitability. Banking and financial services are a demand driven industry, which constitute an important part of the services industry. During the past two decades or so, regulatory, structural and technological factors have significantly changed the banking environment in which banks are expanding across borders, offering a diverse portfolios of competitive services and reengineering their services in order to make use of rapid technology and to meet the changing needs of customers [2]. Banks in all over the world offer similar kinds of services matching with competitors. However, customers perceive quality of services differently. Because of the highly undifferentiated products and services that financial institutions, and specifically banks, offer, service quality becomes a main tool for competing in this marketplace, a way to increase customer satisfaction, intention, and a strategic tool to improve their core competence and business performance. In general, because of the higher profits and higher customer retention to which they lead, high-quality services are believed to provide banks with competitive edge in the marketplace. From the above mentioned, it becomes obvious that high service quality is essential for banks to survive in the highly competitive environment [3]. This leads to the fact that, a good understanding of the attributes that customers use to judge service quality is necessary in order for the banks to be able to monitor and enhance its service performance and improve its overall service quality [4].

There are numerous studies, to date, that identified the key service quality dimensions and its relationship with customer satisfaction and loyalty in the banking environment in the developed countries and in such other emerging developing countries [4], but relatively little literature has investigated service quality and its relationship with customer satisfaction and other factors in the developing countries, especially in Asian markets such as China and service-related issues have long been neglected as well. In this paper a new approach for assessing potential acceptance of banking services is proposed based on the Garbarino and Johnson's model. Garbarino and Johnson's model applied a nonprofit repertory theater [1], this study attempts to test this model's feasibility in profit business. This study also attempts to take all the above-mentioned constructs in consideration and hopes to explain customers' behavioral intentions to acceptance banking services in China.

2 Research Model and Hypotheses

Cronin and Taylor, using a single-item purchase-intention scale, find a positive correlation between service quality and customer satisfaction [5]. Cronin, Brady, and Hult found a positive relationship between customer satisfaction and behavior intentions [6]. The marketing and related literature posits a positive relationship between trust and satisfaction. Garbarino and Johnson find a positive relationship between overall customer satisfaction and trust [1]. Moorman, Zaltman, and Deshpande find that trust by marketing research users in their search providers significantly affected user commitment to the research relationship [7]. Morgan and Hunt find a positive relationship between trust and relationship commitment [8]. Garbarino and Johnson find a positive relationship between commitment and future intentions [1]. For the purpose of conducting an effective and reliable study, we try to build up a conceptual framework for this study, shown in Fig. 1.

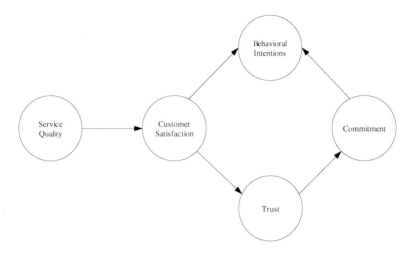

Fig. 1. The Research Model

2.1 Service Quality

Service quality is one of the most dominant themes of research in services. Service quality has been linked with customer satisfaction within the banking industry. It is important that the banks provide customers with high quality services to survive in the highly competitive banking industry [6, 7]. For this, bankers first need to understand the attributes that customers use to judge service quality and then they have to monitor and enhance the service performance. Banking is a high involvement industry and a demand driven industry, which constitutes an important part of the services industry. However, as electronic banking becomes more prevalent, customers still tend to measure a bank's service quality in terms of the personal support they receive, rather than the technical support [8]. For an organization to gain competitive advantage it must use technology to gather information on market demands and exchange it between organizations for the purpose of enhancing the service quality [9].

The quality of the interaction between buyer and supplier (quality of service) has also been identified as a possible and potential antecedent of customer satisfaction [5]. There is ample empirical support for quality as an antecedent of customer satisfaction [10]. Recent research offers some evidence that customer value, satisfaction and/or service quality perceptions positively affect intentions to behave in these ways. Based on this, the following hypotheses are proposed:

H1. Service Quality will have a positive effect on the Customer Satisfaction.

2.2 Customer Satisfaction

From the marketing point of view, satisfied customers will be more likely to continue to purchase the product while dissatisfied customers will be more likely to switch brands and tell others about a bad experience. Customer satisfaction has long been recognized in marketing thought and practice as a central concept as well as an important goal of all business activities [10]. High customer satisfaction has many benefits for the firm, such as increased customer loyalty, enhanced firm reputation, reduced price elasticity's, lower costs of future transactions, and higher employee efficiency [10]. It is believed that customer satisfaction is a fundamental indicator for a firm's future profits, due to the fact that attracting new customers is much more expensive than keeping old ones.

The role of satisfaction in predicting behavioral intention is well established. A study conducted by Woodside, Frey, and Daly uncovers a significant association between overall patient satisfaction and intent to choose the hospital again [11]. Cronin, Brady, and Hult found a positive relationship between customer satisfaction and behavior intentions [8]. Garbarino and Johnson find customer satisfaction have positive relationship between trust and future intentions [1]. Based on this, the following hypotheses are proposed:

H2. Customer Satisfaction will have a positive effect on the Trust.

H3. Customer Satisfaction will have a positive effect on the Behavioral Intentions.

2.3 Trust

The growing importance of relationship marketing has heightened interest in the role of trust in fostering strong relationships [12]. Berry argues that the evidence suggests that relationship marketing is built on a foundation of trust [13]. Trust generally is viewed as to gain the loyalty of customers, your must first gain their trust.

Morgan and Hunt propose dimensions of trust: (1) reliability and (2) integrity [10]. Morgan and Hunt posit that trust is a major determinant of relationship commitment [10]. Based on this, the following hypotheses are proposed:

H4. Trust will have a positive effect on the Commitment.

2.4 Commitment

In the services relationship marketing area, Morgan and Hunt propose that relationship commitment is central to relationship marketing [8]. Similar to trust, commitment is recognized as an essential ingredient for successful long-term relationships. Morgan and Hunt define relationship commitment as an exchange partner believing that an ongoing relationship with another is so important as to warrant maximum efforts at maintaining it; that is, the committed party believes the relationship is worth working on to ensure that it endure indefinitely [8].

Gundlach, Achrol, and Mentzer argue that commitment has three components: (1)an instrumental component of some form of investment, (2) an attitudinal component that may be described as affective commitment or psychological attachment, and (3) a temporal dimension indicating that the relationship exists over time [15].

Garbarino and Johnson find a positive relationship between commitment and future intentions [1]. Based on this, the following hypotheses are proposed:

H5. Commitment will have a positive effect on the Behavioral Intentions.

3 Method and Results

3.1 Questionnaire Design and Data Collection

In order to test the hypotheses, this study relied on five sets of constructs and their indicators. All indicators came from the items in a survey questionnaire designed with a 5-point scale from strongly disagree (1) to strongly agree (5). The questionnaire is designed to analysis the factors affecting consumers to acceptance banking services, all the dimensions included in the questionnaire have been described and used based on the researches of Parasuraman, Zeithaml and Malhotra.

Before the formal distribution questionnaire, we do pre-test to understand subjects whether confused the theme of questionnaire. First of all, there are 50 postgraduates of finance of graduate schools join the pretest. After collected subjects' opinion, we modified these unclear items and adjusted these statistic verification items. Finally, we refined and finalized the appearance and format of the questionnaire.

This study used online questionnaire to collect data. We are not only posted the questionnaires on website, but also used e-mail to send the website of questionnaires. A total of 224 questionnaires were returned, and 18 surveys were unusable due to answering the same scale during February 25 through May 20, 2009. Therefore, the final useful sample contained 216 respondents.

3.2 Measurement Model

A confirmatory factor analysis (CFA) was used to assess the goodness-of-fit of the measurement model, which considering e-service quality as predictor variables, e-customer satisfaction and perceived value construct as mediating variable, and e-loyalty construct as dependent variable. Nine common model-fit measures were used to assess the model's overall goodness of fit: the ratio of χ^2 to degrees of freedom (d.f.), normalized fit index (NFI), non-normalized fit index (NNFI), comparative fit index (CFI), goodness-of-fit index (GFI), adjusted goodness-of-fit index (AGFI), and root mean square error of approximation (RMSEA). As shown in Table 1, all the model-fit indices exceeded their respective common acceptance

Table 1. Fit Indices For Measurement And Structural Models

Fit indices	Recommended value	Measurement model	Structural model
χ^2/df	≤3.00	1.48	1.82
NFI	≥0.90	0.95	0.91
NNFI	≥0.90	0.91	0.90
CFI	≥0.90	0.94	0.92
GFI	≥0.90	0.93	0.94
AGFI	≥0.80	0.88	0.93
RMSEA	≤0.08	0.04	0.05

Table 2. Reliability, average variance extracted and discriminant validity

Factor	CR	1	2	3	4	5
Service quality	0.79	1.00				
Customer Satisfaction	0.80	0.35	1.00			
Trust	0.84	0.34	0.19	1.00		
Commitment	0.79	0.44	0.32	0.06	1.00	
Behavioral Intentions	0.86	0.45	0.38	0.21	0.33	1.00

levels suggested by previous research, thus demonstrating that the measurement model exhibited a fairly good fit with the data collected. Therefore, we could proceed to evaluate the psychometric properties of the measurement model in terms of reliability, convergent validity and discriminant validity.

Reliability and convergent validity of the factors were estimated by composite reliability and average variance extracted (see Table 2). The composite reliabilities can be calculated as follows: (square of the summation of the factor loadings)/{(square of the summation of the factor loadings)+(summation of error variables)}, where the factor loadings are obtained directly from the program output, and the error variables is the measurement error for each indicator. The interpretation of the composite reliability is similar to that of Cronbach's alpha, expect that it also takes into account the actual factor loadings, rather than assuming that each item is equally weighted in the composite load determination.

3.3 Structural Model

This study took a staged approach (i.e., nested models comparisons) to testing hypothetical models that describe the relationship between both observed and unobserved measures. This staged approach, similar to hierarchical regression, allows us to determine if the addition of new set of relationships adds significantly to the explanation of the variation in the data. The result of the best fitting model is shown in figure 2. Goodness-of-fit statistics, indicating the overall acceptability of the structural model analyzed, are acceptable. Most path coefficients are significant ($p<0.05$). The p-values of the estimates for hypotheses testing were determined with two-tailed t tests.

4 Conclusions and Discussions

The objective for this study is to predicting consumer intention to acceptance banking service in China. The above statistical analysis supported all of the hypotheses.

Factors Affecting Consumer Intentions to Acceptance Banking Services in China

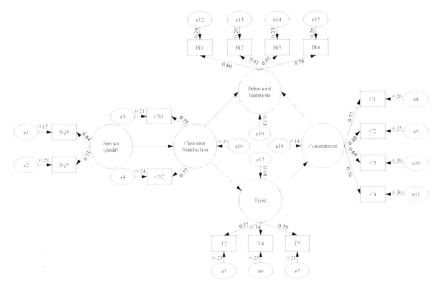

Fig. 2. The best fitting model

In banking, the quality of the core service is difficult for the average customer to judge and he or she has relatively little contact with the service provider. Therefore, an important implication for bank managers is that it is essential to meet customer expectations for the service core. The basic promise or implicit contract must be delivered, as it is a significant driver of customer satisfaction, which directly related to future intentions. This confirms prior research that has identified the importance of delivering the core service or the basic promise to customers. It also points out the importance of "getting it right the first time." Bank managers need to understand what their basic promise is to the customer and deliver on that promise. This promise generates the basic expectations that customers have with respect to the service. The promised could also include the time to complete of the service. Customer will evaluate core service quality based on the promises made, which may include secondary aspects of the core. Thus, the bank manager needs to deliver on all the promised made to meet core expectations. Before bank managers want to acquire customer satisfaction, they focus on service quality.

To gain more positive behavioral intentions from customers, satisfaction and commitment, which directly affected behavioral intentions, should be further considered simultaneously. Although the results emphasize the importance of quality as an operational tactic and strategic objective, the acceptable price range concept should not be ignored.6 That is to say, buyers have a price range that is acceptable for a given purchase, rather than a single price. Besides, bank managers should actively enhance customer satisfaction in many ways, such as service providers' performances and enhancement of corporate image, so as to attract customers.

Firms seek to differentiate themselves from rivals through the utilization of a relationship marketing approach. Relationship marketing is built on long-term relationship between firms and customers. Policies and decisions must favor the development of a customer-oriented culture and relationships characterized by the supplier's in -depth knowledge, commitment, and understanding of customer needs along with an atmosphere of trust. Commitment is based on trust. Hence, bank managers should set trust and commitment as a long-term objective and customer satisfaction as a short-term objective.

Finally, the services manager who only contemplates the possible effect of service quality momentum on his or her customers' behavioral intention may make a mess if he or she does not also ponder over the impact of such a strategy on satisfaction attributed to his or her firm's services. Nowadays, though there are no effective approaches to solving the complicated decision- making process, at least, making efforts to ensure core service quality will pay in terms of customer satisfaction traits.

References

1. Garbarino, E., Johnson, M.S.: The Different Roles of Satisfaction, Trust and Commitment in Customer Relationships. Journal of Marketing 63(2), 70–87 (1999)
2. Yavas, U., Benkenstein, M., Stuhldreier, U.: Relationships between service quality and behavioral outcomes: A study of private bank customers in Germany. The International Journal of Bank Marketing 22(2), 144–157 (2004)
3. Wang, Y., Lo, H.P., Hui, Y.V.: The antecedents of service quality and product quality and their influences on bank reputation: evidence from the banking industry in China. Managing Service Quality 13(1), 72–83 (2003)
4. Arasli, H., Katircioglu, S.T., Mehtap-Smadi, S.: A comparison of Service Quality in the banking industry. International Journal of Bank Marketing 23(7), 508–526 (2005)
5. Cronin Jr., J.J., Taylor, S.A.: Measuring Service Quality: A Reexamination and Extension. Journal of Marketing 56(3), 55–68 (1992)
6. Cronin, J., Brady, M.K., Hult, G.T.: Assessing the Determination of Consumer Behavioral Intentions in Service Environments: An Investigation of A Comprehensive Model of the Effects of Quality, Value, and Satisfaction. Journal of Retailing 76, 193–218 (2000)
7. Moorman, C., Zaltman, G., Deshpande, R.: Relationships between Providers and Users of Market Research: The Dynamics of Trust within and between Organizations. Journal of Marketing Research 29(3), 314–329 (1992)
8. Morgan, R.M., Hunt, S.D.: The Commitment-Trust Theory of Relationship Marketing. Journal of Marketing 58(3), 20–38 (1994)
9. Seth, N., Deshmukh, S.G., Vrat, P.: Service Quality models: A review. International Journal of Quality & Reliability Management 22(9), 913–949 (2005)
10. Anderson, E.W., Sullivan, M.W.: The Antecedents and Consequences of Customer Satisfaction for Firms. Marketing Science 12(2), 125–143 (1993)
11. Woodside, A.G., Frey, L.L., Daly, R.T.: Linking Service Quality, Customer Satisfaction and Behavioral Intentions. Journal of Health Care Marketing 9(4), 5–17 (1989)
12. Sirdeshmukh, D., Singh, J., Sabol, B.: Consumer Trust, Value, and Loyalty in Relational Exchanges. Journal of Marketing 66(1), 15–37 (2002)
13. Berry, L.L.: Relationship Marketing of Services - Growing Interest, Emerging Perspectives. Journal of the Academy of Marketing Science 23(4), 236–245 (1995)
14. Urban, G.L., Sultan, F., Qualls, W.J.: Placing Trust at the Center of Your Internet Strategy. Sloan Management Review 42(1), 39–48 (2000)
15. Gundlach, G.T., Achrol, R.S., Mentzer, J.T.: The Structure of Commitment in Exchange. Journal of Marketing 59(1), 78–92 (1995)

Comparative Study of Workload Control Methods for Autonomic Database System

Yan Qiang, Juan-juan Zhao, and Jun-jie Chen

College of Computer and Software,
Taiyuan University of Technology,
Taiyuan, China
dicom8@yahoo.com.cn

Abstract. Autonomic database system is a performance optimization system, which can improve the resource utilization of database system by effective management of the access order to the workload of the database system. In this paper, the existing workload autonomic framework of database system has been improved, giving emphasis on the workload control component within the framework, and the comparative analysis of the replacement strategy and the NSGA- II algorithm was carried out. The simulation shows that the NSGA- II workload control algorithm can achieve higher accuracy with scarifying longer time to run, whereas the replacement strategy of workloads has lower accuracy with less time-consuming, which is more favorable for a real-time autonomic database system.

Keywords: autonomic database, workload control, NSGA- II algorithm, replacement strategy.

1 Introduction

At present, the database workload autonomic technology is one of the typical computer scientific research objectives in the domestic research community. One of the examples is the university of Huazhong science and technology. Professor Feng Yucai studied the self-adjusting technology of database system by taking examples of the network database DM4 subsystem and I/O subsystem [1]. Professor Wang Yuanzhen from Huazhong University of Science studied the design and implementation of system resources optimization tools of autonomic database system based on the model of feedback-loop circuit [2]. The above studies are mainly focused on the parametric settings of the database systems, without considering the workload characteristics. Abroad, Pang et al [3] put forward the "Priority Adaptation Query Resource Scheduling" algorithm to minimize the number of workload classes exceeding time of queries, and evenly distribute the time exceeding queries proportionally to all the workload classes as defined by the system administrator. The recognized drawbacks of this approach are inaccuracy of the performance prediction, and less supported by workload classes. Queen's University and the IBM Toronto lab in Canada co-operated and presented a basic framework for workload autonomic database system, which will be the basis of this study in this paper [4]. The modified workload autonomic system framework can be illustrated in Fig. 1 as follows.

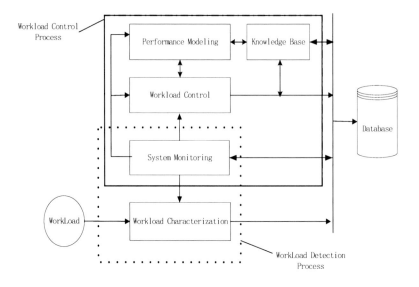

Fig. 1. The autonomic database framework

The framework consists of the workload characterization component, performance-modeling component, workload control component, system-monitoring component and knowledge base system. The functions of each component are as follows: system-monitoring component monitors the resources utilization of the database system, such as CPU, memory and so on; workload characterization component defines the characteristic parameters of the workload, such as workload response time, the necessary system resources consumed by the workload, and workload values; the performance modeling component establishes the system performance model to predict the performance characteristics of the system and application performance; workload Control component controls the workload filtering by the optimal search algorithm, making the workload submitted to the database system in reasonable orders; knowledge base system stores the knowledge to run the various components. Among all of these components, the workload control component is the core component. The other three components lay the foundation for the workload control component and prepare for the autonomic workload control database system. The workload control component controls the filtering of the workload according to the characteristics of the workload parameters and the system resources consumption information returned by the system-monitoring component. This paper will focus on the workload control component.

2 Problem Description

The workload control of the autonomic database system contains the process to filter and control the workloads. It selects optimal workloads according to the characteristic parameters (such as workload response time, CPU consumption, and workload values) provided by the workload characterization component. The less resources consumption and high value workloads are selected as pre-submission objects. In the process of workload filtration, the system uses the current resources consumption as the workload control threshold provided by the system-monitoring component to limit the resources consumption of the pre-submitted workload not exceeding the spare resources available. In the process of workload optimal selection, because the selection is based on at least two workload characteristic parameters such as workload response time, CPU consumption, memory consumption and workload values etc, the workload control of the autonomic database system is a typical multi-objective selection issue. The workload filtration control objective will not be able to achieve by ordinary optimization methods.

Based on the practical application, we use workload response time rt_i, CPU consumption ch_i, workload value val_i and the density of value vd_i as the workload control characteristic parameters. Assuming workload L_i, we may use vector $L_i = (val_i, rt_i, ch_i, vd_i)$ to represent a workload. Parameter vd_i represents the importance factor of the differences between workload L_i and other workloads. rt_i represents the required time from getting the workload to receiving the processing data of the workload. ch_i represents the percentage of CPU consumption with the processing of the workload by the system. vd_i represents the value with a unit consumption of CPU time, thus we have $vd_i = v_i / cpu_i$.

3 Workload Control Algorithm

Presently there are series of algorithms dealing with multi-objective optimization. A popular example is the multi-objective genetic algorithms family, such as MOGA、SPEA、NSGA-II etc., In addition, multi-objective particle swarm optimization (PSO) algorithm, multi-objective ant colony optimization (ACO) algorithm, multi-objective immunization algorithm etc are also well known. However, the NSGA-II [5] algorithm differentiates with other methods by using non-dominated sorting mechanism, which brings this algorithm into the frontier of the best approximation methods such as Pareto optimization. Therefore this approach is widely used in the research field. The above-mentioned multi-objective algorithms are carefully studied with special consideration to the workload control of the autonomic database system. In lieu of their significant advantages, this paper initiates a replacement strategy as the workload optimal selection method as an attractive alternative. A comparison study and experiment was carried out between the replacement strategy and the NSGA-II algorithm in the application to the workload control of the autonomic database system, and conclusions regarding their respective characteristics are also presented.

3.1 Workload Control by Using NSGA-II Algorithm

Workload control using the NSGA-II algorithm starts with considering each workload as a gene, and each workload between the waiting list of control intervals as an individual. For each individual, a group of workloads having the highest values and the shortest responding time will be selected as submission objectives. The processing procedure of the workload control using NSGA-II algorithm takes the following steps [6] :

(1) Define decision variables and constraints. Here the decision variables are workload responding time t_i and value density vd_i, and the constraint is the spare CPU resource of the system.

(2) Establish the optimization model, which includes to define the types of the objective function and to set up it's mathematical descriptions.

(3) Define the coding methods for chromosomes. In the process of workload control of autonomic database system, we use binary 0/1 coding method. Each bit of the binary represents a workload. The value of 1 indicates that the workload is ready for submission, whereas 0 means not ready yet.

(4) Generate initial population. A reasonable size of population S_n can be generated randomly according to practical requirements. We will stipulate here the population size as 50.

(5) Conduct genetic manipulations on S_n by crossover or mutation to generate next generation of population R_n.

(6) Join population S_n together with R_n to generate new population T_n.

(7) Conduct non-dominated sorting operation on T_n to get crowding distance values for each individual.

(8) To inherit the next generation R_{n+1} from the 50% of group T_n, which have better genetic components.

3.2 Workload Control Using the Replacement Strategy

The control procedure of workload control using the replacement strategy can be divided into two steps. First step is to seek a group of pre-submission objects having the highest values within the limitations of the CPU resources of system by using search optimization algorithms. This step will be called search optimization step. The second step involves the replacement of the workloads, which have lower value densities and longer responding times by some urgent workloads. This replacement process is implemented using priority table ranking method.

Search optimization procedure. There are many types of search optimization methods available for use in the literature. They can be divided basically into two groups. One is accurate algorithm, and the other is approximate. The accurate group includes dynamic planning, back searching method etc, whereas the approximate family comprises greedy algorithm, genetic algorithm and ant colony algorithm etc. Based on the practical requirement of autonomic database system, we use greedy algorithm to carry out the search optimization procedure.

Assume s_0 is the workloads waiting list and sort the workloads according to their value densities by descending order. Let s_1 represent the workload pre-submission list, and V represent the spare resources of workloads. Select the group having the highest value densities using greedy algorithm, and put this group into the workload pre-submission list s_1. Let cf represent the left CPU resources after workload submission, thus we have

$$cf = V - \sum_{i \in s_1} ch_i \qquad (1)$$

Replacement Procedure

(a) Priority table ranking. The comprehensive sorting method based on the priority table ranking starts with sorting the workloads of s_1 by their value densities vd_i ascendingly, and their responding time rt_i descendingly. Let i, j represent the position in the list of value densities and responding time, respectively. With parameters i and j, each workload can be distributed with an exclusive priority ranking function P based on the priority ranking distribution formula introduced in [7] :

$$p = (i + j - 1) * (i + j - 2) / 2 + i \qquad (2)$$

The workloads in the list of s_1 with lower value densities and longer responding time will be put in the front of the queue as priority to be replaced based on the priority ranking function P.

(b) Replacement strategy. As mentioned above, the workloads in the list of s_1 with lower value densities and longer responding time will be put in the front of the queue as priority to be replaced based on the priority ranking function P. The next important step is to replace these workloads in the priority list in s_1. The detailed procedure of the replacement strategy is as follows:

(1) The much urgent workloads in the rest of the workloads data set s_0 is put into another data set called s_4. Let the number of these urgent workloads be n. Dataset s_5 is used to collect the workloads to be replaced from dataset s_1.

(2) Let the index of the $i(i=1,2,...,n)$ workload in dataset s_4 be k. The workload L_k is to be replaced into dataset s_1, and the workloads ready to be replaced from dataset s_1 to be put into dataset s_5. The detailed procedures are as follows:

Step 1: Initialize $j=1$, and let s_1^j represents the j workload of dataset s_1.
Step 2: Let

$$cf = cf + ch_{s_1^j} \qquad (3)$$

if $cf < ch_k$, $s_5 = s_5 + \{s_1^j\}$, and $j = j+1$. Otherwise go back to step 3.

Step 3: Conduct replacement manipulation on dataset s_1 according to $s_1 = s_1 + \{k\} - s_5$, and also mark workload k not being replaced. Thus we have $cf = cf - ch_k$.

(3) Clear dataset s_5 to make the next replacement of workload in dataset s_4 ready to go. Using the same logic to replace all the workloads in dataset s_4 into dataset s_1, which eventually becomes the dataset of workloads ready to be submitted to the database system.

Using replacement strategy, the much urgent workloads with shorter responding time in the waiting queue s_0 can be effectively replaced into dataset s_1, which becomes the priority to be submitted to the database system. This approach not only improves the implementation values of the system, but also improves the throughput of the system.

4 Experiment and Discussion

Based on the practical requirement, system implementation value and system throughput will be considered as criteria for the workloads control performance of the autonomic database system. The simulation experiment was also carried out in terms of these two parameters with the system throughput defined as follows:

cnt_i: is the total number of workloads within the i th control interval;

com_i: is the number of workloads submitted within the i th control interval; and

rat_i: is the system throughput, which can be formulated as:

$$rat_i = com_i / cnt_i \qquad (4)$$

In this experiment, the value of workload is a random number from 1-10 generated by random function. CPU consumption by the workload is a random number from

1% - 40% generated also by random function. The responding time is a random number from 1-500 (mSec) generated again by random function. The experiment was carried out with 20 control intervals with each interval as 1 second. The experiment results are as follows:

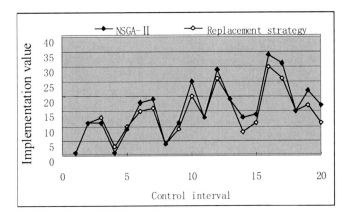

Fig. 2. Comparison of system implementation value

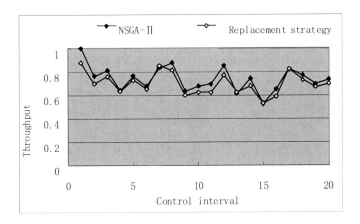

Fig. 3. System throughput

Table 1. Comparison of time consumption

	NSGA-II	Replacement strategy
Time consumption	2549ms	1768ms

It can be seen from Fig. 2 that using NSGA-II algorithm, the implementation value and system throughput of workload control is higher than using replacement strategy. This suggests that the NSGA-II overall yields higher accuracy. However, from table 1, we see that after running 20 intervals, the time consumption for NSGA-II algorithm takes 2549 ms, whereas the time consumption for the replacement strategy takes 1768 ms, which indicates the efficiency of using replacement strategy is quite significant. It is quite obvious that although the NSGA-II algorithm overall has a higher accuracy for workload control of the autonomic database system, it takes much longer time to run, which may impose a significant impact on the real-time performance of the autonomic database system.

5 Conclusion

A comparative study and experiment of workload control of autonomic database system using replacement strategy and NSGA-☐ algorithm has been carried out in this paper. The study shows that the replacement strategy has the merit of less time-consumption and higher efficiency, whereas the NSGA-☐algorithm requires longer time spending with a higher accuracy. Therefore these two methods have different advantages. Taking the strong real-time performance requirement nature of autonomic database system into account, the replacement strategy is more preferred as the study suggests. The limitation of this study is that the replacement strategy is only valid for two objectives in the multi-objective optimization processing, and it will not be suitable to tackle more than two objectives' cases. This could be an interesting scope of study for further research efforts.

References

1. Dong, A.-h.: Research on self-tuning technology of DM4 database (2005) (in Chinese)
2. Wang, Y.-z., Jiang, H., Xie., M.-y.: Design and implementation of self-tuning model of resource for DBMS. Computer Applications 25(9) (2005) (in Chinese)
3. Pang, H., Carey, M.J., Livny, M.: Multiclass Query Scheduling in Real-Time Database Systems. IEEE Transaction on Knowledge and Data Engineering 7(4) (1995)
4. Niu, B., Martin, P., Powley, W., Horman, R., Bird, P.: Workload Adaptation in Autonomic DBMSs. In: Proceedings of CASCON 2006, Toronto, Canada, October 16-19, pp. 161–173 (2006)
5. Deb, K., Pratap, A., Agrawal, S., et al.: A Fast and Elitist Multi-Objective Genetic Algorithm: NSGA-II[R]. KanGAL Report No. 200001, India (2000)
6. Zhou, M., Sun, S.-d.: Genetic algorithm theory and application. National Defense Industry Press, Beijing
7. Wang, Q., Xu, G.: A New Priority Table Based Real-Time Scheduling Algorithm. Acta Electronica Sinica 32(2), 310 (2004) (in Chinese)

Consensus of Mobile Agent Systems Based on Wireless Information Networks and Pinning Control*

Hongyong Yang, Lan Lu, and Xiao Li

School of Information Science and Engineering, Ludong University,
Yantai 264025, China
hyyang_ld@yahoo.com.cn

Abstract. In this paper, consensus of mobile multi-agent systems with wireless information networks and pinning control is studied. Suppose mobile multi-agent systems consist of n agents, there is a directed interconnected graph with a globally reachable node. Based on the Lypunov stability theory, the consensus of multi-agent systems with pinning control is analyzed. By Applying generalized Nyquist criterion and Greshgorin's disc theorem, the consensus of delayed algorithm with pinning control is analyzed. Finally, computer simulations are used to show the validity of the result.

Keywords: Mobile agent systems, wireless networks, consensus, pinning control, communication delays.

1 Introduction

In recent years, there has been an increasing research interest in the control design of multi-agent systems. Many results have been obtained with local rules applied to each agent in a considered multi-agent system. For example, Jadbabaie et al. [1] demonstrated that a simple neighbor rule makes all agents eventually move in the same direction despite the absence of centralized coordination and each agent's set of neighbors changing with times as the system evolves under a joint connection condition. Also, with similar technique, Lin et al. [2] studied three formation strategies for groups of mobile autonomous agents. The stability analysis of multi-vehicle formations was given with a Nyquist-type criterion in (Fax & Murry[3]). Moreover, by a Lypunov-based approach, Olfati-Saber et al.[4] solved the average-consensus problem with directed interconnection graphs or time-delays.

The collective behavior of multiple agents with leaders is one of the most interesting topics in the motion control of multiple agents. The coordinated motion of a group of motile particles with a leader has been analyzed in Mu et al. [5] and Wang et al [6]. Recently, Olfati-Saber [7] has introduced a theoretical framework including a virtual

---

* This research was supported in part by Chinese National Natural Science Foundation (under the grant 60774016, 60875039, 60904022, 60805039), and the Science Foundation of Education Office of Shandong Province of China (under the grant J08LJ01).

leader/follower architecture. Hong et al.[8] and Cheng et al.[9] proposed an observer to estimate the variable leader's velocity, and analyzed the stability with time delays.

In this paper, we apply the method of pinning control to stabilize the states of multiple agents. Feedback pinning has been a common technique for the control of synchronization in a complex dynamical network. Grigoriev et al. [10] studied the pinning control of spatiotemporal chaos. Wang and Chen [11] proposed an effective measure to pin a scale-free dynamical network to its equilibrium. Li et al. [12] used virtual control for microscopic dynamics in different network models. Zhou, et al. [13,14] presented a pinning adaptive synchronization of a general complex dynamical network.

This paper is organized as follows: In Section 2, a problem description of mobile agent systems is presented. The consensus of mobile agent systems with pinning control is studied in Section 3. In Section 4, the communication delays are considered in mobile agent systems. Computer simulations are used to verify in Section 5. Conclusions are drawn in Section 6.

2 Problem Formulations

Let $G=(V,E,A)$ be a weighted digraph of order n with the set of nodes $V=\{1,2,\cdots,n\}$, set of arcs $E \subseteq V \times V$ and a weighted adjacency matrix $A=[a_{ij}] \in R^{n \times n}$ with nonnegative elements. The node indexes belong to a finite index set $I=\{1,2,\cdots,n\}$. The element a_{ij} associated with the arc of the digraph is positive, i.e., $a_{ij} > 0$ if there is an edge between agent i and agent j. Moreover, we assume $a_{ij} = 0$ for all $i \in I$. The set of neighbors of node i is denoted by $N_i = \{j \in V : (i,j) \in E\}$.

A diagonal matrix $D = diag\{d_1, d_2, ..., d_n\} \in R^{n \times n}$ is a degree matrix of G, whose diagonal elements $d_i = \sum_{j \in N_i} a_{ij}$ for $i=1,2,...,n$. Then the Laplacian of the weighted digraph G is defined as $L=D-A$.

In the graph G, if there is a directed path beginning with i and ending with j, i.e. a sequence of edges $a_{ij_1} > 0$, $a_{j_1 j_2} > 0$, ..., $a_{j_m j} > 0$, then node j is said to be reachable from node i. If there is a node which is reachable from every other node in the graph, then the node is said to be globally reachable.

Consider a communication network with n agents, which update their states based on information exchange with their neighbors. Olfati-saber [4,15] studied the following consensus algorithm of multi-agent systems

$$\dot{x}_i(t) = \sum_{j \in N_i} a_{ij}[x_j(t) - x_i(t)], \qquad (1)$$

where $x_i(t) \in R^m$ is the state of agent i, a_{ij} is the link weight between agent i and agent j. Suppose the graph G is strong connected and equilibrium, the consensus of system (1) can be reached, i.e., for all $i \in I$, $x_i(t) \to \alpha$, where

$$\alpha = \sum_{i=1}^{n} x_i(0)/n .$$

The consensus state of system (1) is affected by the initialized values of multi-agent systems. If the initialized values are set at random, the consensus state can not be converged to an expected value. In this paper, we apply pinning control to bring the consensus state of multiple agents to an expected value.

3 Mobile Agent Systems with Pinning Control

Suppose mobile agent systems are composed of n agents with directed connected topology and have a globally reachable node i. Now, we add a pinning control on the node i

$$u = -k(x_i - \alpha_0), \qquad (2)$$

where control gain k>0, α_0 is the expected consensus state of multi-agent system.

Without loss of generality, rearranges the order of the nodes in the network, and let first agent is a globally reachable. Thus, pinning control (2) can be rewritten as

$$u = -k(x_1 - \alpha_0).$$

The pinning controlled network can be described as:

$$\dot{x}_1(t) = \sum_{j \in N_1} a_{ij}[x_j(t) - x_1(t)] - k(x_1 - \alpha_0), \qquad (3)$$

$$\dot{x}_i(t) = \sum_{j \in N_i} a_{ij}[x_j(t) - x_i(t)], \ i = 2,..., n \qquad (4)$$

Lemma 1[16]. Assume the interconnection topology graph of n agents with wireless information networks has a node as a globally reachable node. Then, the matrix L+B has no zero eigenvalues, where L is the Laplacian matrix of the interconnection topology of n agents, and $B = diag(b_i, i \in I)$, b_i is nonnegative real number and $b_{i_0} \neq 0$ for $i_0 \in I$.

Theorem 1. Assume mobile agent systems are composed of n agents with wireless information networks and the interconnection topology graph has a node as a globally reachable node. Then, the multi-agent systems (3-4) with pinning control (2) can asymptotically reach the expected value.

Proof. In order to prove conveniently, suppose the state of agent is one dimension, i.e., $x_i \in R$. Let $\delta x_i = x_i - \alpha_0$, system (3-4) can be changed as:

$$\delta \dot{x}_1(t) = \sum_{j \in N_1} a_{ij}[\delta x_j(t) - \delta x_1(t)] - k\delta x_1, \qquad (5)$$

$$\delta \dot{x}_i(t) = \sum_{j \in N_i} a_{ij}[\delta x_j(t) - \delta x_i(t)], \ i = 2,...,n, \quad (6)$$

Let $\delta x(t) = [\delta x_1, ..., \delta x_n]^T$, system (5-6) can be described as:

$$\delta \dot{x}(t) = -(L+K)\delta x(t), \quad (7)$$

where $K = diag(k_1, k_2, ..., k_n)$, $k_1 = k > 0$, $k_i = 0$ ($i = 2,...,n$). Based on Lemma 1, matrix $L+K$ is positive definition matrix. Let Lyapunov function $V = \delta x^T \delta x$, the derivative of $V(t)$ along the trajectories of (7) gives

$$\dot{V}(t) = -2\delta x^T(L+K)\delta x \le 0.$$

It is easy to see that system (7) is asymptotically stable because of the positive definition matrix $L+K$. i.e., for all $i \in I$,

$$\lim_{t \to \infty} \delta x_i(t) = 0.$$

Then, we have $\lim_{t \to \infty} x_i(t) = \alpha_0$. The proof is completed.

4 Mobile Agent Systems with Communication Delays

Due to the finite speeds of transmission and spreading as well as traffic congestions, there are usually time delays in spreading and communication in reality. Therefore, it is very important to study the delay effect on convergence of consensus protocols. Suppose the communication delays are considered between the agents, the multi-agent systems apply following dynamical equation:

$$\dot{x}_1(t) = \sum_{j \in N_1} a_{ij}[x_j(t - T_{1j}) - x_1(t)] - k(x_1 - \alpha_0), \quad (8)$$

$$\dot{x}_i(t) = \sum_{j \in N_i} a_{ij}[x_j(t - T_{ij}) - x_i(t)], \ i = 2,...,n \quad (9)$$

where T_{ij} represents the communication delay from agent j to agent i.

Theorem 2. Assume mobile agent systems are composed of n agents and the interconnection topology graph has a node as a globally reachable node. Then, the delayed multi-agent systems (8-9) with pinning control (2) can asymptotically reach the expected value.

Proof. Let $\hat{x}_i(t) = x_i(t) - \alpha_0$, submitted into Eq.(8-9), we have,

$$\delta \dot{x}_1(t) = \sum_{j \in N_1} a_{ij}[\delta x_j(t - T_{1j}) - \delta x_1(t)] - k\delta x_1, \quad (10)$$

$$\delta\ddot{x}_i(t) = \sum_{j\in N_i} a_{ij}[\delta x_j(t-T_{ij}) - \delta x_i(t)], i = 2,...,n \qquad (11)$$

Applying the Laplace transform, let $X_i(s) = L(\hat{x}_i(t))$, we obtain

$$sX_1(s) - X_1(0) = \sum_{j\in N_1} a_{ij}[e^{-sT_{1j}}X_j(s) - X_1(s)] - kX_1(s), \qquad (12)$$

$$sX_i(s) - X_i(0) = \sum_{j\in N_i} a_{ij}[e^{-sT_{ij}}X_j(s) - X_i(s)], i = 2,...,n \qquad (13)$$

Let $X(s) = [X_1(s),\cdots,X_n(s)]^T$, we get

$$sX(s) - X(0) = -[L(s) + K]X(s), \qquad (14)$$

where $K = diag(k_1, k_2,...,k_n)$, and $L(s) = [l_{ij}(s)]$ satisfying

$$l_{ij}(s) = \begin{cases} -a_{ij}e^{-sT_{ij}}, & j \in N_i \\ \sum_{j\in N_i} a_{ij}, & j = i \\ 0, & otherwise \end{cases}$$

It is easy to know $L(0) = L$. We get the characteristic equation of the system (14)

$$\det(sI + L(s) + K) = 0. \qquad (15)$$

When $s = 0$, $\det(L(0) + K) = \det(L + K)$. Since L+K is positive definition matrix from Lemma1, s=0 is not the zero of the characteristic equation. Let

$$D(s) = \det(I + (L(s) + K)/s). \qquad (16)$$

If we prove the zeros of D(s) are on the open left complex plane, which is equivalent to all zeros of (15) with negative real parts, the consensus of multi-agent delayed system will be obtained. Let

$$G(s) = (L(s) + K)/s, \qquad (17)$$

and $s = j\omega$, we get

$$G(j\omega) = (L(j\omega) + K)/(j\omega). \qquad (18)$$

Based on the general Nyquist stability criterion, the zeros of D(s) lie on the open left half complex plane, if the trajectory-tracking of $\lambda(G(j\omega))$ does not enclose the point $-1 + j0$ for all $\omega \in R$, where $\lambda(G(j\omega))$ is the eigenvalue of matrix $G(j\omega)$.

We use the Greshgorin disc theorem to estimate the matrix eigenvalue $\lambda(G(j\omega))$, i.e.

$$\lambda(G(j\omega)) \in \bigcup_{i \in I} G_i, \tag{19}$$

where $G_i = \{\xi : \xi \in C, |\xi - (\sum_{k=1}^{n} a_{ik} + k_i)/(j\omega)| \leq \sum_{k=1}^{n} |a_{ik} \frac{e^{-j\omega T_{ik}}}{j\omega}|\}$, C is complex number field. We can obtain by simple calculating

$$G_i = \{\xi : \xi \in C, |\xi - (d_i + k_i)\frac{1}{j\omega}| \leq |\frac{d_i}{j\omega}|\}. \tag{20}$$

where $d_i = \sum_{k=1}^{n} a_{ik}$. The center can be denoted by

$$G_{i0}(j\omega) = j(d_i + k_i)/\omega, \tag{21}$$

The radius of the disc is $d_i / |\omega|$. When the point in the Nyquist plot of $G_{i0}(j\omega)$ changes as $\omega \in R$, the disc G_i changes correspondingly. Since the disc G_1 has none point across the real axis for $i=1$ with $k_1 = k \neq 0$, $-1 + j0$ can not be contained in G_1. Since disc G_i has one point across the real axis at the origin for $i = 2, ..., n$ with $k_i = 0$, $-1 + j0$ can not be contained in G_i. Therefore, we have

$$(-1, j0) \notin \bigcup_{i \in I} G_i.$$

Since $\lambda(G(j\omega)) \in \bigcup_{i \in I} G_i$ by Greshgorin's disc theorem, the track of $\lambda(G(j\omega))$ does not contain the point $-1 + j0$. Theorem 1 is proved.

5 Computer Simulations

Consider mobile agent systems of 4 agents with the directed connect topology graph in Fig. 1. It is obvious that the agent 1 is globally reachable. Let the expected consensus state is a virtual agent 0. Suppose the weights of the directed edges are a12=0.4, a23=0.5, a31=0.6, a41=0.7, based on the network topology, we choose the pinning control $u = -k(x_1 - \alpha_0)$ where k=1 so that the agents converge to the expected value $\alpha_0 = 2$ asymptotically.

Case 1. Firstly, we consider the multi-agent systems without communication delays. With the initial states generated randomly, the agents in the system asymptotically converge to the expected state (Fig.2) with the pinning control. The result of the simulation consists with that of Theorem 1, and the consensus of multi-agent systems is reached.

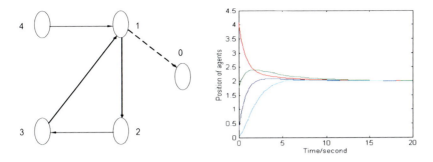

Fig. 1. Topology of Multi-agent systems   **Fig. 2.** Systems without communication delays

Case 2. Secondly, we consider the multi-agent systems with communication delays. Let the communication delays be $T_{12} = 0.17$ s, $T_{23} = 0.15$ s, $T_{31} = 0.13$ s, $T_{41} = 0.11$ s. With the parameters set as Case 1 and the initial states generated randomly, the agents in the system asymptotically converge to the expected state (Fig.3, (a)) with the pinning control. The result of the simulation consists with that of Theorem 2, and the consensus of multi-agent systems is reached.

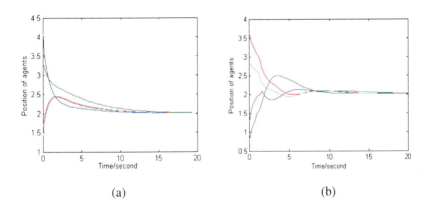

(a)   (b)

Fig. 3. Mobile agent systems with communication delays

Case 3. The effects of communication delays on the convergence of the system. With the parameters set as Case 1 and the initial states generated randomly, the pinning control is added the agent 1. Let $T_{12} = 1.7$ s, $T_{23} = 1.5$ s, $T_{31} = 1.3$ s, $T_{41} = 1.1$ s, the agents in the system asymptotically converge to the expected state in Fig.3 (b). Although the communication delays have effects on the convergence of the systems, the expected state can be reached asymptotically in multi-agent systems.

6 Conclusion

In this paper, consensus of mobile multi-agents system with pinning control is studied. Suppose multi-agent systems consist of n agents with wireless information networks, there is a static directed interconnected graph with a globally reachable node. Based on the Lypunov stability theory, the consensus of multi-agent systems without pinning control is analyzed. By Applying generalized Nyquist criterion and Greshgorin's disc theorem, the consensus of delayed algorithm with pinning control is analyzed, and a decentralized convergence condition is obtained to ensure the consensus of multi-agent systems. This consensus condition is independent of communication delays. Finally, computer simulations are used to show the validity of the result. Future work will consider the effects of time-delays on the consensus algorithms with time-variable position of the expected state and switching interaction graphs.

References

1. Jadbabaie, A., Lin, J., Morse, A.: Coordination of groups of mobile autonomous agents using nearest neighbor rules. IEEE Trans. Automatic Control 48(6), 998–1001 (2003)
2. Lin, Z., Broucker, M., Francis, B.: Local control strategies for groups of mobile autonomous agents. IEEE Trans. Automatic Control 49(4), 622–629 (2004)
3. Fax, A., Murray, R.M.: Information flow and cooperative control of vehicle formation. IEEE Trans. Automatic Control 49(9), 1453–1464 (2004)
4. Olfati-Saber, R., Murray, R.M.: Consensus problem in networks of agents with switching topology and time-delays. IEEE Trans. Automatic Control 49(9), 1520–1533 (2004)
5. Mu, S., Chu, T., Wang, L.: Coordinated collective motion in a motile particle group with a leader. Physica A 351, 211–226 (2005)
6. Wang, W., Slotine, J.J.E.: On partial contraction analysis for coupled nonlinear oscillators. Biol. Cybern. 92(1), 38–53 (2005)
7. Olfati-Saber, R.: Flocking for multi-agent dynamic systems: algorithms and theory. IEEE Trans. Automatic Control 51(3), 401–420 (2006)
8. Hong, Y., Hu, J., Gao, L.: Tracking control for multi-agent consensus with an activeleader and variable topology. Automatica 42(7), 1177–1182 (2006)
9. Yang, Z., Liu, Z., Chen, Z., Yuan, Z.: Tracking Control for Multi-agent Consensus with an Active Leader and Directed Topology. In: Proceeding of the 7th World Congress on Intelligent Control and Automation, Chongqing, China, June 25-27, pp. 1037–1041 (2008)
10. Grigoriev, R.O., Cross, M.C., Schuster, H.G.: Pinning Control of spatiotemporal chaos. Physical Review Letters 79(15), 2795–2798 (1997)
11. Wang, X., Chen, G.: Pinning control of scal-free dynamical networks. Physica A 310(3-4), 521–531 (2002)

12. Li, X., Wang, X., Chen, G.: Pinning a complex dynamical network to its equilibrium. IEEE Trans. Circuits and Systems I 51(12), 2074–2087 (2004)
13. Zhou, J., Lu, J., Lv, J.: Pinning adaptive synchronization of a general complex dynamical network. Automatica 44, 996–1003 (2008)
14. Yu, W., Chen, G., Lv, J.: On pinning synchronization of complex dynamical networks. Automatica 45, 429–435 (2009)
15. Olfati-Saber, R., Fax, J.A., Murray, R.M.: Consensus and cooperation in networked Multi-agent systems. Proceedings of the IEEE 95(1), 215–233 (2007)
16. Wei, R., Beard, R.W., Atkins, E.: Information Consensus in Multivehicle Cooperative Control: Collective Group Behavior through Local Interaction. IEEE Control Systems Magazine 27(2), 71–82 (2007)

Dynamic Evolution in System Modeling of Knowledge-Intensive Business Services' Organizational Inertia

Min Liu[1,2]

[1] Glorious Sun School of Business and Management,
DongHua University,
Shanghai, China, 200051
[2] School of International Business,
Tianjin Foreign Studies University,
Tianjin, China, 300204
7718cathy@sohu.com

Abstract. This research analyzes inertia as a metaphor to explain organizational change and innovation by adopting system dynamics modeling of the organizational inertia of knowledge-intensive business services (KIBS). KIBS are regarded as the main study object in virtue of more uncertainties and complicatedly ever-changing environments and their winning inertia for the survival is a buzzword. This study aims at making out the specific component of organizational inertia of KIBS, which have the prominently different characteristics from the traditional industries. It highlights the level and rate model to illustrate the dynamic evolutionary processes of organizational inertia and works out the input and output graphs of organizational inertia in Vensim that demands the relevant system modeling foundations in the empirical practice.

Keywords: Organizational Inertia, System Modeling, Knowledge-intensive Business Services, Level and Rate Model, Input and Output Model.

1 Introduction

The relationship among the countries is becoming much stronger with the help of modern transportation and advanced communication technologies in the 21st century. And the world's economy is becoming a network in which each economic event must be impacted by the others or would influence on others. This is a nonliner, multi-feedback complex system. Under this background, the research aims at changing its previous content and method by considering the complex science.

Successful innovation of KIBS can be the chief determinant for the survival of an enterprise in a knowledge-based economy. Enterprises are encouraged to adopt novel ideas while reforming old operational procedures and creating new ones (Nonaka, 1994). Hence, to weaken the negative impact of organizational inertia,

KIBS should seek ways to strengthen the research and development of knowledge and comprehending the essence of inertia in modeling.

This study frames inertia in organizations with respect to the laws of motion (Newton, 1995/1687). The fresh insights are stressed to use the principles of inertia in physics to the study of KIBS innovation management. The inertia perspective has been imported into organizational theory from physics study and this view plays an essential role in solving hinders of organizational innovation (Hannan and Freeman, 1984).

2 Theoretical Framework

2.1 Conceptional Explanation of KIBS

Generally speaking, KIBS are concerned with providing knowledge-intensive inputs to business process of other enterprises to the full extent, including private and public sector clients. Miles et al. (2005) explained the principal characteristics of KIBS: they rely heavily on professional knowledge; they either are themselves primary sources of information and knowledge or they use knowledge to produce intermediate services for their clients' production processes; they are of competitive importance and supplied primarily to business. Miles defined KIBS more precisely as "services that involved economic activities which are intended to result in the creation, accumulation or dissemination of knowledge" and classified the two kinds of KIBS such as "traditional professional services" (P-KIBS) and "new technology-based services" (T-KIBS).

"In many ways, what they are doing is locating, developing, combining and applying various types of generic knowledge about technologies and application to the local and specific problems, issues and contexts of their clients. They are involved in a process of fusing generic and local knowledge together (Miles, 2005)". This sentence summarizes more than a decade of research on the topic of knowledge business intensive services (KIBS), which have recently become an important subject of study, both conceptually and empirically.

One essential point derived from the analysis of definitions is that KIBS refer to service firms that are characterized by high knowledge intensity and services to other firms and organizations, services which are predominantly non-routine (Muller, 2009).

All contemporary literature review of KIBS contains at least three key elements of the definition:

- The term "business services" or specialized services, which are demanded by firms and public organizations and are not produced for private consumption (Strambach, 2001);
- The phrase "knowledge intensive" which can be indicated either in terms of labor qualification or in terms of the conditions for transactions between the service provider and the service user or procurer (Hauknes, 1999);

- The term "knowledge intensive firms" which refers to firms that undertake complex operations of an intellectual nature where human capital is the dominant factor (Alvesson, 1995).

Table 1. KIBS Sectors and Sub-sectors

NACE	DESCRIPTION
72	Computer and related activities
721	Hardware consultancy
722	Software consultancy and supply
723	Data processing
724	Database activities
725	Maintenance and repair of office, accounting and computing machinery
726	Other computer-related activities
73	Research and development
7310	Research and experimental development in natural sciences and engineering
7320	Research and experimental development in social sciences and humanities
74	Other business activities
741	Legal, accounting, book-keeping and auditing activities; tax consultancy; market research and public opinion polling; business and management consultancy; holdings
742	Architectural and engineering activities and related technical consultancy
743	Technical testing and analysis
744	Advertising
7484	Other business activities

To sum up, there is no standard approach and universally accepted explanation of KIBS. However, a certain consensus exists about the branches and organizations that comprise the KIBS sector. NACE (a European Classification of Economic Activities) provides the nomenclature with prevalent power to identify KIBS in Europe. Table I displays that each category contains several subcategories in detail.

2.2 Organizational Inertia

In physics, the principle of inertia states that objects continue in a state of rest or uniform motion unless acted upon by forces. Unless interrupted, an object's motion is subject to physical constraints and objects will move in the predicted trajectory. Three implications are emphasized to be derived from this statement:

- Prediction is on the basis of the comprehension of the previous trajectory or trend which can be followed according to the inertia;
- External force plays an important role in changing the previous trajectory;
- Enforcement of change from the outside force is meaningful but not instinctive.

A large body of diverse literature and research dedicate the apprehension of organizational inertia. The concept of organizational inertia is very broad; consequently, researchers and scholars have looked at it from a variety of perspectives. For example, Hannan and Freeman (1984), in a population ecology perspective, proposed the structural inertia as a starting point for this discussion. They claimed that "high levels of structural inertia in organizational populations can be explained as an outcome of an ecological-evolutionary process" (1984). The theory underlines reliability and accountability to be the primary sources of survival advantage for the complex organizations. The frame of organizations are as the studied object and it describes that structures of organizations have high level of inertia pressure if the speed of reorganization is much lower than the rate at which environmental conditions change. Thus the concept of inertia refers to a correspondence between the behavioral capabilities of a class of organizations and their environments (Hannan & Freeman, 1984).

In the context of strategic change, Huff et al. (1992) defined the inertia as an "overarching concept that encompasses personal commitments, financial investment sand institutional mechanisms supporting the current ways of doing things... Inertia describes the tendency to remain with the status quo and the resistance to strategic renewal outside the frame of current strategy". Everything stemming from past experience and knowledge without revision and updating would imply predictable management behavior and problem-solving strategy of an enterprises (Liao, 2002).

The theory of knowledge inertia proposed by Liao in 2002 and then was tested empirically in 2008, which he established the constructs of knowledge inertia using principal analysis and examines the relationships between knowledge inertia, organizational learning and organizational innovation with structural equation modeling approach.

Shull (1999) discussed a hot issue that why a lot of famous fig firms behaved more badly and the answer he gave is action inertia constrains the further development when changes happened on the occasion of previous successful pattern of implement in management activities.

An organization is continuously subjected to the need for change in its strategy and structure to align with a dynamic environment. Otherwise, routines and established patterns of thinking and activities can lead to organizational inertia, producing resistance to change (Hannan & Freeman, 1984; Nelson & Winter, 1982).

Considered sunk costs, KIBS are changed slowly although their structures are not of a deadweight quality unlike manufacturing industries. There are still large lags in response to environmental changes and to attempts by decision makers to implement change (Hannan & Freeman, 1984). Since lags can be longer than typical environmental fluctuations and longer than the attention spans of decision makers and outside authorities, inertia often blocks organizational innovation. Structural inertia is not constant over the organizational life-course, but varies systematically with age and size.

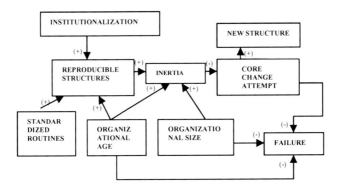

Fig. 1. A Basic View of Structural Inertia Theory (Dawn Kelly, 1991)

Dawn Kelly (1991) proposed Fig. 1 to demonstrate a basic view of structural inertia theory. It is the first time that the inertia theory has been demonstrated in light of the dynamic model including positive and negative feedbacks. However, there are so many limitations in this model. Some researchers regard organizational type as a moderator variable. Organizational type is a determinant factor in modern knowledge economic tide to reveal the inimitable features of organizational change. For example, KIBS are totally different from other types of industries, especially the traditional manufacturing industry.

Erik Larsen et al. (2002) redesign Kelly's model in Fig. 2 to demonstrate the ecological model of organizational inertia and change.

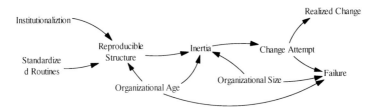

Fig. 2. Ecological Model of Organizational Inertia and Chang (Erik Larsen, 2002)

Erik Larsen's model is contributed to his further investigation of the relationship of inertia, capabilities and changes. It makes much progress to design an incommensurable empirical models but it does not pinpoint the specific dimensions of inertia in light of organizational innovation. This paper aims to provide a tentative study of inertia elements and system modeling to exert the inertia theory investigation.

3 Research Method

System dynamics (SD) theory is proposed by Jay W. Forrester in 1956. It is remarkable as "the complexity, nonlinearity, and feedback loop structures that are inherent in social and physical systems" (Forrester, 1994). Assumptions about how organizational inertia and organizational innovations illustrate the evolutionary dynamics of corporations and other institutions are frequently taken as primitive terms in empirical studies. Limited researches are available that examine these assumptions directly by representing inertia and its factors as fundamentally interdependent modeling processes.

KIBS regard the knowledge and knowledge workers as the key competitive advantages in response to more uncertainty and more complicated ever-changing environment. As described above, although the core concerns addressed by these mutually contentious perspectives on organizational innovation are a little bit different, there also exist domains of substantial overlap that could be practically investigated.

Against the background of this general discussion, accumulation processes of organizational inertia are developed in the context of level and rate representation of a model of inertia inspired by ecological theories. This paper uses system dynamics (SD) to build the model, test its internal accumulator variability, and explore the full dynamic implications of a theory with delay that relates to dynamic evolution of organizational inertia.

Fig. 3. Inertia Level and Rate Modeling

As regards Fig.3, organizational inertia (I) is represented as a stock (or an accumulator) variable that integrates the corresponding net flow, defined as the balance between increase in inertia ($I^{(+)}$) and decrease in inertia ($I^{(-)}$).

$$I_{t_i} = \int_{t_0}^{t_i} [I^{(+)}(t_i) - I^{(-)}(t_0)] dt + I(t_0) \tag{1}$$

As noted above, the initial value of inertia is 0 and its maximum is 1000 in Fig.4. The first step is to identify and represent dynamic elements that are telegraphed through the described above and the second objective is to explore conceptual connections between core features of inertia theories of organizational innovation that are typically believed to form incommensurable empirical models. In this regard, innovation is considered to change for enterprises.

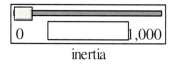

Fig. 4. Inertia Input Graph

In view of KIBS' special elements, such as knowledge worker, flexible organization, special business cultures, and more determinant factors should be designed. The diagram identifies the nonlinear and complicated dynamic system of organizational inertia.

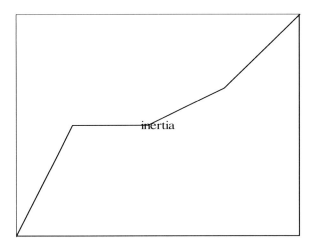

Fig. 5. Inertia Output Graph

Fig. 5 illustrates the output graph of inertia which is so strong at beginning and at last because its functions are positive and negative respectively at these two periods of time in organizational innovation and in the middle period of time inertia remains like a line because it meets a balance. In evolutionary perspective there is stronger inertia at some period of the life-cycle and sometimes weaker inertia. Thus, what we are eager to do is that no matter what kind of inertia will be expected not to block the process of organizational innovation.

4 Implications and Conclusions

It is the first time to make KIBS as the main research object which is always regarded as the modulator of inertia in the previous study; this paper has established a system dynamic model of KIBS' organizational inertia, namely the level and rate process; it enlarges the apprehension of inertia from its original theoretical foundation (structural inertia) to organizational inertia by considering features of KIBS with uncertainty and ever-changing knowledge. Inertia exists not only in structures, but also in human cognition and behavior in the process of learning knowledge and innovation activities.

The dynamic evolutionary study of organizations directly deals with organizational inertia and change. It reformulates some of the central assumptions and propositions in terms of System Dynamics and is combined the features of KIBS and inertia. By the input and output graphs, it is available to explore more specific dynamic processes implicit in the original formulation to test its dynamic consistency and the link among inertia level variables.

The current modeling has two limitations: the first one is that absence of necessary data which makes the dynamic simulation model unavailable; the second is that some feedback elements and their corresponding functional relations as well should be explored more elaborately, which is the next step in the further study.

References

1. Nonaka: A dynamic theory of organizational knowledge creation. J. Organization Science 15(1), 14–37 (1994)
2. Newton: The Principia, Translated by Motte, A. Prometheus Books, Amherst (1995) Original work published in 1687
3. Hannan, M.T., Freeman, J.: Structural Inertia and Organizational Change. J. American Sociological Review. 49, 149–164 (1984)
4. Miles, I.: Knowledge Intensive Business Services: Prospects and Policies. J. Foresight – The Journal of Future Studies, Strategic Thinking and Policy 7, 39–63 (2005)
5. Muller, E., Doloreux, D.: What we should Know about Knowledge-intensive Business Services. J. Technology in Society 31, 64–72 (2009)
6. Strambach, S.: Innovation Processes and the Role of Knowledge-intensive Business Services. In: Innovation Networks: Concepts and Challenges in the European Perspectives. Physica-Verlag, Heidelberg (2001)
7. Hauknes, J.: Knowledge Intensive Services – What is Their Role. OECD Forum on Realizing the Potential of the Service Economy, Paris (1999)
8. Alvesson, M.: Management of Knowledge-Intensive Companies. de Gruyter, Berlin and New York (1995)
9. Huff, J.O., Huff, A.S., Thomas, H.: Strategic Renewal and the Interaction of Cumulative Stress and Inertia. J. Strategic Management Journal 13 (1992)
10. Liao, S.H., Fei, W.C., Liu, C.T.: Relationships Between Knowledge Inertia, Organizational Learning and Organization Innovation. J. Technovation 28, 183–195 (2008)
11. Shull: Why does a Good Firm Become so Bad. J. Foreign Scial Science 2 (1999)
12. Nelson, R., Winter, S.: An Evolutionary Theory of Economic Change. Belknap, Cambridge (1982)
13. Forrester, J.W.: System Dynamics, Systems Thainking, and Soft OR. System Dyanmics Review 10(2), 1 (1994)
14. Kelly, D., Amburgey, T.: Organizational Inertia and Momentum: a Dynamic Model of Strategic Change. J. Academy of Management Journal 34, 591–612 (1991)
15. Larsen, E., Lomi, A.: Representing Change: a System Model of Organizational Inertia and Capabilities as Dynamic Accumulation Processes. J. Simulation Modelling Proactice and Theory 10, 271–296 (2002)

Research on Java Imaging Technology and Its Programming Framework

M.A. Weifeng[1] and Mao Keji[2]

[1] School of Information and Electronic Engineering,
Zhejiang University of Science and Technology
310023 Hangzhou, China
mawf@lreis.ac.cn
[2] College of Information Engineering, Zhejiang University of Technology
310023 Hangzhou, China

Abstract. With the rapid increase on requirement of the cross-platform distributed imaging technology, and the common, stable imaging lib, SUN provided a complete solution based Java platform technology. This paper analyzed the evolution of java imaging technology such as Java AWT, Java 2D and Java Advanced Imaging. Especially, it details the image operators, the core classes and the programming framework of Java Advanced Imaging. At last, the paper gave out two imaging examples included the programming codes and the result images, which can be used in project.

Keywords: Java Imaging, Java AWT, Java 2D, JAI.

1 Introduction

For the characteristics of storing conveniently, saving a long time, transporting easily on network and containing rich information, digital image has occupied an important position on the scientific research, engineering applications and modern life, and it has been widely applied to various fields. In recent years, the progress of digital image acquisition technology and the rapid increase in demand for imaging applications, greatly promoted the development of digital imaging technology.

At present, most of the image processing systems implemented with C/C++, these systems are very difficult to upgrade. The requirement of the cross-platform distributed image processing technology, and the common, stable imaging lib, are proposed to us. Fortunately, SUN attached importance to the image and image processing technology all the time. A complete solution on Java platform-based image processing technology was provided by SUN. This paper focused on the Java Imaging Technology.

2 The Evolution of Imaging in Java

The Java language which developed by SUN supports cross-platform imaging from its formal launch, and it developed through three phases (Table 1): Java AWT, Java 2D API and Java JAI (Java Advanced Imaging) [1].

Table 1. Comparison of three stages of Java imaging technology[1]

Content	Java AWT	Java 2D API	JAI
Model	Push mode	Immediate mode	Pull mode
Major feature	Simple image rendering. Almost no processing function.	Enhanced image rendering. Basic image processing.	Advanced image processing, supporting distributed, scalability, robustness, performance, etc.
Major interface and class	Image ImageProducer ImageConsumer ImageObserver	BufferedImage Raster BufferedImageOp RasterOp	JAI RenderableImage TiledImage ParameterBlock PlanarImage

2.1 Java AWT Technology

Java AWT is the earliest technology that supports graphics and image, it releases with the first version of java in 1991. The earlier version only provides a simple rendering package for general HTML page rendering, for example, render a number of simple graphic images which are line-type or geometric, also, it can support simple bitmap images such as GIF, JPEG and so on through TOOKIT object, but do not provide a number of complex image processing functions.

Java AWT mainly samples the Producers/ Consumer "push" model. Java AWT achieves basic image processing through providing the package of java.awt.image, mainly uses three interfaces and one class: ImageProducer interface, ImageConsumer interface, ImageObserver interface and Image class. In the producer/consumer model of the image, it is easy to appear the following problems:

- **Problem of image data loss**
- **Problem of image data overlap**

Java AWT image processing is not suited to the development of high-performance image processing code. It does not have a persistent image data object. Its model of the image filter is limited, only two kinds of image data format. All these are unable to fit uses' needs.

2.2 Java 2D API

Because of the limitations of the Java AWT and people's demands of improving image processing, Java 2D technology appears. Compared with the Java AWT, Java 2D API retains and extends the Producer/Consumer model, and increase the lasting image data which supported by the memory. It extends the image processing filters, various types of image data formats and the color models, as well as the more complex output device expression. Java 2D is a immediate mode imaging model, six of the classes or interfaces play the key role. Figure 1 describes the organizational structure of the Java 2D image data model[1][2]:

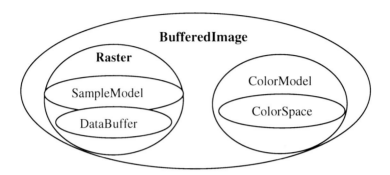

Fig. 1. The image organizational structure of Java 2D

The real-time model imaging of Java 2D is mainly implemented by the BufferedImage class. This class is mainly used to deal with the image of memory, file and URL path, provides image data management, color rendering, etc. The former one implemented by the Raster class, and the latter one implemented by the ColorModel class. At the same time, the Raster class includes the DataBuffer class which used to manage image data and the sample model classes SampleModel which used to interpret the image data value. The operation of the image mainly depends on two interfaces BufferdImageOp and RasterOp. The image processing is encapsulated in a class, this class implements these two interfaces or their interfaces.

3 Java Advanced Imaging

Java AWT and java 2DAPI have considerable limitations in image processing. They can not content the specific users' image processing needs. Therefore, SUN joints many companies and proposes JAI high-level image processing as an extension of the JDK class package which is based on AWT, 2D API. JAI technology has many characteristics such as cross-platform, distributed, device independence, robustness and so on, it has very high performance[1]. JAI provides very powerful image processing API, including image manipulation, image enhancement and image analysis, etc. (see in table2).

Table 2. JAI Imaging Function Summary[1]

Category	Functions
Image Manipulation	Region of interest (ROI) control; Relational operators; Logical operators; Arithmetic operators; Dithering; Clamping pixel values; Band copy
Image Enhancement	Adding borders, Cropping an image, Amplitude rescaling, Histogram equalization, Lookup table modification, Convolution filtering, Median filtering, Frequency domain processing, Pixel point processing, Thresholding (binary contrast enhancement)
Geometric Manipulation	Geometric transformation (Translate, Scale, Rotate, and Affine), Perspective transformation (PerspectiveTransform), Transposing (Transpose), Shearing (Shear), Warping (Warp, WarpAffine, WarpPerspective, WarpPolynomial, WarpGeneralPolynomial, WarpQuadratic, and WarpOpImage)
Image Analysis	Finding the mean value of an image region, Finding the minimum and maximum values in an image (extrema), Producing a histogram of an image, Detecting edges in an image, Performing statistical operations

JAI's program implementation is also based entirely on object-oriented. In order to implement the image processing functions, JAI defined some core classes as follows[1,2,3].

JAI is a final class from the javax.media.jai package. It cannot be instantiated and contains only static methods. It is used to setup operators for execution in the imaging chain. The primary static method of create() returns a RenderedOp encapsulating the operation name, parameter block, and rendering hints. The most common form of this method takes the string name of the operator, such as RenderedOp im1 = JAI.create(opName, paramBlock, renderHints), a list of parameters provided by an instance of a ParameterBlock class, and optional rendering hints provided by an instance of the RenderingHints class.

ParameterBlock class encapsulates all the information about sources and parameters required by a RenderableImageOp, or other classes that process images. It provides a parameter list for imaging used in JAI.Create() method.

RenderableImage is a common interface defined as a rendering-independent image object, so it can be rendered for various specific devices such as screen, printer, etc.. All of the classes implementing RenderableImage interface must have a data buffer of the image and a sample model object, which describes how the pixels are arranged in the image.

PlanarImage class implements the RenderedImage interface, so it can contains the complete information such as the image's layout, sources, properties, etc., and contain or produce image data in form of Rasters. RenderedOp is the subclass of PlanarImage which can be returned by JAI.create() method. Also, it serves as the parent for TiledImage classes.

TiledImage class is a subclass of PlanarImage and implements the WriteableRenderedImage interface to provide manipulation of its buffered image. Moreover, it is very important that TiledImage represents images containing multiple tiles arranged into a grid. The tiles form a regular grid, and the size of tile can be defined by yourself via the constructor method. Because of each tile's independence, large images can be processed with tiles partly and high-efficiency.

4 Programming Framework on JAI

4.1 Graphs Programming and Imaging Steps

In JAI, it provides so many imaging operations and any operation is defined as an object. An operator object is instantiated with zero or more image sources and other parameters that define the operation[1]. Its programming framework likes a graph or chain through linking one operator to another as in figure2.

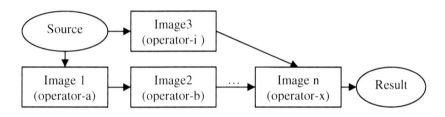

Fig. 2. The imaging graph[1]

In figure2, two or more operators may be strung together, and the above operator (operator-a) can become an image source to the next operator(operator-b).When we want to do an imaging operation within JAI, there are about three steps as follows[1]:

(1) Obtain the source image or images.
(2) Define the imaging graph.
(3) Run and Process the result, such as save the image in a file or display the image on the screen, etc..

4.2 Examples

4.2.1 Image Manipulation

JAI provides many image operators based on pixel manipulation. For example, if we want to make dynamic monitoring of traffic conditions, we can use the same location and make the arithmetic subtraction operation of the current time picture and the background picture. The implemented codes are shown in Figure 3 (a), the result is shown in Figure 3 (b).

```
RenderedImage result;
String filename☐ "test/交通图11.bmp";
RenderedOp image1 = JAI.create("fileload", fileName);
filename☐ "images/data/image2.bmp";
RenderedOp image2 = JAI.create("fileload", fileName);
result=JAI.create("subtract",image1,image2);
```

(a) The codes

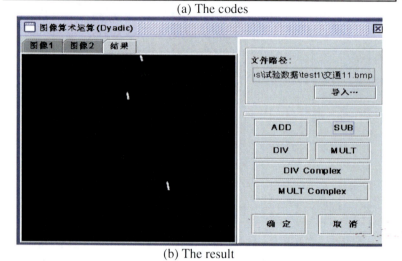

(b) The result

Fig. 3. Image Manipulation

4.2.2 Image Analysis

Edge detection is one of simple and effective image analysis method, and can locate the boundaries of objects within an image in effect. As an edge detector, the GradientMagnitude operation computes the magnite of the image gradient vector in horizontal and vertical directions.

GradientMagnitude operation may be defined as[1]:

$$dst[x][y][b] = \sqrt{(SH(x,y,b))^2 + (SV(x,y,b))^2} \qquad (1)$$

Through correlating the band b of the source image with orthogonal gradient masks, the horizontal and vertical gradient images will be generated as SH(x,y,b) and SV(x,y,b), and be construted to new result image dst[x][y][b].

In JAI, the default is the Sobel Edge Enhancement Masks(in two direcitons): KernelJAI.GRADIENT_MASK_SOBEL_VERTICAL and KernelJAI.GRADIENT_MASK_SOBEL._HORIZONTAL. The implemented codes of an edge detection operation used in Sobel masks, are shown in Figure 4 (a), the result is shown in Figure 4 (b).

```
// Load the image.
PlanarImage im0 = (PlanarImage)JAI.create("fileload", filename);
// Create the two kernels.
KernelJAI kern_h = KernelJAI.GRADIENT_MASK_SOBEL_HORIZONTAL
KernelJAI kern_v = KernelJAI.GRADIENT_MASK_SOBEL_VERTICAL.
// Create the Gradient operation.
PlanarImage im1 = (PlanarImage)JAI.create("gradientmagnitude", im0, kern_h, kern_v);
```

(a) The codes

(b) The result

Fig. 4. Edge detection used in Sobel masks

5 Conclusion

With the development of network technology, it enabled digital imaging applications over on the network environment. The paper discusses the Java-based image processing technology, and this technology is very suitable for digital image applications under network environments. It has broad prospects. In the future works, we will focus on research of web image processing, and do some related experiments.

Acknowledgements. The Project Supported by Zhejiang Provincial Natural Science Foundation of China (Y1090387).

References

1. Sun Microsystems. Programming in Java Advanced Imaging, Release 1.0.1, Palo Alto, CA (November 1999)
2. Santos, R.: Java Advanced Imaging API: A Tutorial. RITA XI(1) (2004), http://www.inf.ufrgs.br/~revista/docs/rita11/rita_v11_n1_p93a124.pdf
3. Sun Microsystems. JAI API, http://java.sun.com/products/java-media/jai/forDevelopers/jai-apidocs/index.html
4. Weifeng, M., Keji, M.: Development and Application of Online Image Processing System based on Applet and JAI. In: ESIAT 2009, July 2009, pp. 382–385 (2009)

The Performance of SFN in Multi-Media System

Zhang Naiqian and Jin Libiao

Information Engineering Department, Communication University of China
Beijing, China, 100024
{jlb,ddznq}@cuc.edu.cn

Abstract. This paper describes the principles of SFN (Signal Frequency Network) in details and analyses the affection of time-delay of multi-transmitters in SFN. This paper also presents the insufficiency of transmitters' maximal distance designed in traditional literatures based on the transmission circumstance of big city. At last, some simulations are made to validate the results.

Keywords: SFN, OFDM, TMMB, CMMB.

1 Introduction

OFDM (Orthogonal Frequency Division Multiplexing) system has been fully used in digital TV, including TMMB (Terrestrial Mobile Multimedia Broadcasting) and CMMB (China Mobile Multimedia Broadcasting). The using of OFDM transmission system makes it possible to build SFN. SFN means that all the transmitters in the network work on the same frequency (block) and transmit same bits in the same time. SFN brings the advantages, such as saving transmission power and frequency resource, due to robustness to the severe multi-path fading[1][2]. There are also some key problems in SFN, such as transmitters' position and transmission power adjustment. These problems always determine the performance of the SFN. So, in this paper, we study on these problems and make some simulations to prove the results

2 Theory of SFN

In traditional single carrier system, transmitters, which beside each other, using the same frequency will lead to a severe co-channel interference. But the using of OFDM transmission method makes the SFN become possible. Differing from single carrier system; OFDM is a multi-carrier transmission system. It uses many sub-carriers, which are orthogonal to each other, to form a broad band system. For example, in the 1.7MHz mode of TMMB, the OFDM system is made up from 1536 sub-carriers. These huge numbers of sub-carriers are generated by IFFT as following formula[3]:

$$s_i = \frac{1}{\sqrt{N}} \sum_{k=0}^{N-1} S_k \cdot e^{j2\pi \frac{k}{N} i} \qquad i = 0,1,\cdots,N-1 \qquad (1)$$

where S_k is the QAM symbol transmitted on the kth sub-carrier, N is the number of IFFT point, and s_i is the OFDM signal after IFFT. At the receiver end, we can use FFT to demodulate the OFDM signals. From (1) we can see that OFDM makes a parallel transmission of signals in frequency domain. Thus, the duration of a single symbol will be greatly enlarged. For example, the symbol duration is 1536 times of single carrier system in the same transmission speed. The longer duration means more robustness in multi-path transmission.

For one transmitter in SFN, the signals from other transmitters are just like the multi-path signals. These signals meeting at the receiver antenna will cause ISI (Inter-Symbol Interference) which is harmful to the signal receiving. And in the system which symbol duration is very short, ISI can damage the signals severely. But in OFDM system, much longer symbol duration can resolve this problem perfectly. So, in SFN, if all the distances of transmitters are not very far which can generate large time-delay, the damage of ISI will be limited in a reasonable scope and the signals still can be demodulated accurately.

In addition, a guard time is added in the front of each OFDM symbol in order to enhance the resistance of multi-path fading even more greatly. The guard time is filled with CP (Cyclic Prefix), which is to repeat the last part of an OFDM signal which has the same duration as guard time in the front of this OFDM signal. The reason of using CP is to maintain the orthogonality of sub-carriers. The duration of CP is determined by the max time-delay of the transmission system. For the transmitters in SFN, the ISI will never happen if the distances between each other are within the scope that the guard time provides. Suppose the SFN is composed of L+1 transmitters, so if one of them is considered as main transmitter which provide most energy of the received signals of a receiver, then other L transmitters will be considered as vice transmitter. And the signals come from vice transmitters are just like multi-path signals to the main signals. For convenience, we ignore the Doppler Effect in SFN then we get following formula[4]:

$$r(t) = \sqrt{1-p^2}\, s(t) + \frac{p}{\sqrt{L}} \sum_{l=1}^{L} e^{j\theta_l} s(t-\tau_l) \qquad (2)$$

where p is the relative power of vice transmitters to the main transmitter, θ_l is the carrier phase of lth vice transmitter and τ_l is the time delay of the lth vice transmitter. When all the τ_l are less than the guard time, each OFDM symbol is only affected by itself. Because of the orthogonality, changing one of the sub-carriers gives no affection on other sub-carriers. So, we can analyze the whole OFDM symbol from one sub-carrier. We can change formula (2) to formula (3) and deduct following:

$$\sqrt{1-p^2}\cos\omega_k t + \frac{p}{\sqrt{L}}\sum_{l=1}^{L} S_k \cos\omega_k (t-\tau_l+\theta_l) \qquad (3)$$

$$=\sqrt{1-p^2}\cos\omega_k t + \frac{p}{\sqrt{L}}\sum_{l=1}^{L} S_k \cos[\omega_k t-\omega_k(\tau_l-\theta_l)]$$

$$=\sqrt{1-p^2}\cos\omega_k t + \frac{p}{\sqrt{L}}\sum_{l=1}^{L} S_k [\cos\omega_k t \cdot \cos\omega_k(\tau_l-\theta_l)]$$

$$+\frac{p}{\sqrt{L}}\sum_{l=1}^{L} S_k [\sin\omega_k t \cdot \sin\omega_k(\tau_l-\theta_l)]$$

$$=\cos\omega_k t \cdot \left[\sqrt{1+p^2}+\frac{p}{\sqrt{L}}\sum_{l=1}^{L} S_k \cos\omega_k(\tau_l-\theta_l)\right]$$

$$+\sin\omega_k t \cdot \left[\frac{p}{\sqrt{L}}\sum_{l=1}^{L} S_k \sin\omega_k(\tau_l-\theta_l)\right] \qquad (4)$$

For a certain receiving position in SFN, the time-delay (τ_l) and carrier phase (θ_l) of each vice transmitter are certain values. So, now we make following definitions:

$$a=\sqrt{1+p^2}+\frac{p}{\sqrt{L}}\sum_{l=1}^{L} S_k \cos\omega_k(\tau_l-\theta_l)$$

$$b=\frac{p}{\sqrt{L}}\sum_{l=1}^{L} S_k \sin\omega_k(\tau_l-\theta_l)$$

and $c=\frac{b}{a}$,

so (4) and be written as:

$$\sqrt{a^2+b^2}\cdot\sin(\omega_k t+c) \qquad (5)$$

From (5) we can see that all the signals from vice transmitters can be seen as a fixed phase shift to the signals from main transmitter. In the practical system such as TMMB and CMMB, the symbol of each sub-carrier always is modulated as differential modulation such as differential PSK. So the shift of phase is easy to remove. Thus the signals will never be interfered when the distances of all the transmitters are in the scope which is prescribed by guard time. Further more, because the signals meet at receive antenna are all themselves, so the signals in SFN can get a extra network gain, thus can save transmission power remarkably. Experimentation shows that transmission power can be 10dB lower[5] in SFN than in MFN (Multiple Frequency Network). But this is only happened when the distances of all the transmitters are in the certain scope, otherwise the different OFDM symbols will be overlapped together which causes the co-channel interference and even ISI in some severe situations. Although the symbol duration of OFDM is much longer than single carrier system and the resistance to multi-path is much stronger, co-channel interference is still happened when max time-delay is longer than guard time and the quality of the signals is getting worse when time-delay becomes longer. So, when we build a SFN, we must carefully design the distances between every two transmitters based on the duration of guard time. Keep the max distance within the value which equal to max time-delay plus radio transmission speed. This is the basic design criterion of SFN.

3 Performance of SFN

In all kinds of wireless multi-media transmission system, SFN mode is always defined. In traditional documents, the max distance between two transmitters is usually defined as max time-delay plus radio transmission speed. But this max distance can not always ensure the accurate receiving. Then we analyze the performance of SFN by simulation.

The simulation bases on the 8MHz system of CMMB[6] which adopts 16QAM on each sub-carrier. For the sake of describing the affection of the transmitters' distance more efficiently, we have not added the channel noise in the simulation. We assume that there are four transmitters in the SFN, one is the main transmitter, and the other are the vice transmitters. All the four transmitters sent same bit in the same time. For the convenience of comparing, all the results come from a same series of original signals. Figure 1 shows the signal constellation of one sub-carrier when the distances are shorter than the value gives from guard time. From the figure, we can see that the signals can be demodulated perfectly although there is a little constellation extension. We can recover the 16QAM constellation with signal standardization.

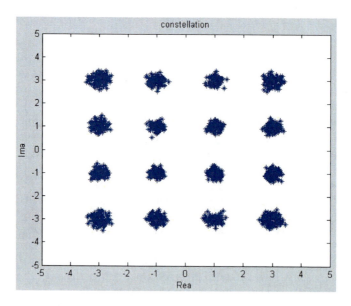

Fig. 1. Constellation when transmitters' distances are within the guard time restriction

When the distances become longer than the guard time restriction, ISI will be happened. This is also called co-channel interference in traditional single carrier system. Figure 2 describes this situation. In the simulation of figure 2, the distances between main transmitter to each vice transmitter is longer than the restriction of guard time. The max time-delay is as much as guard time pluses 10% of the OFDM symbol duration. In this case, we can see the constellation extension obviously. And some signals even can not be modulated. This causes the rapidly increasing of BER (Bit Error Rate). And this situation becomes worse when the distances become longer.

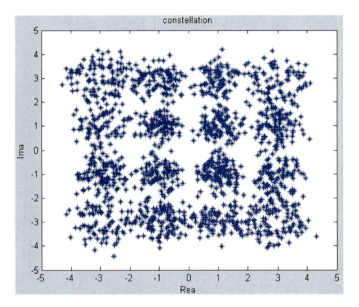

Fig. 2. Constellation when transmitters' distances are beyond the guard time restriction

However, figure 2 is made under the situation that the channel noise and the multi-path transmission are ignored. This is far from the real transmission circumstance of multi-media broadcasting. Currently, mobile multi-media broadcasting systems are usually applied in big city which has numbers of tall buildings. So the situation of multi-path transmission is much severe. Even some place can not see the transmitter antenna directly because of the shelters. In this circumstance, the existing of multi-path transmission means to extend the time-delay of the system. So, the ISI still emerges even if the designed distances of transmitters are within the restriction of guard time. The following picture shows this situation. In this simulation, we suppose that the max distance between main and vice transmitters is equal to the guard time restriction. And each LOS (Line Of Sight) signal has two multi-paths which max time-delay is equal to the 10% of OFDM symbol duration.

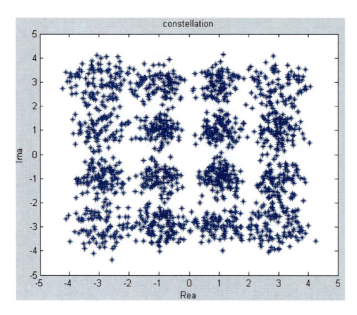

Fig. 3. Constellation under multi-path circumstance when transmitters' max distance is equal to the guard time restriction

Comparing to figure 2, we find that the constellation of figure 3 does not become better. So, when we design a SFN, guard time shall not be considered as the only criterion. The max distance shall be shorter than the restriction of guard time, although more transmitters will cost more investment. The max distance must contain some redundancy for the multi-path transmission. In practical application, the power of most multi-path signals is very low that can not disturb the main signals severely. But for the wireless communication in such complex circumstance as big city, especially for the mobile multi-media broadcasting, we should always consider about the worst situation.

In traditional broadcasting system, when the signals are not good enough for receiving, people usually enhance the transmission power. But this is not always beneficial to receiving in SFN. In figure 4, we maintain the transmitters' position and transmission circumstance just the same as figure 3. Then we raise the power of the vice transmitters and the constellation shows as following:

After carefully comparing, we find that the constellation in figure 4 is a little worse than in figure 3. This illuminates that raising transmission power only can make signals even worse when the distances of transmitters are beyond the guard time restriction or the max time-delay of the SFN system is longer than the guard time.

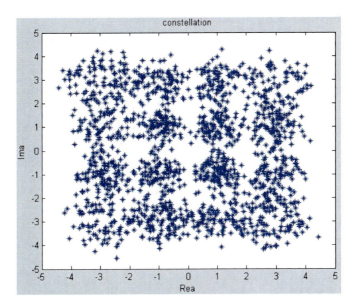

Fig. 4. Constellation after raising the power of vice transmitters when the max time-delay is longer than the guard time

3 Conclusions

This paper analyzes the performance of SFN in detail and provides some simulations on signal constellation based on the CMMB system. The simulation results show that SFN can work well and even provide network gain when the distances of transmitters are within the restriction of guard time. But the max distance should not be considered only by the value that guard time plus radio transmission speed. There must be some redundancy in the design for the multi-path transmission in the network. And when ISI, which cause a severe signal quality decline, happens, raising transmission power may cause the constellation extension and a even higher BER.

References

1. Hikmit, S., Georges, K., Jsabelle, J.: Transmission techniques for digital terrestrial TV broadcasting. IEEE Comm. Magazine, 100–109 (February 1995)
2. Zou, W.Y., Wu, Y.: COFDM: an overview. IEEE Trans., On Broadcasting 41, 1–8 (1995)
3. Yumin, W.: Application and Key Problem on OFDM. China Machine Press, Beijing (2007)
4. Tao, H., Chaowei, Y.: Technology and Application of MIMO. China Machine Press, Beijing (2007)
5. Dong, L.: Digital Sound Broadcasting. Beijing Broadcasting Institute Press, Beijing (2001)
6. GY/T 200.1-2006, Mobile Multimedia Broadcasting Part I: Framing Structure, Channel Coding and Modulation for Broadcasting Channel, The State Administration of Radio Film and Television (November 2006)

A Domain-Oriented Goal Elaborate Method

Yonghua Li and Yingjie Wu

Computer school, Wuhan University of Technology, Wuhan, China
liyonghua@whut.edu.cn

Abstract. Most traditional methods elaborate goals by answering "HOW" and "WHY" questions. There may be some defects in elaborate results by using traditional methods because different analyzers usually have different views on the goals and turn to different elaborate results. This article quotes domain rules to instruct goals elaborate process upon the traditional methods. Analyzers' "HOW" and "WHY" questions in turn are based on domain rules, and the sequence of questions is determined by the relationship between rules and goals. Analyzers answer such questions which can instruct the elaborate process. In elaborate process, analyzers should make the most of formal goal refinement patterns, because those patterns can reduce human's influence to the elaborate results.

Keywords: We would like to encourage you to list your keywords in this section.

1 Introduction

Requirements engineering is the branch of software engineering concerned with the real-world goals for, functions of, and constraints on software systems. It is also concerned with the relationship of these factors to precise specifications of software behavior, and to their evolution over time and across software families.

Goal-oriented requirements engineering is concerned with the identification of goals to be achieved by the envisioned system, the refinement of such goals and their operationalization into specifications of services and constraints, and the assignment of responsibilities for the resulting requirements to agents such as humans, devices and software [1,2]. It is also help to the evolution of requirements engineering [3]. To sum up, requirements "implement" goals much the same way as programs implement design specifications [4,5].

There are some defects in traditional goal elaborate method. Firstly, different analyzers may have different refine results[6]. Secondly, the process of elaborate is lack of instruction, so some analyzers may neglect some connotative domain restricts that must match the goals. Lastly, there are short of methods to eliminate conflicts of goals among users. So it is important to set up criteria to instruct analyzers to refine the goals.

2 The Background Picture

A goal-based elaboration typically consists of a hybrid of top-down and bottom-up processes [1], plus additional processes driven by the handling of possible abnormal agent behaviors, the management of conflicting goals, the recognition of analogical situations from which specification can be transposed, and so forth. However, that for explanatory purpose the resulting requirements document is in general better presented in a top-down way.

Nowadays, an obvious informal technique for finding out subgoals and requirements is to keep asking HOW questions about the goals already identified. The method is effective to goal refinement, but there is an unresolved problem that different goal analyzers have different refined result. In order to solve this question and keep the consistent and complement of subgoals, Formal goal refinement patterns are put forward [6]. Such patterns are proved correct and complete once for all; refinements in the goal graph are then verified by matching them to one applicable pattern form the library. A frequently used pattern is the decomposition-by-milestone pattern that refines a parent **Achieve** goal:

$$P \Rightarrow \Diamond Q$$

Into two subgoals:

$$P \Rightarrow \Diamond R, R \Rightarrow \Diamond Q$$

The techniques above redound to different goal analyzers to get the same result for the same system. But formalized refinement patterns handicap normal analyzers to catch on the method. It is useful for normal analyzers to set up a series rules to use refinement patterns. Requirements often based on domain knowledge and goals based on domain knowledge too. So domain knowledge can instruct the refinement process of goal and we can set up above rules from domain knowledge.

Reference [7] gives us a series steps to achieve domain knowledge:

- Get domain intrinsic description
- Get support technologies
- Get the management and organization of domain
- Get rules and regulations of domain

Get descriptions of human behavior of domain

3 The KAOS Language

The specification language provides constructs for capturing various kinds of concepts that appear during requirements elaboration, namely, goals, constraints, agents, entities, relationships, events, actions, views, and scenarios.

Each construct in the KAOS language has a two-level generic structure: an outer semantic net layer for declaring a concept, its attributes and its various links to other concepts; an inner formal assertion layer for formally defining the concept. The generic structure is instantiated to specific types of links and assertion languages according to the specific type the concept is an instance of.

A Domain-Oriented Goal Elaborate Method 79

For example, a goal that appears at some stage of the elaboration of a meeting scheduler system is to get all participants' constraints known to the scheduler. The concept **ParticipantsConstraints — Known** is of type "Goal"; it could be partially declared and formally defined as follows:

Goal *Achieve*[ParticipantsConstraintsKnown]
  InstanceOf InformationGoal
  Concerns Meeting, Participant, Scheduler,...
  RefinedTo ConstraintRequested, ConstraintProvided
  InformalDef *A meeting scheduler should know the constraints of the various participants invited to the meeting within C days after appointment*
  FormalDef $\forall m$: Meeting, p : Participant, s : Sheduler
     Invited(p,m) $\wedge$ Scheduling(s,m) $\Rightarrow \Diamond_{\leq cd}$ Knows(s,p.Constraints)

The declaration part of the specification above states that the goal **ParticipantsConstraints- Known** (i) is concerned with keeping agents informed about object states, (ii) refers to objects such as Participant or Scheduler, and (iii) is refined into two subgoals. (The latter turn to be constraints assignable to single agents; the fact that the **ConstraintRequested** constraint is assignable to the Scheduler agent would be stated in the declaration of that constraint.)

4 Goal Elaborate Method Based on Domain Knowledge

Experience with Formal goal refinement patterns has revealed that appropriate patterns are often hard to find for normal analyzers; in other way, manual goal-decompositions are usually incomplete even sometimes inconsistent. Domain knowledge is relatively more steady than requirements. We can use domain properties to help analyzers to select appropriate patterns from formal goal refinement patterns library.

We can refine the parent goal by following:

- To Achieve passengers' initialized goals and to confirm the domain of the goals;
- According to the way of the domain's management and organization, we should ask "HOW" question that how the initialized goals are managed and organized in the domain? We can refine the initialized goals by "AND" link according to this question.
- To score each rule in the domain to estimate the correlation for those goals, and to insert the rules whose score are nonzero to a list descendingly.
- We ask "HOW" questions based on the rules of the rule list seriatim that how each rule should meet subgoals, and refine them by "AND" link or "OR" link. We should try our best to refine those subgoals by formal goal refinement patterns that can be proved correct and complete once for all.

- According to the restriction of human's behavior in the domain, we take the assignment of subgoals' agents into account, and optimize those refinement steps whose assignment of agents are irrationality.

The refinement of goal is for supporting requirement engineering. When do we consider that the refinement process can be finished? There is a criterion to end the refinement that if a goal only describes operation, we can consider that the refinement of the goal is finished. For example, if a goal describes that system transmits the email at once when system receives an email, we can consider that the goal need not be refined.

We can apply the method as following.

Now we should construct a query system of scheduled flight. Passengers require that system should return all results to passengers and system should deal with multi-requirement at the same time, when they provide the start, destination and leave time. So we can describe the initialized goal as below:

Goal Achieve[TicketOfficeSystem]

UnformDef : *When passengers submit a query, system should give passengers a result table sooner or later.*

FormalDef : $\forall start, destination : Address, d : Date$

$Inquire(start, destination, d) \Rightarrow \Diamond ResultTable$

Refining the initialized goal by domain management and organizing:
From the requirement of passengers, we can draw a conclusion that the requirement belongs to airplane circulating domain. In the domain, we find that the query disposal needs both sides participation. On one side system needs passengers to submit queries and wants to get the results as soon as possible, on the other hand when system receives a query, it should give passengers results during acceptable time according to the schedule of airplane. We pose the question that how the initialized goals are managed and organized in the domain. We can get the answer that there is a system dealing process after passengers submit a query. So we can refine the initialized goal as following:

Goal Achieve[GetResultWhenProduce]

UnFormalDef : *When passengers submit a query, if there is a system dealing process, passengers will get a result table sooner or later.*

FormalDef : $\forall start, destination : Address, d : Date$

$Inquire(start, destination, d) \wedge SystemProcess(start, destination, d) \Rightarrow$
$\Diamond ResultTable$

Goal Achieve[ProcessWillDo]

UnFormalDef : *If passengers submit a query, there is a system dealing process sooner or later.*

FormalDef : $\forall start, destination : Address, d : Date$

$Inquire(start, destination, d) \Rightarrow \Diamond SystemProcess(start, destination, d)$

A Domain-Oriented Goal Elaborate Method

Goal Maintain[UserWaiting]

UnFormalDef : *There is not other state between submitting query and getting the result table in the system.*

FormalDef ∀start,destination : Address,d : Date

Inquire(start,destination,d) ⇒ Inquire(start,destination,d)WResultTable

In this step, we refine the initialized goal by using Fig 1 refinement pattern. So we can insure that the refinement in this step is correct and complete.

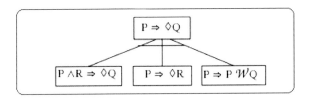

Fig. 1. A refinement pattern in patterns library

Refining the goals by domain rules:
From the subgoals above, we get that some rules in the domain are related to the subgoals. For example, there is a rule that passengers may arrive in destination by nonstop flight or change flight. This rule will instruct "SystemProcess" action, so it is related to subgoal "**ProcessWillDo**". There are other rules to be associated with subgoals above, but the rule of arriving way is most relative to the subgoal in all rules, so the score of this rule is highest. We will use the rule to instruct the refinement process.

According to the rule of arriving way, we pose the question that how we get the result table. The answer can be that we get the result tables of nonstop flight and change flight respectively, and we combine the two tables into a large one for passengers. According to the answer, we can refine the subgoals "**ProcessWillDo**" into following subgoals:

Goal Achieve[ProcessNonstop]

UnformalDef : *When passengers submit a query, system will get nonstop flight result table sooner or later.*

FormalDef : ∀start,destination : Address,d : Date

Inquire(start,destination,d) ⇒ ◊GetNonStop(start,destination,d)

Goal Achieve[ProcessChangeFlight]

UnformalDef : *When passengers submit a query, system will get change flight result table sooner or later.*

FormalDef : ∀start,destination : Address,d : Date

Inquire(start,destination,d) ⇒ ◊GetChangeFlight(start,destination,d)

Goal Achieve[ProcessInfo]

　UnformDef : *System dealing process includes to get nonstop flight result and change flight result.*

　FormalDef : ∀start,destination : Address,d : Date
　　　　SystemProcess(start,destination,d) ⇒ ◊GetNonStop(start,destination,d) ∧
　　　　　　　　　　　　　　　　　　　　　◊GetChangeFlight(start,destination,d)

In this step, we refine the initialized goal by using Fig 2 refinement pattern.

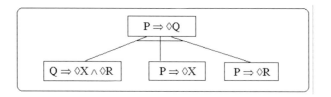

Fig. 2. A refinement pattern in patterns library

We can find that "SystemProcess" include "GetNonStop" and "GetChange-Flight" from the refinement. In terms of nonstop flight, according to domain rule, system can get the result by matching start, destination and start time with the query.

Goal Achieve[NonstopResult]

　UnformalDef : *When system get nonstop result,if start, destination and start time of a flight match that of the query, we add the flight No. to the result table.*

　FormalDef : ∀start,destination : Address,d : Date, result : the set of Flight
　　　　GetNonStop(start,destination,d) ⇔ ∀f ∈ Flight((f.start = start ∧ f.destination
　　　　　　　　　　　　　　　　　= destination ∧ f.date = d) → f ∈ result)

The subgoal describes the seeking criterion of nonstop flight, and it is the operation of system. So we can draw the conclusion that the subgoal needn't be refined any more.

As far as change flight, there are three domain rules which are relevant to **"ProcessChangeFlight"** subgoal. They are (1)Change flight can't be more than 3 times, otherwise passengers can't accept it;(2) The free time between two adjacent flights is more than half an hour and less than 6 hours;(3)Circular change flights aren't permitted. Change flights are restricted with all those 3 rules, according to the three rules how can we get the right change flights result? We refine the subgoal **"ProcessConnect- Flight"** according to the question as following:

Goal Achieve[ChangeFlighResult]

　UnformalDef : *When system deal with change filght operation,system can get a final rusult.*

　FormalDef: ∀start,destination : Address,d : Date
　　　　GetChangeFlight(start,destination,d) ⇒ ◊ChangeResult

A Domain-Oriented Goal Elaborate Method

Goal Maintain[ChangeResultStrict]

    UnformalDef : *Change flight can't be more than 3 times, and the interval between two adjacent flights is more than half an hour and less than 6 hours, and circular change flights aren't permitted.*

    FormalDef : $\forall$start, destination : Address, takeofftime, putdowntime : Time
        $\forall$result $\in$ ChangeResult(NUM(result.planelines)
        $\leq 3 \wedge \forall a,b \in$ result.planelines((a!= b)
        $\rightarrow$ ((a.start != b.start) $\wedge$ (a.start !=
        result.start $\rightarrow \exists!c \in$ result.planelines
        (c.destination = a.start $\wedge$ c.putdowntime
        + 30 < a.takeofftime $\wedge$
        c.putdowntime + 360 > a.takeofftime)))))

The subgoal "**ConnectResultStrict**" gives the limit to change flights according to domain rules. In the subgoal "a.start!=b.start" make the restrict to circular change flights, and it can limit change flights linking up by adding "c".

Optimize goals by human action and assignment of agents:
According to the restriction of human's behavior in the domain, taking passengers into account, we can draw the conclusion that nonstop flights result should be got at first because passengers need lower flights fee and less flight time. From the domain knowledge, we can see that change flights seeking is very complex, so the agent of subgoal "**ChangeResultStrict**" tends to deal with the nonstop flight seeking in order to lighten the burden of system. According to the analysis, we optimize the subgoal "**ProcessInfo**" as below:

Goal Achieve[ProcessInfo]

    UnformDef : *System dealing process includes getting nonstop flight result and change flight result.*

    FormalDef $\forall$start, destination : Address, d : Date
        SystemProcess(start, destination, d) $\Rightarrow$
            $\Diamond$GetNonStop(start, destination, d) $\wedge$
            $\Diamond$GetChangeFlight(start, destination, d) $\wedge$
            (GetChangeFlight(start, destination, d) $\rightarrow$
                • GetNonStop(start, destination, d))

Refinement graph:
The goal of flights query system elaboration process is ended hereto. Fig 3 is the refinement tree.

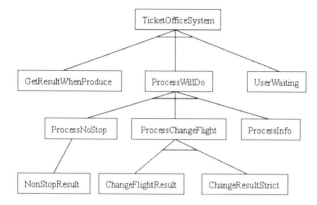

Fig. 3. Flights query system goal refinement tree

5 Discussion

System goals should satisfy domain knowledge. In elaborate process domain knowledge can suggest and remind analyzers, because some analyzers may not know what goals they need in the process. And domain knowledge is relatively steady, so it can eliminate conflicts of goals among users.

The method is based on the domain knowledge, so it has above advantages of domain. Setting up all-around domain knowledge is too complicated and burdensome for analyzers, but the knowledge can instruct not only this software requirement but also that of other kind software. That is to say the method does not add any burden to analyzers and gives a road to reuse software requirements.

There are still lots of work to improve the method. Such work include: (i) Setting up a strict criteria to judge the degree of relationship between rules and goals; (ii) Giving a method to judge the quality of subgoals.

References

1. Lamsweerde, A.V.: Goal-Oriented Requirements Engineering: A Guided Tour. In: Proceedings RE 2001, 5th IEEE International Symposium on Requirements Engineering, Toronto, pp. 249–263 (2001)
2. Letier, E., Lamsweerde, A.V.: Agent-Based Tactics for Goal-Oriented Requirements Elaboration. In: Proc. ICSE 2002: 24th Intl. Conf. on Software Engineering, pp. 83–93. IEEE Press, Orlando (2002)
3. Feather, M.S., Cornford, S.L., Dunphy, J., Hicks, K.: A QuantitativeRisk Model for Early Lifecycle Decision Making. In: Proceedings of the Conference on Integrated Design and Process Technology, Pasadena, California (2002)
4. Lamsweerde, A.V., Letier, E.: From Object Orientation to Goal Orientation: A Paradigm Shift for Requirements Engineering. In: Wirsing, M., Knapp, A., Balsamo, S. (eds.) RISSEF 2002. LNCS, vol. 2941, pp. 325–340. Springer, Heidelberg (2003)

5. Letier, E., Lamsweerde, A.V.: Reasoning about Partial Goal Satisfaction for Requirements and Design Engineering. In: SIGSOFT 2004/FSE-12, pp. 53–62. ACM, CA (2004)
6. Darimont, R., Lamsweerde, A.V.: Formal Refinement Patterns for Goal-Driven Requirements Elaboration. In: Proceedings 4th ACM Symposium on the Foundations of Software Engineering (FSE4), pp. 179–190. ACM, San Francisco (1996)
7. Bjørner, D.: Domain Engineering: A "Radical Innovation" for Software and Systems Engineering? A Biased Account. In: Wirsing, M., Knapp, A., Balsamo, S. (eds.) RISSEF 2002. LNCS, vol. 2941, pp. 436–439. Springer, Heidelberg (2002)

A Hybrid Genetic Routing Algorithm in Wireless Sensor Networks

Lejiang Guo, Bengwen Wang, and Qian Tang

Department of Control Science and Engineering
Huazhong University of Science and Technology
Wuhan,Hubei Province, China
radar_boss@163.com

Abstract. Wireless Sensor Networks (WSN) represent a new dimension in the field of networking. Through learning from the energy multi-path routing protocol of WSN and the hybrid genetic algorithm, this paper presents a novel routing protocol to find the optimal path. The Algorithm consists of two stages: single-parent evolution and population evolution. The initial population is formed in the stage of single-parent evolution by using gene pool, then the algorithm continues to the further evolution process, finally the best solution will be generated and saved in the population. The simulation results show that the algorithm is effective. It can optimize the network path, balance energy consumption of the network and extend the network life cycle.

Keywords: wireless sensor networks, routing protocol, network lifetime.

1 Introduction

In Wireless Sensor Networks, the energy of the node is limited, the maximum of the lifetime of WSN becomes an important goal to the design of the routing protocol [1]. However, the traditional routing protocol is little regard for the energy consumption, such as Flooding, Gossiping, and SPIN. Shah proposes an energy-aware routing protocol that improves the energy efficiency of the Directed Diffusion protocol, but it selected a path depending on the probability of random which lead to reliability decreased. Therefore, the routing protocols must consider not only the reliability of the optimum path, but also the energy consumption of the whole network.

In 1975, John Holland proposed a global optimization algorithm Genetic Algorithm (GA). In recent years, based on genetic algorithm in the WSN, routing optimization research is also very active [2]. For path optimization, genetic algorithms have shown a tremendous advantage. Based on the model of the energy multi-path routing protocol, this paper presents a new algorithm which abandons the randomness in the generation of initial population and replaces the gene fragments by gene pool. From simulation, the algorithm is effective and extends the network lifetime.

2 Hybrid Genetic Algorithm

Nowadays, in the stage of the whole evolution, the genes involved in genetic operator are mostly from the individual itself [3]. The quality level of the individual determines the efficiency of the algorithm. If the fitness of all individuals is poor, the algorithm performance will be affected. In order to overcome these weaknesses, this paper sorts n points, constructs a $n*n$ matrix of the gene pool, prepares for the genetic operators. This method greatly improves the efficiency of the algorithm.

2.1 Network Model

The network models as an undirected connection diagram $G(S,V,P)$, where S on behalf of the node of the sink, V presents the nodes. Set the nodes numbered 1, 2, 3... n, $P_k = V_1^k V_2^k ... V_n^k$ for a feasible path, the first k nodes of the path starting point is V_1^k, the aggregation node is V_n^k, then the total length can be expressed as:

$$f(P_k) = \sum_{i=1}^{n-1} PE(V_i^k, V_{i+1}^k) + PE(V_n^k, V_1^k)$$

$PE(V_i^k, V_j^k)$ is the energy consumption between the nodes, $f(P_k)$ is evaluated the individual's good or bad by using P_k.

2.2 Construction of Gene Pool

According to the cost between the nodes $PE(i,j)$, $PE(i,j)$ construct a $n*n$ matrix $D = \{PE(i,j)\}$, $num[i,j] = \begin{cases} j+1 & j \geq i \\ j & j < i \end{cases}$, $i = 1,2,3,...n$, $j = 1,2,3,...n$ then defines a $n*(n-1)$ matrix $num = \{num[i,j]\}$.

For each node i, according to the size of $PE(i,j)$, the num in the first line of the corresponding i elements in accordance with the order from small to large order, and before all the elements i come in, so will the expanded num (increase in first column) to get a $n*n$ of the square, gene pool is formed in this paper. Each row element in gene pool is the problem a feasible solution, the start node of the feasible solution is the line number of the gene pool [4]. Solving the problem is divided into single-parent evolution and the evolution of population. By single-parent evolution, the initial population $P = \{P_1, P_2, ..., P_n\}$ is generated.

2.3 The Single-Parent Evolution Algorithm

(1) Randomly generate one individual V, calculate fitness value $f(V)$;

    Step1-1: Randomly generate gene length $l \leq len \leq maxlen$;

    Step1-2: Randomly generating a gene location $l \leq pos \leq n - len$;

    Step1-3: The gene pool in the pos line of len elements replace the current total of the individual V in the genes, their location from the beginning of pos, get a new individual $Vnew$;

    Step1-4: Calculation $f(Vnew)$;

    Step1-5: If $f(Vnew) < f(V)$ then $V := Vnew$;

    Step1-6: If (no end) then goto Step1-1;

(2) Randomly generate two integers $l \leq pos\ 1, pos\ 2 \leq n$, $pos\ 1 \neq pos\ 2$

    Step2-1: The gene segment between $pos\ 1$ and $pos\ 2$ of the individual V is reversed in order to be new individual $Vnew$;

    Step2-2: Calculation $f(Vnew)$;

    Step2-3: If $f(Vnew) < f(V)$ then $V := Vnew$;

    Step2-4: If (no end) then goto (2);

(3) Randomly generate two integers $l \leq pos\ 1, pos\ 2 \leq n$, $pos\ 1 \neq pos\ 2$

    Step3-1: The genome $Vpos\ 2$ of individual genes V will be inserted into the genome $Vpos\ 1$ to be new individual $Vnew$;

    Step3-2: Calculation $f(Vnew)$;

    Step3-3: If $f(Vnew) < f(V)$ then $V := Vnew$;

    Step3-4: If (no end) then goto (3);

    Step3-5: Output.

In the single-parent evolution algorithm, only a single individual is evaluated. The speed of evolution which produces a good general is very fast. At the same time, a global optimal path is generated.

2.4 The Population Evolution Algorithm

The population evolution algorithm presents only as an amendment to the role, and the purpose is to improve the solution quality [5]. After producing the initial population p with the single-parent evolution algorithm, select two individuals pr_1, pr_2 randomly to hybrid operation, if the new individual is better than the mother, the new individual replace the mother directly.

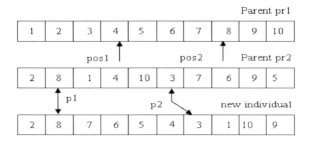

Fig. 1. The procession of Hybrid gene operator

The evolution which repeats the hybrid process will not be terminated until the best individual is generated [6]. In another the mother pr_2 the gene position which is equal to p_1 and p_2 in the gene pr_1 is found. The same value of pr_2 between pos_1 and pos_2 from the gene pr_1 is deleted in its entirety, the genes of p_1 and p_2 will be moved to the adjacent location. Then the genes between pos_1 and pos_2 will be cut and inserted pr_2 between the individual p_1 and p_2 according to the order of the position of p_1 and p_2, as shown in Fig.1.

3 Simulation

3.1 Simulation Environment

Under the Matlab Environment, simulation environment for wireless sensor networks are established [7]. The number of sensor nodes is 210, communication radius is 20. The sensors randomly are distributed in the area of 100 × 100, the initial energy of sensor nodes are uniformly distributed between 200J and 400J, data packet collected by the sensor nodes is 512byte. Each node sends a packet consumes 0.2J; each node receiving a packet consumes 0.01J. The network lifetime will be terminated when the number of nodes in the network is below 85%.

3.2 Simulation Result

In this paper, the largest energy path routing mechanism (MCP) and the maximum energy path switching routing mechanism (MCP-PS) are also simulated. Three kinds of protocol simulation results is shown in Fig.2, 3.In Fig.2, When the node density is small, the network work cycle of three kinds of algorithms is similar. Compared with MCP and MCP-PS, MCP-GEN algorithm has more work-cycle when the number of nodes is increased. This algorithm achieves the requirements of real-time path, the network lifetime of MCP-GEN is longer than the other two algorithms.

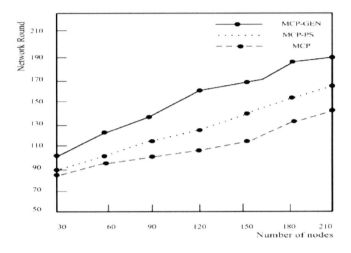

Fig. 2. Network round with number of nodes

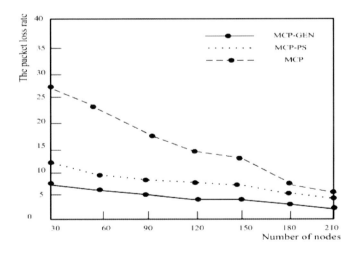

Fig. 3. The packet loss rate with number of nodes

From Fig.3, it shows the packet loss rate of three kinds of protocol. As can be seen from the figure, when the node density is small, the packet loss rate of the MCP protocol is higher than the protocol of MPC-GEN or MCP-PS. When the node density is large, three kinds of protocol packet loss rates are very similar and below 10%, and the change range of MCP-GEN is lower slightly than the other two kinds of routing protocol in the whole procession.

4 Conclusions

In this paper, we build a gene pool for the design of genetic operators to avoid the algorithm to a part optimal solution which results in premature convergence problem. At the same time, the algorithm balances energy consumption and extends the network lifetime, finally the network efficiency of WSN is improved. In future, we will focus on how to build a high-quality gene pool.

References

1. Younism, A.: An Energy-aware QoS Routing Protocol for Wireless Sensor Networks. In: Proceedings of the 23rd International Conference on Distributed Computing Systems Workshops, pp. 710–715. IEEE Computer Society, Los Alamitos (2003)
2. Perris, A., Szewczyk, R., Wen, V., et al.: SPINS. Security Protocols for Sensor Networks. In: Proceedings of the 7thAnnual International Conference on Mobile Computing and Networking, pp. 189–199. ACM, USA (2001)
3. Hou, X.B., Tipper, D., Kabara, J.: Simulation based on Multi-path Routing in Wireless Sensor Networks. In: The Advanced Radio Technologies (ISART), Boulder, CO (2004)
4. Ying, L., Hai-bin, Y., Peng, Z.: Optimization of Clusterbased Routing Protocols in Wireless Sensor Network PSO. Control and Decision 21(4), 453–456, 461 (2006)
5. Akyildiz, I., Su, W., Sankara Subramanian, Y., et al.: Wireless sensor networks: a survey. Computer Networks 38(4), 393–422 (2002)
6. Tian, D., Georganas, N.D.: Energy Efficient Routing with Guaranteed Delivery in Wireless Sensor Networks. In: 1 WCNC 2003, New Orleans, USA (2003)
7. Bhatia, A.K., Basu, S.K.: Tackling 0/1 Knapsack Problem with Gene Induction. Soft Computing (8), 1–9 (2003)

The Application of Fast Multipole-BEM for 3-D Elastic Contact Problem with Friction

Gui Hai-lian[1,*] and Huang Qing-xue[2]

[1] Mechanical & Electronic Engineering College
Taiyuan University of Science and Technology
Taiyuan, Shanxi, 030024, P.R. China
guihl2000@yahoo.com.cn
[2] Material Science & Engineering Science College
Taiyuan University of Science and Technology
Taiyuan, Shanxi, 030024, P.R. China

Abstract. In this paper, three-dimensional elastic contact problem is discussed. Based on traditional boundary element method (BEM), mixed boundary integral equation is presented to solving several bodies contact problem with friction. It is not only reducing the dimension from domain to boundary, but also solving the displacement and stress independently at the same time. A new judgement criterion of contact state is established to judge the contact state of separate, adhere and slip. Through numerical analysis, the displacement and stress obtained by FM-BEM is better than traditional BEM in calculation time and accuracy. It illustrates that FM-BEM is better than traditional BEM for solving contact problem with friction.

Keywords: elastic contact problem with friction, mixed boundary integral equation (MBIE), boundary element method (BEM), fast multipole boundary element method (FM-BEM).

1 Introduction

In the elastic contact problem with frictional, the displacement and stress in the contact area are most concerned. The boundary element method (BEM) is divided elements in the surface; this merit is justly suit for contact problem. So BEM was applied as simple use, small iterative calculations requirement, saving computation memory. In addition, BEM can solve directly the couple of law stress and tangential stress in the edge, eliminating the iterative process for determining the law stress and tangential stress.

With the using of the large mechanical device in engineering, the contact problem is becoming huge. In BEM, the matrix is full rank, this is very inconvenient in

---

[*] The support of Innovative projects of outstanding graduate students in Shanxi Province (20081082) and the Returned Overseas Students to Scientific Research Foundation of Shanxi Province in China (20081037) is gratefully acknowledged.

numerical calculation. Fortunately the fast multipole method was introduced in BEM; this new method could enhance the speed of solving the large-scale numerical calculation in engineering.

In this paper, firstly the boundary mixed integral formulation is founded, this formulation is not only reducing the dimension from domain to boundary, but also solving the displacement and stress at the same time. In several bodies contact problem, a unitary matrix is founded by boundary relativity; Then researching the three dimensional elastic contact problem, it is simulated by fast multipole boundary element method (FM-BEM); At last, a numerical example is given about three bodies elastic frictional contact problem and the displacement and stress are analyzed in contact areas by FM-BEM. It illustrates that FM-BEM is better than traditional BEM in solving contact problem.

2 Establish Mixed Boundary Integral Equtaion (MBIE)

2.1 Boundary Mixed Integral Formulation in 3-D Elastic Contact Problem

Consider the two bodies Ω^A and Ω^B, Suppose the boundary of A body is $\Gamma^A = \Gamma^{AC} \cup \Gamma^{AD}$, the boundary of B body is $\Gamma^B = \Gamma^{BC} \cup \Gamma^{BD}$, where Γ^{AC} and Γ^{BC} represent the contact boundary between A body and B body. So the boundary integral formulation is written as follow:

$$c_{ij}u_j = \int_{\Gamma^A} u_{ij}^{K*} p_j d\Gamma - \int_{\Gamma^A} p_{ij}^{K*} u_j d\Gamma \qquad K = A, B \qquad (1)$$

Matrix formulation is established as:

$$\begin{pmatrix} \mathbf{H}^{KD} & \mathbf{H}^{KC} \end{pmatrix} \begin{bmatrix} \mathbf{U}^{KD} \\ \mathbf{U}^{KC} \end{bmatrix} = \begin{pmatrix} \mathbf{G}^{KD} & \mathbf{G}^{KC} \end{pmatrix} \begin{bmatrix} \mathbf{P}^{KD} \\ \mathbf{P}^{KC} \end{bmatrix} \qquad K = A, B \qquad (2)$$

Using of the boundary condition in contact areas Γ^{AC} and Γ^{BC}:

$$u_{ij}^A = u_{ij}^B \text{ and } p_{ij}^A + p_{ij}^B = 0$$

The boundary integral formulation is coupled as follow:

$$\begin{pmatrix} \mathbf{H}^{AD} & \mathbf{H}^{AC} & 0 \\ 0 & \mathbf{H}^{BD} & \mathbf{H}^{BC} \end{pmatrix} \begin{bmatrix} \mathbf{U}^{AD} \\ \mathbf{U}^{C} \\ \mathbf{U}^{BD} \end{bmatrix} = \begin{pmatrix} \mathbf{G}^{AD} & \mathbf{G}^{AC} & 0 \\ 0 & -\mathbf{G}^{BC} & \mathbf{G}^{BD} \end{pmatrix} \begin{bmatrix} \mathbf{P}^{AD} \\ \mathbf{P}^{C} \\ \mathbf{P}^{BD} \end{bmatrix}$$

2.2 Expansion the Fundamental Solution

At first, the fundamental solutions are given and divided into dependent parts:

$$u_{ij}^* = \frac{1}{16\pi\mu(1-v)r}\{(3-4v)\delta_{ij} - r_{,i}r_{,j}\} = P_{ij}(x)(\frac{1}{r}) + Q_i(x)(\frac{1}{r}y_j) \quad (3)$$

$$p_{ij}^* = \frac{-1}{8\pi(1-v)r^2}\left[\frac{\partial r}{\partial n}\left[(1-2v)\delta_{ij} + 3r_{,i}r_{,j}\right] + (1-2v)\left(r_{,i}n_j - r_{,j}n_i\right)\right] \quad (4)$$

$$= W_{ijm}(x)(\frac{1}{r}n_m(y)) + S_{im}(x)(\frac{1}{r}n_m(y)y_j)$$

where: $P_{ij}(x) = \dfrac{\alpha}{2\mu}\left[(3-4v)\delta_{ij} - x_j\partial_i\right]$

$Q_i(x) = \dfrac{\alpha}{2\mu}\partial_i$, $S_{im}(x) = -\alpha\partial_i\partial_m$

$W_{ijm}(x) = \alpha[(1-2v)(\delta_{jm}\partial_i - \delta_{ij}\partial_m) - 2(1-v)\delta_{im}\partial_j + x_j\partial_i\partial_m]$

$i, j, m = 1, 2, 3$, ∂_i is partial derivative about x_i, $\alpha = 1/8\pi(1-v)$.

3 Key Technique in Calculation

3.1 Establish the Judgement Criterion of Contact State

The contact boundary is changing with step loading, some nodes and elements are contact state in n step, but in $n+1$ step, these maybe become slip or separate state, so it is necessary to establish the judgement criterion of contact state. In different state, the calculation formulas of displacement increment and stress increment are different.

3.1.1 Initial Contact State

At first, it must satisfy the condition: $\Delta\hat{t}_3 < 0$

(1) adhere state(Γ_{ca})

$$\Delta\hat{t}_j^A - \Delta\hat{t}_j^B = 0, \quad j = 1, 2, 3$$

$$\Delta\hat{u}_k^A + \Delta\hat{u}_k^B = 0, \quad k = 1, 2, 3$$

$$\Delta\hat{u}_3^A + \Delta\hat{u}_3^B = \delta_0^{n-1}, \quad \delta_0^{n-1} \text{ is clearance in currently step}$$

(2) slip state(Γ_{cs})

$$\Delta\hat{t}_j^A - \Delta\hat{t}_j^B = 0, \quad j = 1, 2, 3$$

$$\Delta\hat{u}_k^A + \Delta\hat{u}_k^B = \Delta u_{AB}, \quad k = 1, 2, 3$$

$$t_1^A = -\mu\cos\phi\,|\hat{t}_3^A|, \quad t_2^A = -\mu\sin\phi\,|\hat{t}_3^A|,$$

$$\phi = \cos^{-1}\left(-\Delta\hat{t}_1^A \Big/ \sqrt{(\Delta\hat{t}_1^A)^2 + (\Delta\hat{t}_2^A)^2}\right)$$

3.1.2 Distinguish Dynamic Contact State ($n-1$ Step as Reference State)

(1) adhere state ($n-1$ step)

Separate: $\hat{t}_3^A > 0$

Adhere: $\hat{t}_3^A \leq 0$, $\sqrt{\hat{t}_1^2 + \hat{t}_2^2} \leq \mu\,|\hat{t}_3|$

Slip: $\hat{t}_3^A \leq 0$, $\sqrt{\hat{t}_1^2 + \hat{t}_2^2} > \mu\,|\hat{t}_3|$, $\phi = \cos^{-1}\left(-\hat{t}_1^A \Big/ \sqrt{(\hat{t}_1^A)^2 + (\hat{t}_2^A)^2}\right)$

(2) slip sate (n-1 step)

Separate: $\hat{t}_3^A > 0$

Adhere: $\sum \Delta\hat{t}_K^A \Delta\hat{u}_{ABK} \geq 0$, $K = 1, 2$

Slip: $\sum \Delta\hat{t}_K^A \Delta\hat{u}_{ABK} < 0$, $K = 1, 2$, $\phi = \cos^{-1}\left(-\hat{u}_{AB1} \Big/ \sqrt{(\Delta\hat{u}_1)^2 + (\Delta\hat{u}_2)^2}\right)$

(3) separate state (n-1 step):

Contact: $\Delta\hat{u}_3^A + \Delta\hat{u}_3^B = \delta_0^{n-1}$

Separate: $\Delta\hat{u}_3^A + \Delta\hat{u}_3^B < \delta_0^{n-1}$

Residual frictional force:

$$\Delta\hat{t}_1^{\varepsilon\cdot n} = \mu\cos\phi\,|\Delta t_3^{n-1}| - \Delta t_1^{n-1}, \quad \Delta\hat{t}_2^{\varepsilon\cdot n} = \mu\sin\phi\,|\Delta t_3^{n-1}| - \Delta t_2^{n-1}$$

3.2 Processing Load Step

Calculating the elastic frictional contact problem, contact state is judged in every add load step. Suppose contact tolerate is δ. In add load process, penetrate phenomena maybe happen, so checking penetrate to every element is necessary in add load step. Suppose time change from t to $t+\Delta t$, point P move from $P(t)$ to $P(t+\Delta t)$, penetrate will happen when $P(t+\Delta t)$ is larger than contact tolerance. In this case, we must subdivision the add load step in order to avoid penetrate phenomena.

4 Numerical Analysis

Above the discussion, the program of three-dimensional elastic contact problem with friction is compiled by FORTRAN. The computation will run on a Windows XP computer equipped with one 2.8 GHz Intel Pentium 4 unit and 2.0GMB of core memory.

Consider the elastic contact frictional problem about three cubes, the computing model shows as Fig.1. Supposed Poisson's ratio is 0.3, elastic module is 210 GPa, the frictional coefficient is 0.25, the contact limit is 0.001mm. There is 100 MPa pressure acting on upper surface of body C and body A is fixed constraint on the bottom surface. In discrete model, the number of elements and nodes show in Tab.1.

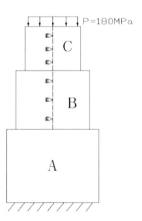

Fig. 1. Computing model

Using the judgement criterion of contact state, the displacement and stress are discussed in contact area. Solving this problem spend 163 seconds using traditional BEM and 49 seconds using FM-BEM, but The displacement distribution

Table 1. The number of elements and nodes

body	Contact nodes	Contact elements	Total nodes	Total elements
A	36	25	152	150
B	98	72	218	216
C	25	16	98	96

shows as Fig. 2 and Fig. 3. From the figures, we know that the maximum displacement appear in the edge and the displacement is very small in the center of contact area. This phenomenon is called "cat ear phenomenon". The continuity of displacement is well because that the displacement is solved independent in MBIE.

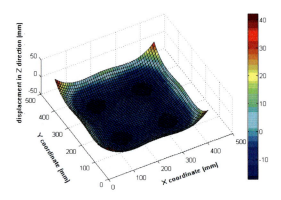

Fig. 2. Displacement of body B in contact area

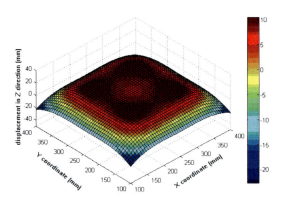

Fig. 3. Displacement of body C in contact area

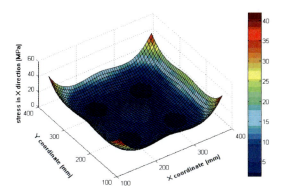

Fig. 4. Stress distribution in contact area of upper surface

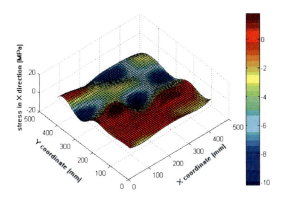

Fig. 5. Stress distribution in contact area of bottom surface

Fig. 4 and Fig. 5 shows the stress distribution in contact area of body B. In the contact area of upper surface, the change trend is similar to the displacement, the stress change mainly happen in edge, but in the low contact area, the stress change is sharply. This is because the force and friction influence in this area.

5 Conclusion

In this paper, FM-BEM is applied to solve the elastic contact problem with frictional. Through establishing the mixed boundary integral equation (MBIE), the displacement and stress are solved independent, this is benefit for ensure continuity. To judge the contact state in every step loading, the judgement criterion of contact area is given. This criterion respectively discuss the state of sperate, adhere and slip and give different calculation formula of displacement increment and stress increment. Solved the contact problem of three cubes with friction by FM-BEM, the results of displacement and stress distribution is better than tradition in calculation time and accuracy. It illustrates that FM-BEM is better than traditional BEM in solving contact problem with friction.

References

[1] Liu, D.: Three dimensional multipole BEM for elastic-plastic contact with friction and rolling simulation of four-high mill, Yanshan University (2003)
[2] Junjie, J., Qingxue, H.: Frictional contact multipole-BEM analysis of traction field in screw pairs. Heavy Machinery 1, 10–14 (2007)
[3] Wang, H., Yao, Z.: Application of Fast Multipole BEM for Simulation of 2D Elastic Body with Large Number of Inclusions. In: Proceedings of the Third International Conference on Boundary Element Techniques, pp. 77–82. Tsinghua University Press, Beijing (2002)
[4] Huang, N.J., Deng, C.X.: Auxiliary Principle and Iterative Algorithms for Generalized Set- valued Strongly Nonlinear Mixed Variational-like Inequalities. J. Math. Anal. Appl. 256, 345–359 (2001)
[5] Qingdong, Z., Xiangming, S., Jian, B.: Analysis of Rolls' Elastic Deformation on CVC 6-h Mill by FEM. China Mechanical Engineering 7, 789–892 (2007)
[6] Zhao, L., Yao, Z.: Fast Multipole BEM for 3-D Elastostatic Problems with Application for Thin Structures. Tsinghua Science and Technology 10, 67–75 (2005)
[7] Guangxian, S., Deyi, L., Chunxiao, Y.: Multipole boundary element method and rolling engineering. Science Press, Beijing (2005)
[8] Yu, C., Shen, G., Liu, D.: Mathematical Foundation of the Fast Multipole-BEM for 3-D structural Objects. Computational Engineering, 51–60 (2004)
[9] Hui, S., Mingfu, F., Guotai, Y.: Generalized Variational Inequality Principle of Rate from in Elastic-plastic Contact Problem with Finite Displacement and Friction. Chinese Journal of Mechanical Engineering 41, 38–41 (2005)
[10] Ting, L., Zhenhan, Y., Haitao, W.: High performance parallel computations of 3-D fast multipole boundary element method. Journal of Tsinghua University (Science and Technology) 42, 280–283 (2007)
[11] De-zhi, N., Bin, T., Gou, Y.: Application of fast multipole boundary element method to 3-D potential flow problem. Journal of Dalian University of Technology 45, 243–247 (2005)
[12] De-zhi, N., Bin, T., Ying, G.: Implementation of the fast multipole expansion technique in the higher order BEM. Chinese Journal of Computational Mechanics 22, 700–704 (2005)
[13] Linbin, Z., Zhenhan, Y.: The fast multipole-BEM about 3-D elastic problem applied in thin shell structure, pp. 20–43 (2003)
[14] Liu, Y.J., Shen, L.: A dual BIE approach for large-scale modeling of 3-D electrostatic problems with the fast multipole boundary element method. Int. J. Numer. Meth. Engng. 71, 837–855 (2007)
[15] Chun-Xiao, Y., Guang-Xian, S.: Program-iteration pattern Fast Multipole BEM for elasto-plastic contact with friction. Chinese Journal of Computational Mechanics 25(1), 65–71 (2008)
[16] Giner, E., Tur, M., Vercher, A., Fuenmayor, F.J.: Numerical modeling of crack-contact interaction in 2D incomplete fretting contacts using X-FEM. Tribology International 42(9), 1269–1275 (2009)
[17] Blázquez, A., París, F.: On the necessity of non-conforming algorithms for 'small displacement' contact problems and conforming discretizations by BEM. Engineering Analysis with Boundary Elements 33(2), 184–190 (2009)
[18] Gill, J., Divo, E., Kassab, A.J.: Estimating thermal contact resistance using sensitivity analysis and regularization. Engineering Analysis with Boundary Elements 33(1), 54–62 (2009)

Simulation Method Research of Ground Target IR Scene Based on Aerospace Information

Chen Shan and Sun Ji-yin

Research Inst. of High-tech Hongqing Town
Xi'an, Shanxi, China, 710025
chenshan1223@126.com

Abstract. Infrared imaging guidance is a guidance system with strong anti-interference ability. It is a technology with high efficiency-cost ratio which increases the missile power. It also is one of precision guided munitions development directions nowadays. It is first step that studies on the infrared radiation characteristics of target region to study on infrared imaging guidance, in which the key is how to obtain the infrared information of target region. This paper presents a kind of target IR scene simulation methods based on aerospace information. Firstly, three-dimensional scene model is generated based on terrain data, target 3D data and remote sensing image data. Secondly, the value of target and scene surface temperature is computed with different methods. Finally, target region IR scene 24h image sequence is produced. After comparative analysis, the IR scene image sequence produced by this method reflects target region infrared radiation characteristics in truth, and has a good fidelity.

Keywords: Infrared imaging guidance; aerospace information; IR scene simulation.

1 Introduction

The precision guided weapon will be the information warfare's main weapons in the condition of high-tech. improving its guidance accuracy is the most effective means to strengthen attack effect. At present, in the aspect of terminal guidance, passive infrared imaging terminal guidance system has been more and more paid attention, which realizes the precision guidance by using the difference between target and background to form the thermal image of target and its surrounding scene. It indicates from all sorts of data that infrared imaging guidance is a kind of guidance system with strong anti-interference ability, is a technology with high efficiency-cost ratio, and makes the missile much more powerful. It now is one of development directions of precision guided weapons.

It is impossible to be the same thermal image to different targets in the infrared imaging terminal guidance system, so the ability to identify the target is unique. Before study infrared imaging guidance, infrared characteristics of target region must be studied, and the key problem is how to gain target region's infrared information. There are many reconnaissance techniques at present, such as obtaining from satellite remote sensing images and obtaining from reconnoiters on the spot. Because of the uncertainty of infrared characteristics, the target region's infrared characteristics are different in different weather condition or different time of day. It is impossible to gain infrared information in different weather condition. It also is difficult to meet the requirements of precision-guided to satellite remote sensing images for its low accuracy at infrared band. Therefore, how to simulate target region 3D infrared scene by using the aerospace information such as satellite remote sensing images, target region terrain data, target 3D data from reconnoiters on the spot, etc. to provide data guarantee for the precision guidance, becomes the research hotspot at present.

In view of this, this paper presents a kind of ground target infrared scene generation method based on aerospace information, gives infrared simulation image sequence one day of 24 hours. It can be seen from the contraction between simulation images and real-time images that simulation image sequence not only reflect the real temperature (gray) value of target and background, but also have a good verisimilitude.

2 The Overall Design of Scene Simulation

To realistically simulate target region infrared scene, infrared characteristics of target region must be studied firstly. The general approach is to build 3D scene model, calculate surface temperature by establishing conductivity equation and at last obtain the infrared characteristics. As for the research of target's infrared radiation characteristics, this approach can accurately predict each surface temperature value while it is high-computational. As for the research of background, this approach is much more time-consuming and high-computational with high errors. Therefore, this paper firstly divides the target region into two parts. One is background such as ground surface, water body, vegetation, etc. The other is target such as tall building, inhabited area, factory, etc. different part uses different simulation method: for background, uses the simplified simulation method for its large data and rich infrared characteristics; for target, uses complex simulation model to obtain more precise infrared characteristic.

As for target and background, the difference is the complex degree of simulation model; therefore, the overall program of 3D infrared scene simulation is shown in Figure 1.

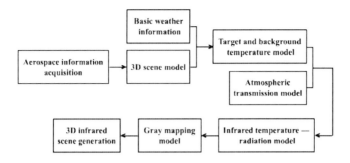

Fig. 1. Overall program of 3D infrared scene simulation

3 3D Scene Model Generation of Target Region

There usually are two kinds of ideas to establish 3D scene model of target region. One is reconstructing according to the principles of photogrammetry by using many stereo pairs of different viewpoints. The other is obtaining manually by using 3D protracting software such as 3Dmax, Maya, etc.

According to the terrain data, target 3D data and remote sensing image data obtained, this paper adopts the second approach. The background part is produced by terrain software using terrain data such as DEM, DSM, etc. while the target part is produced by 3D protracting software such as 3Dmax using target 3D data, remote sensing image data, etc. as the basic model of 3D scene simulation, background and target use different algorithm. Figure 2 is the 3D scene model produced by using aerospace information.

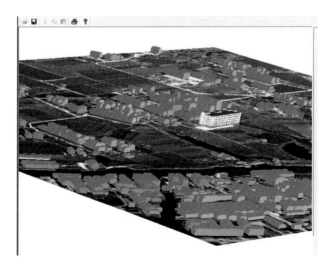

Fig. 2. 3D scene model (protracted based on VC platform)

4 Surface Temperature Calculation of Target and Background

It can be seen from the basic law of heat radiation that temperature is the key factor of infrared radiation characteristics. To generate the target 3D infrared scene, firstly the temperature equilibrium model of target and background must be established and the surface temperature value of target and background must be calculated.

4.1 External Factors Affecting the Temperature

Object surface temperature mainly affected by the background and internal heat source. Ground target locates in a very complex surrounding. How to effectively make clear various factors and their processes has an important influence to the computational accuracy. The heat exchange between object surface and external environment, namely the external factors influencing object surface temperature, mainly are: solar radiation, ambient temperature (low altitude atmospheric temperature), sky atmospheric radiation, wind speed, etc. which is closely related to season, weather and geographical position.

Solar radiation is a function related to time, latitude, atmospheric characteristics, etc. It mainly has three parts: direct solar radiation, scattering solar radiation and ground reflect solar radiation. The distance between the Sun and the Earth is very far, so the sunlight reaching the ground can be as parallel light. The solar radiation received by object surface can be expressed as follows.

$$Q_{sun} = CCF \cdot (Q_{p,b} + Q_{p,d} + Q_{p,\rho}) \tag{1}$$

$Q_{p,b}$ is direct solar radiation, $Q_{p,d}$ is scattering solar radiation, $Q_{p,\rho}$ is ground reflect solar radiation and CCF is cloud cover factor.

Ambient temperature is low altitude atmospheric temperature, which is a function related to season, geographical location, altitude and time. Ambient temperature has a periodicity change daily or yearly. Usually, ambient temperature reaches the highest-point at about 2 pm to 3 pm, and reaches the lowest-point at 5 am to 6 am. It has a change form like harmonic wave but not strictly tally with sine or cosine. According to relevant theory, ambient temperature can be expressed as the form of second-order Fourier series.

$$T_a(t) = g_0 + g_1 \cos\omega(t-b_1) + g_2 \cos 2\omega(t-b_2) \tag{2}$$

t is time, g_0, g_1, g_2, b_1, b_2 is constant.

Sky atmospheric radiations also a factor affecting object temperature, which mainly is a kind of long-wave radiation. The atmosphere has certain temperature after absorbing certain solar heat and Earth's heat; therefore it also can give radiation to the object. The expression is:

$$Q_{sky} = C.C \cdot \varepsilon_{sky} \sigma T_{sky}^4 F_{sky} \tag{3}$$

T_{sky} is sky temperature, F_{sky} is radiation angle coefficient of object surface to atmospheric equivalent gray body plane, ε_{sky} is atmospheric equivalent emissivity, $C.C$ is the coefficients related to the cloud type.

Wind is caused by low-level atmospheric convection, which produces convection heat transfer with object surface. The exchange heat caused by the relative movement between object surface and the air can be expressed as follows.

$$Q_{conv} = H(T_{air} - T) \tag{4}$$

H is convection heat transfer coefficient, T_{air} is low altitude atmospheric temperature, T is object surface temperature.

In addition, as to the surface with water and water vapor, latent heat is also an important attribute of affecting object surface temperature. Latent heat plays the role of adjusting surface temperature. When the surface absorbs the heat in air, the water on surface will evaporate into water vapor to delay the hoist of surface temperature. When the air temperature drops, the water vapor nearby the surface will freeze into water to delay the decline of surface temperature.

4.2 Calculation of Object Surface Temperature

Calculating object surface temperature must establish the object temperature equilibrium model. In general, object 3D transient heat equilibrium equation is shown as follows.

$$\rho c \frac{\partial T}{\partial \tau} = \frac{\partial}{\partial x}\left(k\frac{\partial T}{\partial x}\right) + \frac{\partial}{\partial y}\left(k\frac{\partial T}{\partial y}\right) + \frac{\partial}{\partial z}\left(k\frac{\partial T}{\partial z}\right) + S \tag{5}$$

ρ is density, c is specific heat capacity, T is temperature, τ is time, k is thermal conductivity and S is inner heat source.

To different object, equation (5) can be transformed into equation under the coordinate system like cylindrical coordinate or spherical coordinate, even can be transformed into two-dimensional or one-dimensional equation.

Boundary conditions refer to the relation or interaction between object's boundaries and external environment in the heat change. To transient heat conduction, it usually is the external driving to force making the process happen and develop.

The boundary condition of object outer surface belongs to third-boundary conditions, which mainly is the radiation heat transfer and convection heat transfer from external environment factors to object. The boundary condition of object inner surface is mainly caused by the inner heat sources, for buildings it can be considered as room temperature.

4.3 Calculation of Background Surface Temperature

Compared with the target, the background (like ground surface) has less survey information (mainly terrain information) such as DEM/DSM data and large dimensionality. It is high-computational to obtain surface temperature by establish 3D heat conduction equation. Because the target infrared radiation is paid more attention and it is unnecessary to obtain the precise temperature of background like ground surface, water body and sky, this paper simplifies background temperature balance equation into one-dimensional equation to obtain the average temperature value of surface, and then simulates infrared background combining the surface texture gained from terrain information.

Take ground surface for example, the analysis of calculating surface temperature is shown in Figure 3.

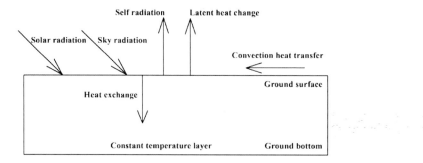

Fig. 3. Ground surface temperature model

In the Figure, the heat change between ground surface and external environment and between ground surface and the inner is analyzed, in order to simplify the calculation, one-dimensional heat conduction equation in vertical direction is established as follows.

$$\rho c \frac{\partial T}{\partial \tau} = \frac{\partial}{\partial z}\left(k \frac{\partial T}{\partial z}\right) \tag{6}$$

ρ is density, c is specific heat capacity, T is temperature, τ is time, k is thermal conductivity and z coordinate in vertical direction.

On the bottom boundary, $T|_{bottom} = const.$

On the top boundary, $T|_{surface} = Q$, Q is the summation of various heat change on surface.

5 Infrared Scene Generation

After 3D scene modeling and temperature – key factor of infrared radiation characteristics – calculation, infrared scene can be produced by simulating infrared thermal imaging sensor.

5.1 Infrared Temperature – Radiation Model

Target surface infrared radiation received by sensor mainly is three parts: target self radiation, target reflects radiation and environmental radiation, which is shown as follows.

$$E_{\det ector} = \varepsilon\overline{\tau} \cdot E_{self} + (1-\varepsilon)\overline{\tau} \cdot E_{reflect} + (1-\overline{\tau}) \cdot E_{envi} \tag{7}$$

$\overline{\tau}$ is atmospheric average transmittance, ε is target surface emissivity. In the formula:

◇ E_{self} is self radiation, $E_{self} = \int_{\lambda_1}^{\lambda_2} \frac{C_1}{\lambda^5 (e^{C_2/\lambda T} - 1)} d\lambda$, λ_1, λ_2 is range of sensor wavelength;

◇ $E_{reflect}$ is reflect radiation, mainly is reflect solar radiation, ground radiation and sky atmospheric radiation;

◇ E_{envi} is infrared environment radiation received by detector, $E_{envi} = \int_{\lambda_1}^{\lambda_2} \frac{C_1}{\lambda^5 (e^{C_2/\lambda T_{atm}} - 1)} d\lambda$, T_{atm} is air temperature.

5.2 Infrared Radiation Gray Mapping Model

Sensor transfer function is a major indicator of sensors. Different sensor has different transfer function. The infrared radiation gray mapping model is established as follows.

$$G = [r + (1-r) \cdot \frac{E - E_{\min}}{E_{\max} - E_{\min}}] \times 255 \tag{8}$$

[] is get full function, G is the gray value corresponding to the radiation value E. $E_{\min}$ and $E_{\max}$ respectively represent the maximum and minimum of radiation in infrared scene. $r (0 \leq r < 1)$ is a fixed value to a given scene, which is equivalent to the value of infrared ambient light.

5.3 Infrared Scene Texture Generation

As for the background, the gray value is relatively simple for its simple model. In order to achieve a more lifelike simulation effect, this paper makes a mapping from the surface texture obtained by terrain information to the infrared scene produced. This method not only retains the infrared radiation information truly, but also adds texture information to enhance the simulation fidelity.

Figure 4 is the infrared scene image sequence of target region one day of 24 hours. It can be seen by comparatively analyzing the image sequence that the

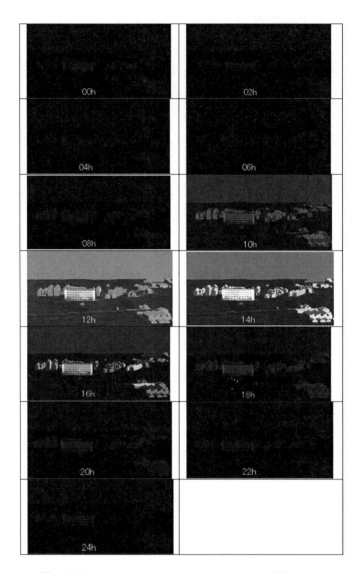

Fig. 4. Infrared scene image sequence (one day of 24 hours)

infrared scene sequence produced by this method truly reflects the infrared radiation characteristics, the gray (temperature) gradually increase from night to day until reaches its highest point in the afternoon, and then gradually decreases until reaches its lowest point in midnight, the change rule is consistent with the actual situation. In addition, through adding texture information, it predigests the calculation of background temperature, also has the fidelity of scene simulation.

6 Conclusion

This paper presents a kind of infrared scene simulation method. Infrared simulation image sequence is obtained by using satellite remote sensing data, which not only reflects target region variation rule in the infrared band but also has good fidelity. Next the algorithm of target model will be improved to reach a higher fidelity.

References

1. Xiang-Yin, L., Ling, Y.-S., Huang, C.-c.: Calculation of surface temperature and infrared radiation for ground target. Infrared and Laser Engineering 35(5), 563–567 (2006)
2. Zhang, J.-Q., Fang, X.-P.: Infrared Physics. Xidian University Press, Xi'an (2004)
3. Zhang, Y.-S., Gong, Z.-D., et al.: High resolution remote sensing satellite application – imaging model, processing algorithm and applying technology. Science Press, Beijing (2005)
4. Repasi, E., Greif, H.J.: Generation of dynamic IR-Scenes for ground-based systems and missile applications. Proceeding of the SPIE 3436, 460–461 (1998)
5. Cao, Z., Chen, H.: Quick 3D reconstruction of the IKONOS-satellite stereo images. Computer and Digital Engineering 34(11), 9–11 (2006)
6. Ben-Yosel, N., Rahat, B., Feigin, G.: Simulation of IR images of natural backgrounds. Applied Optics 22(1), 190–193 (1983)
7. Wollenwebe, F.G.: Weather impact on background temperatures as predicted by an IR background model. SPIE 1311, 119–128 (1990)
8. Akiyama, T., Tamagawa, Y., Yanagisawa, T.: Simulation of visible/infrared sensor images. SPIE 2744, 61–67 (1996)
9. Hahn, V.: Scene generation integration into a common simulation framework. SPIE 6237, 23–31 (2006)

Research on Role-Based Agent Collaboration in WSN

Lei Yan, Xinying Wang, and Dongyang Zhang

Department of Computer, North China Electric Power University,
Yonghuabei Street.619, 071003 Baoding, Hebei, P. R. China
leio@live.com, wangxinying@126.com, zdynew@163.com

Abstract. The sensor nodes in wireless sensor network have limited capacities, complete complex tasks through collaboration. In multi-agent system, the agents complete their work through collaboration. Collaboration is one of the key issues in multi-agent system research. Taking into account the similarities between multi-agent systems and wireless sensor networks, it is possible to use agent collaboration in wireless sensor network. In this paper, we use the role-based agent collaboration to help sensor nodes, and construct a Wireless Data Acquisition and Analysis System with role-based agent collaboration. Simulation experiment results show that, when the number of nodes is large, the use of role-based agent collaboration can help it to effectively improve the system load, and reduce channel congestion. However, in the area of role definitions, roles dynamically added and so on, there are great challenges.

Keywords: wireless sensor network; agent; collaboration; role; data analysis.

1 Introduction

WSN(Wireless Sensor Network, WSN) contains a large number of intelligent sensor nodes, distributed in a wide range of geographical areas, quasi real-time monitoring, sensing and collecting data of environment or objects monitored in the region, process data, get accurate detailed information , and deliver it to users [1]. WSN has features such as high-precision monitoring, flexible arrangement, low cost etc., and illustrates a good prospect of application in military surveillance, industrial control, traffic monitoring, environmental monitoring, etc.[2][3]

Since the communication, processing and sensing capability of a single sensor node is limited, it is unable to handle large-scale complex problems. In most cases sensor nodes can not access global information around whole network, and require collaborative communication capabilities [4]. WSN collaboration mainly refers to the collaboration on resource, collaboration on task, collaboration on Signal and Information [5]. Collaboration on resource, signal and information services for the collaboration on task in deed [6][7].

Agent and MAS (Multi-Agent System , MAS) can be regarded as the further development of a distributed problem solving[8].Agent has a different problem-solving ability each, Agents can communicate and collaborate with each other in accordance with agreed protocols, making the whole system a superior performance. The collaboration between multiple Agents can improve the performance; enhance the ability of solving problems. Collaboration among the MAS is the core issue of research on MAS, and it is the key to make it working together over Agents, but also one of the key concepts to distinguish MAS from other related research areas (such as distributed computing, object-oriented systems, expert systems, etc.) [9]. Considering the similarity between WSN and MAS, arranging agent to resolve the issue of collaboration between sensor nodes is possible.

The remainder of the paper is organized as follows. Section 2 describes the role-based agent collaboration used in WSN in this role-based agent collaboration. Section 3 introduces the structure of sensor node and structure of agent in this system. Section 4 presents the role definition and the overall of the system function. Finally, Section 5 concludes providing final remarks.

2 Role-Based Agent Coordination

When one agent plays a role, it will provide specific services and request the some corresponding services. the role defines that the user can request the service from others, on the other hand defines that the user must provide the service to others. The role becomes the key medium between users` interaction [10]. As the collaborative dynamical and variable environment, the process need to be adjusted dynamically, so the requests of users change dynamically [11]. The definitions of services and roles are as follows:

2.1 Service

Service is a group of independent feature set, defined as a five-tuple s:: = <I, n, d, In, Out>, where: I is the identity of the service; n is the name of the service; d is a description of function provided by the service; In is the service input; Out is output of the service.

2.2 Role

The role defines description of services that the user must provide and it can request, defined as a five-tuple r:: = <I, Sp, Sr, Na, Ds>, where, I is the identity of the role; Sp is a set of Services provided, Sr is a set of services that may request, Na is a group of agent ID who are playing the role; Ds is a description of the role.

2.3 A Collaborative System

A collaborative system can be defined as a 9-tuple.:: = <C, O, A, M, R, E, G, s0, H>, where C is a set of classes, O is a set of objects, A is a set of agents, M is a set of messages, R is a set of roles, E is a set of environments, G is a set of groups, s0 is the initial state of a collaborative system, and H is a set of agent developers. Every group has an environment in which agents play roles.

The main tasks developing a role-based system are specifying roles and the relationships among roles, specifying role players and assign roles to agent.

3 Node Architecture

3.1 Sensor Node Architecture

Sensor node architecture is shown in Fig.1. the hardware part includes CPU, main memory, Flash memory, sensors, wireless communication module, power supply ,etc. The software part is set of agents with roles. Software programs is stored in Flash memory, and will not be lost even the sensor node loses power. When the system gets normal power supply, the program will be carried into the main memory to run.

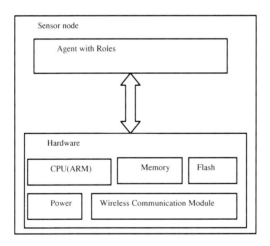

Fig. 1. Sensor Node Architecture

3.2 Agent Architecture

Agent consists of a number of knowledge modules, and executable modules. Knowledge module contains the information about WSN environment, the number of close nodes, the roles acting, and the goals to accomplish. The problem solver module is responsible for selecting the appropriate solution. Planning module is to perform the corresponding actions in accordance with the planning table. Communication module is responsible for sending and receiving information through a wireless module, part of the hardware. MOBILE agent module will be deployed when necessary to the appropriate node, process the data, carry the result back. Agent Architecture is shown in Fig.2.

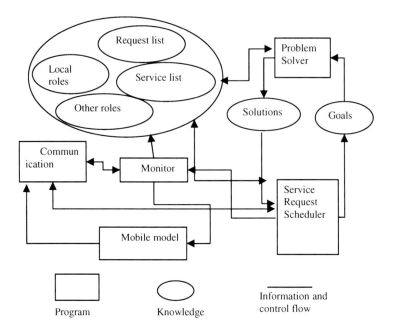

Fig. 2. Agent Architecture

4 Role Definition

There are a number of roles in the system. They are defined as follows:

4.2 *Data Acquisition Node*

The role is responsible for collecting a variety of data from the field, determining data simply, storing data, and transmitting part of the data via wireless paths to the agents playing role of fault analysis node nearby. Such roles can be deployed in large numbers to achieve a good coverage, its main functions are: to provide the original sampling data inquiry service, simple data fault diagnosis service, and to request collecting data from others.

4.3 Data Analysis Node

The role is responsible for analysis the data received, resulting the initial judgment including whether the failure occur of the general type of failure. Based on results of the analysis, it send mobile agent to the Data Acquisition Node sending fault data back and request details of the original sampled data. Based on judgment of mobile agent, and detailed sampling data, it analyzes data and makes judgments. It makes it as a service provided to other roles. Such node has both response structure and deliberative structure, and higher bandwidth wireless transmission. But there is not many agents playing this role. Such roles can provide services as: data analysis services, mobile agent deployment services, providing data analysis, routing and other services. Node requested service as: query original sampled data, query the results of simple data failure judgment.

4.4 Fault Diagnosis Node

The role detailed analysis results and related sample data get from Data analysis node, using its powerful computing capabilities and a more detailed expert knowledge base to determine the detailed type of failure to record the original sampled data. And the results will be provided as a service to interface node. Such roles can provide services as: a detailed analysis of fault data, fault determination. Nodes request services as: request data analysis, request for the original sampling data, and request the use of expert knowledge base.

4.5 Expert Knowledge Base Node

This role is responsible for maintaining a detailed knowledge base of experts, making knowledge base up to date through learning content information. Such roles can provide services: using expert knowledge base. Node requested service: query of the original sample data, query the results of Fault diagnosis.

4.6 Interface Role

The role is responsible for providing interface between fault monitoring system and fault diagnosis nodes. Through this interface a fault monitoring system running on the PC can receive the running conditions of all sensor nodes; fault conditions, detailed original sample data, etc., and thus demonstrate them to the technical staff. Such roles are able to provide services as follows: providing system running condition, providing results of fault determines and interaction with technical staff. Node requested service: query the original sample data, query the results of failure determine.

5 System Overall

5.1 Choosing the Roles

Agents and roles constitute the role net which is shown in Fig.3

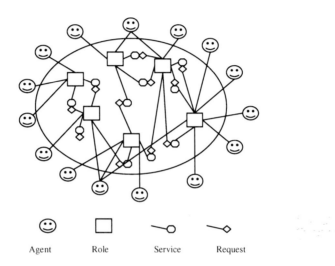

Fig. 3. The Roles Net

After deployment, Agent must choose roles to play. Each agent must submit its chosen roles to the role registry agent. An agent can choose a number of roles, but is not allowed choosing none of the roles. Just System deployment begins, agent selects roles randomly. As the system running, the agent who is not busy can choose more roles, and relatively busy agents can refused to play some certain roles any more. The role registry agent knows all the changes about agent playing roles. The role registry agent can assign roles to any particular agent by sending a mobile agent to it. After choosing the roles, agents begin to accomplish different tasks based on the roles chosen.

5.2 Data Sampling

Large amounts of Data acquisition node collect their own data using a variety of high-speed sensing devices. Data Acquisition node packages data and mark as "normal" "abnormal" or other marks, then brought data into their own FLASH storage devices. It reports anomalies.

5.3 Data Analysis

After receiving an "exception" report, Data analysis nodes send mobile agent to the data acquisition node who submitted the exception report. When reaching the data acquisition node, Mobile agent get control of the node, analysis data, return carrying the analysis results and detailed original sample data analysis to the node. Data analysis nodes analysis the results of some mobile agents and the detailed raw data, and determine the result. By sharing of the results concluded by mobile agents, the arbitration mechanism will work and obtained the results of the analysis.

5.4 Faults Identify

Fault identify node requests the original sampled data and coordinates analysis results of Data analysis nodes, and requests for expert knowledge base and fault type library providing knowledge to determine the fault type and maintenance programs.

5.5 Interact with the Technical Staff

Interface Node requests for the results of fault diagnosis. Monitoring software will show them to the technical staff. Technical staff can interact with it. Through the interface node, technical staff can query fault conditions, and the raw data.

6 Simulation

System-level simulation experiments were carried out on the OMNET + +. The simulation result is shown in figure 4. When the number of nodes is large, the use of role-based agent collaboration can help WSN effectively improve the system load, and reduce channel congestion.

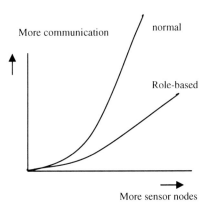

Fig. 4. Simulation Result

7 Conclusion

Agent collaboration applied to WSN nodes can make nodes more collaborative to accomplish complex task only done by the centralized system in the past. The role-based agent collaboration helps the system designer to complete the system design through role definition and definition of the relationship between the roles. The designer can be very flexible to deploy new sensor nodes in the system, in order to enhance the performance of WSN. Good role definitions help the agent to perform effectively to enhance the performance of WSN. However, there is no common approach in the role definition. How to add a new role in the running system dynamically is well worth to study.

Acknowledgment. The research is supported by the youth research fund of North China Electric Power University (200811017).

References

1. Culler, D., Estrin, D., Srivastava, M.: Overview of sensor networks. IEEE Computer 37, 41–49 (2004)
2. Limin, S., Jian, L., Chen, Y.: Wireless Sensor Networks. Tsinghua University Press, Beijing (2005)
3. Haibin, Y., Peng, Z.: Intelligent wireless sensor network systems. Science Press, Beijing (2006)
4. Akyildiz, F., Su, W., Sankarasubramaniam, Y., Cayirci, E.: Wireless Sensor Networks: A Survey. Computer Networks 38(4), 393–422 (2002)
5. Estrin, D., Govindan, R., Heidemann, J., Kumar, S.: Next century challenges: scalable coordination in sensor networks. In: Proceedings of International Conference on Mobile Computing and Networks (MobiCom 1999), Seattle, Washington, pp. 263–270 (1999)
6. Kumar, S., Zhao, F., Shepherd, D.: Collaborative signal and informasion, processing in microsensor networks. IEEE Signal Processing Magazine 19(2), 13–14 (2002)
7. Shi, H.-s., Yang, S.-J., Fai, R.H.: Wireless sensor networks, collaborative information processing Signal Research. Information and Control 35(2), 225–232 (2006)
8. Wooldridge, M.: An Introduction to MultiAgent Systems. John Wiley & Sons Ltd., West Sussex (2002)
9. Hussain, Y.S., Shakshuki, E., Matin, W.: Agent-base System Architecture for Wireless Sensor Networks. In: The Proceedings of The Second International Workshop on Heterogeneous Wireless Sensor Networks (HWISE 2006), in conjunction with the 20th International Conference on Advanced Information Networking and Applications, Vienna, Austria, April 18-20. IEEE Computer Society, Los Alamitos (2006) (accepted for presentation)
10. Zhu, H.: From WYSIWIS to WYSINWIS: Role-Based Collaboration. In: Proceedings of the IEEE International Conference on Systems, Man, and Cybernetics, The Hague, Netherlands, pp. 5441–5446 (2004)
11. Liu, L.y., Zhu, H.-b., Shang, W.-q.: A Role-Based Research on Dynamic Agent Service Coordination Mechanisms. Journal of System Simulation 19(1), 38–43 (2007)

Spectral Matting Based on Color Information of Matting Components

Jia-zhuo Wang and Cui-hua Li

Image and Video Processing Laboratory,
Computer Science Department,
Xiamen University, China
wangjiazhuo777@gmail, com chli@xmu.edu.cn

Abstract. Color information carried by the matting components in spectral matting is used in our method to solve the matting problem. By comparing the color similarity of each matting component, the matting components were separated into two groups: the foreground and the background. There are two advantages of doing so. Firstly, there is no need to use the trimap which is a necessary tool in traditional matting solving methods to accomplish the process of semi-supervised learning. Instead, the color similarity information could be used as constrain or heuristic cue in finishing the process of unsupervised learning to lower the blindness of automatically extracting the foreground object. Secondly, combining the color similarity information with a little user provided foreground and background cue could further reduce the number of matting components whose group have not been known yet, and thus a semi-supervised learning process which could only be done by using a lot user provided cue in traditional matting solving methods could be accomplished.

Keywords: spectral method; matting; color histogram; unsupervised learning; semi-supervised learning.

1 Introduction

How to extract the foreground object from an image is a main problem in image editing. Many matting algorithms typically assume that each pixel I_i in an input image I is a linear combination of a foreground color F_i and a background color B_i:

$$I_i = \alpha_i F_i + (1-\alpha_i) B_i \qquad (1)$$

α_i is used to depict the opacity of foreground object in that pixel, i=1,2,.....,N(N is the total number of pixels in input image I).

Traditional matting algorithms, such as [1][2][3] typically use an auxiliary tool, trimap, to solve this problem. As illustrated in Fig. 1, a trimap is used to point the

input image   trimap

Fig. 1. An image and it's trimap

definite foreground region (white region), definite background region (black region), and unknown region (gray region).

However, it requires a large amount of user effort to draw a trimap. To eliminate this disadvantage, [4] introduced a method to automatically generate a trimap. But, it used a device that can automatically extract the depth information of an input image in the process of image collection. In our work, we do not consider such requirement for special hardware device.

So how to reduce or even eliminate the user effort becomes a main problem in matting algorithms. [5] introduced a concept of spectral matting, and we manage to solve the problem by adopting color information in spectral matting. The basic idea of spectral matting in [5] is to generalize equation (1) by assuming that each pixel is a convex combination of K image layers $F^1, \ldots, F^k$

$$I_i = \sum_{k=1}^{K} \alpha_i^k F_i^k \qquad (2)$$

The K vectors α^k (k=1,2,......,K) are the matting components of input image I ($\alpha^k \geq 0$ and $\sum \alpha_i^k = 1$). The matting components are grouped into two classes: the foreground matting components and the background matting components. And the foreground object is extracted by adding all foreground matting components. This process is illustrated in Fig. 2: the foreground matting components are framed in red, and combining all these foreground matting components yields the foreground object, framed in blue. Please note that images in Fig. 2 are gained from our algorithm, not from [5].

Our method is to use color information carried by matting components to instruct the grouping process. Our idea is based on the assumption that matting components with similar color information should be grouped into the same class. The process is: firstly, the color information carried by a matting component is presented by the corresponding color histogram of that matting component; then color similarity between different matting components are measured by the distance between the color histograms of different matting components; finally, the gained color similarity between different matting components is used to instruct the grouping process. There are two different ways to utilize the gained color similarity. The first

Spectral Matting Based on Color Information of Matting Components

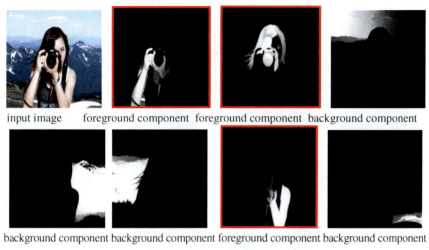

input image   foreground component   foreground component   background component

background component background component foreground component background component

background component   foreground object

Fig. 2. Foreground object gained by adding all matting components

one is an unsupervised learning way which means that under the situation without any user provided foreground and background cues, the color similarity between different matting components could be used as either the constrains of the graph-cut algorithm introduced in [5] or heuristic information during the exhaustive algorithm, reducing the blindness of graph-cut and exhaustive algorithm, and thus the foreground object could be extracted more efficiently. The second one is a semi-supervised learning way: a little amount of user provided cues are used to make sure the class of a portion of matting components, then by utilizing the color similarity between different matting components, the class of some other matting components could be known, so the range of matting components with known class could be gradually enlarged, finally the grouping process will be finished and the foreground object could be extracted by adding all foreground matting components. As illustrated in Fig. 3 the user provided cues in the semi-supervised learning do not involve as much user effort as the trimap.

The process of computing color similarity between matting components will be presented in Section 2. In Section 3, we will explain how to apply the color similarity into the unsupervised and semi-supervised learning, instructing the

Fig. 3. Foreground and background cue

grouping process and thus extracting the foreground object. Experiment results are shown in Section 4.

2 Color Information of Matting Components

There are some pixels in a matting component where α_i^k are near 1, this means these pixels look more light than the other pixels. Then there are some corresponding pixels in the input image I. The color information of the matting component is the color information carried by the corresponding pixels in the input image I. So when generating the color histogram of a matting component, we only need to count the color information of such pixels in the input image I that the gray value of the corresponding pixels in that matting component are above a threshold. Counting only the color information of these light pixels is based on an assumption that the amount of color information carried by these pixels are much enough to represent the color information of the matting component, this means that the other dark pixels are neglected in the process of depicting the color information because they are less representative than the light pixels. This assumption is reasonable in the sense that counting too much the color information of dark pixels is actually disturbance to the most representative color of that matting component and therefore impacts the color similarity measurement between different matting components in the next step. The above concept is illustrated in Fig. 4 where an example of light pixels in a matting component, the corresponding pixels in the input image and the color histogram of that matting component are given. Next we will expound the process of generating the color histogram and the measurement of color similarity between different matting components.

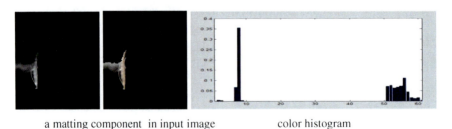

a matting component in input image   color histogram

Fig. 4. A matting component and it's color histogram

2.1 Generation of Color Histogram

In this paper, the HSV color space is adopted to generate the color histogram of matting components. Typically the color space is divided into several different regions, and it will be counted that how many pixels are in each region. However, how to divide the color space to better represent the color distribution information is a problem. [6] gave us a hint about this by analyzing the feature of the HSV color space: for low values of Saturation, a color can be approximated by a gray value specified by the Intensity level while for higher Saturation, the color can be approximated by its Hue. The Saturation threshold that determines this transition is once again dependent on the Intensity. For low intensities, even for a high Saturation, a color is close to the gray value and vice versa. So it is reasonable to use the Saturation value of a pixel to determine whether the Hue or the Intensity is more pertinent to human visual perception of the color of that pixel, and we use the following function to determine the threshold of the Saturation value:

$$T = \frac{1}{1 + \alpha \cdot Value} \quad \alpha \in [1, 4] \quad (3)$$

Value represents the Intensity.

When Saturation value is above T, we use Hue. When Saturation value is under T, we use Intensity.

[7] gave us a further hint about that into how many regions the color space should be divided: the colors that can be perceived by human visual systems generally include red, yellow, green, cyan, blue and magenta, and these colors are not uniform distributed according to the Hue subspace, so the Hue subspace could be divided into 6 unequal regions corresponding to 6 colors mentioned above. And the Intensity subspace could be divided into 3 unequal regions, black, gray, and white. So the HSV color space is divided into 9 different unequal regions.

In this paper, to better distinguish different colors, we further divide the Hue subspace into 50 unequal regions, and Intensity subspace into 11 unequal regions. So finally the HSV color space is divided into 61 different unequal regions.

2.2 Measurement of Color Similarity between Matting Components

The color histograms of matting components are used to measure the color similarity between matting components. We could use the distance between the color histograms to represent the color similarity between matting components: the smaller the distance is, the higher the similarity is and vice versa.

Before computing the distance between the color histograms, the color histogram should be normalized at first: n_i counts the pixels in the ith region, n counts the total number of pixels, after normalization, n_i should be:

$$n_i = \frac{n_i}{n} \qquad (4)$$

[8] introduced four different distance measurement functions of color histograms. Given that n_i, m_i respectively count the pixels in the ith region of two different color histograms, the four kinds of distance measurement functions are:

$$D_I = \sum_i \min(n_i, m_i) \qquad (5)$$

$$D_{L_1} = \sum_i |n_i - m_i| \qquad (6)$$

$$D_{L_2} = \sqrt{\sum_i (n_i - m_i)^2} \qquad (7)$$

$$D_{\chi^2} = \sum_i ((n_i - m_i)^2 / (n_i + m_i)) \qquad (8)$$

In this paper, we adopt equation (8) as our color histogram distance function.

3 Application of Color Information into Unsupervised and Semi-supervised Learning

The color similarity between different matting components could be used to instruct the process of grouping the matting components into two classes. Generally speaking, there are two aspects about the color similarity. On one hand, matting components with high similarity tend to be in the same class. On the other hand, matting components with low similarity tend to be in different classes. Next we will expound how to apply the above two aspects about color similarity into both unsupervised and semi-supervised learning.

3.1 Unsupervised Learning

Among all the matting algorithms, the method in [5] tried to automatically extract the foreground object for the first time without any user effort. It used a balance-cut algorithm to minimize the cost $J(\alpha)$ function to classify the matting components. In our paper, we apply the color similarity information into the graph-cut algorithm, and make an adjustment about the generation of an edge's weight, which represents the cost of grouping the corresponding two matting components, represented by the two vertices of that edge, into the same class.

The adjustment process is that firstly, there are two similarity controlling thresholds, the low similarity threshold and the high similarity threshold (the high one is above the low one): on one aspect, if the color similarity between two matting components is above the high similarity threshold, which signals that there is a high possibility that the two matting components belong to the same class, then we should adjust the corresponding edge's weight into as α times as the original weight (0<α<1), which means that we are making the cost of grouping the two matting components into the same class lower than before; on the other aspect, if the color similarity between two matting components is under the low similarity threshold, which signals that there is a high possibility that the two matting components belong to different classes, then we should adjust the corresponding edge's weight into infinity. After such an adjustment, the edge's weight represents more accurately the cost of grouping the corresponding two matting components into the same class.

The solution of the cost function $J(\alpha)$ gained from the graph-cut algorithm is not optimum, it's just an approximate solution. In fact, when there are a small number of matting components, the exhaustive algorithm could be adopted. We could specify each matting component into either foreground class or background class one by one, in this way we get a combination about the grouping way, then we could see the costs of all possible combinations and find one or several combinations with minimum cost as our candidate grouping way. In this process, we could use similar method as mentioned in the above paragraph to generate the value of the cost function by using the color similarity information between matting components.

3.2 Semi-supervised Learning

In our work, the difference between unsupervised learning and semi-supervised learning is that under during the process of semi-supervised learning, there is a little amount of user provided cues by which we could group in advance a portion of matting components into either foreground or background matting components.

We could enlarge the range of matting components with known class by utilizing the color similarity information between matting components. This process is a four-steps algorithm. The flow diagram of the algorithm is illustrated in Fig. 5.

A more specific version of this algorithm is:

(1). Classifying a portion of matting components by utilizing user provided cues. After this first step, each matting component should be labeled as one of the three states, 'foreground', 'background', and 'unknown'.

(2). Computing the similarity matrix S between matting components. A element of the matrix S, S(i, j), is the color similarity between matting component i and matting component j, computed from equation (8).

(3). Setting a similarity threshold. By utilizing the similarity matrix S, we could find all pairs of matting components whose color similarity is above the similarity threshold, such a pair of matting components is called a candidate pair. In the next step, we will deal with some matting components in 'unknown' state through these candidate pairs, trying to label these matting components as either 'foreground' or 'background' state.

(4). For each candidate pair of matting components, we will make the following judgment process: A. if one matting component of the candidate pair, call it component 1, has been labeled as 'foreground' or 'background' state, and another matting component of the candidate pair, call it component 2, has been label as 'unknown' state, then we should have make component 2's label the same as component 1's label (assuming that this label is 'foreground'), however, if in the judgment process of another candidate pair, there is a situation that we should change component 2's label into 'background', we say that an "conflict" emerges. When we found a "conflict", component 2's label should maintain as 'unknown' state. B. For other situations not appeared in A, the two matting components of the candidate pair should maintain their original state.

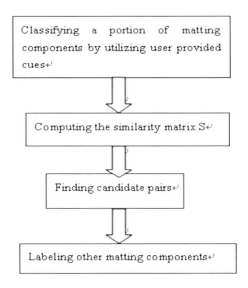

Fig. 5. flow diagram of our semi-supervised learning algorithm

Through the above four-steps algorithm, we enlarge the range of matting components with known class. However, this algorithm can not guaranty that all the 'unknown' matting components could be labeled as 'foreground' or 'background', which means there may exist some matting components that we can't make sure to which class they belong. Fortunately, the number of such left 'unknown' matting components is not too much. So we can get an optimum solution of the cost function $J(\alpha)$ by using an exhaustive algorithm. Until now, we have finished a process of semi-supervised learning to group the matting components, and extract the foreground object.

4 Results and Comparisons

We will present our results under both unsupervised learning and semi-supervised learning.

4.1 Comparison under Unsupervised Learning

Traditional matting algorithms typically use an auxiliary tool trimap, so they can not work with no user input. The method in [5] is the only algorithm before this paper that tries to automatically extract the foreground object. So under the unsupervised learning situation, we will compare our results with that of [5]. The results of [5] under unsupervised learning are presented in Fig. 6, showing us several candidate result effect. We can easily identify that a large portion of background has been recognized as foreground in [5]. Our results are shown in Fig. 7, we apply the color similarity information into the graph-cut algorithm. The quality of the results has been improved compared with that of [5], without recognizing a large portion of background as foreground.

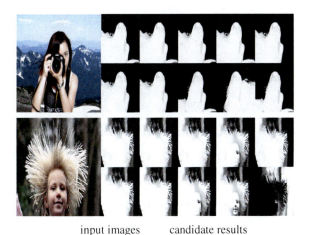

input images   candidate results

Fig. 6. Results of [5] under unsupervised learning

Fig. 7. Our results under unsupervised learning

4.2 Comparison under Semi-supervised Learning

We have compared several matting algorithms with our method under the situation that there is only a little amount of user provided cues, results are shown is Fig. 8. We can tell from the results that other algorithms can not work normally under such a situation.

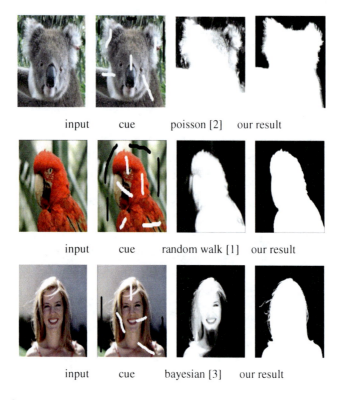

Fig. 8. Results comparison of several methods under semi-supervised learning

4.3 Complexity Analysis of Exhaustive Algorithm

The complexity of exhaustive algorithm is directly associated with the number of matting components whose classes are unknown. Given that this number is M, then the complexity of exhaustive algorithm is $O(2^M)$. Next we will give how many this number has decreased by if we apply the color information into the grouping process.

Under unsupervised learning, the number of matting components whose classes are unknown is initialized as 30, for the two input images in Fig. 6, we run our method 10 times. For the first input image (the photographer) the average number of matting components whose classes are still unknown is 10 during the 10 times run; for the other input image the average number of matting components whose classes are still unknown is 13 during the 10 times run. This means that the number of matting components whose classes are unknown decreases respectively by 20 and 17.

Under semi-supervised learning, the number of matting components whose classes are unknown is initialized as 40, for the three input images in Fig. 8, we run our method 10 times. For the first input image (koala), if we only use the user provided cues about foreground and background, the average number of matting components whose classes are still unknown is 28 during the 10 times run, if we add the color information into this process, the average number of matting components whose classes are still unknown is 14 during the 10 times run; for the second image (parrot), the corresponding two numbers are respectively 27 and 12; for the last image, the corresponding two numbers are respectively 23 and 8. So after using the color information to instruct the grouping process, the number of matting components whose classes are still unknown has further decreased respectively by 14, 15 and 15 for the three input images.

5 Conclusion

In this paper, we introduce the color information of matting components into spectral matting algorithm to instruct the process of grouping the matting components for the first time. Besides, we have specifically illustrated how to apply the color similarity information between matting components into both unsupervised and semi-supervised learning, and thus extract the foreground object. Generally speaking, our method has minimized the user effort under the premise of having good extracting result which is proven by our experiments results under both unsupervised and semi-supervised learning.

References

1. Grady, L., Schiwietz, T., Aharon, S., et al.: Random walks for interactive alpha-matting. In: Proceedings of VIIP 2005, pp. 423–429. Oacta Press, Benidorm (2005)
2. Sun, J., Jia, J., Tang, C.-K., Shum, H.-Y.: Poisson matting. In: Proceedings of ACM SIGGRAPH 2004, Los Angeles, California, USA, pp. 315–321 (2004)
3. Chuang, Y.-Y., Curless, B., Salesin, D.H., Szeliski, R.: A bayesian approach to digital matting. In: Proceedings of CVPR 2001, pp. 264–271. IEEE, Hawaii (2001)
4. Wang, O., Finger, J.: Automatic Natural Video Matting with Depth. In: Proceedings of Pacific Conference on Computer Graphics and Applications, Pacific Graphics, Maui, Hawaii, USA (2007)
5. Levin, A., Rav-Acha, A., Lischinski, D.: Spectral Matting. In: Proceedings of CVPR 2007, pp. 1–8. IEEE, Minneapolis (2007)

6. Sural, S., Qian, G., Pramanik, S.: Segmentation and Histogram Generation using the HSV Color Space for Image Retrieval. In: Proceedings of the 2002 International Conference on Image Processing, pp. 589–592. IEEE, Rochester (2002)
7. Zhengjun, L., Shuwu, Z.: An Improved Image Retrieval Method Based on the Color Histogram. Control and Automation Publication Group 24(2-1) (2008)
8. Jou, F.-D., Fan, K.-C., Chang, Y.-L.: Efficient matching of large-size histograms. Pattern Recognition Letters 25(3), 277–286 (2004)

The Correlation of Conditional Time Series of Sunspot Series

Wenguo Li[1], Haikun Zhou[2], and Hong Zhang[1]

[1] Department of Mathematics, Hebei University of Engineering,
Handan, P.R. of China
[2] Logistics Management Department, Hebei University of Engineering,
P.R. of China

Abstract. This paper investigates the correlation properties of original sunspot series and their conditional time series. Appling the R/S method, we obtain the correlated exponent for original series and the conditional time series. The conclusions indicate that the Hurst exponents value H decrease from 1.1264 to 0.9896, with the increasing of q from $0.15\overline{x}$ to $0.90\overline{x}$.

Keywords: rescaled range method, correlation, conditional time series, Hurst exponent, sunspot time series.

1 Introduction

In recent years the Hurst exponent H invented by H. E. Hurst has been established as an important tool for the detection of long-range correlations in time series. Although its application was to model the fluctuations of Nile River in 1951[1], subsequently the applications were spread to diverse disciplines and problems including traffic science [2], economics time series [3, 4], and geology [5], as well as other fields. For estimating the Hurst exponent, Mandelbrot and Wallis further developed the rescaled range statistic and introduced a graphical technique [6].

For the furtherly detection of long-range correlations, we introduce the concept of the conditional correlation for the physical phenomenon, which are the correlations of the conditional time series following some threshold q. obviously, the conditional correlation of a certain random event depends on the history or previous event.

In this paper, we study systematically how different threshold q affects the correlation of the sunspot time series.

The paper is organized as follows: Firstly, a brief review of the rescaled range analysis method is given in Section 2. Section 3 is about the details of conditional time series. In Section 4, we discuss the conditional correlation properties of the sunspot time series. Finally, the important conclusions of this paper are presented in Section 5.

2 Methodology

We consider a record $x(k), k = 1, 2, 3, \cdots, n$, in most applications, the index k will correspond to the time of the measurements. We are interested in the correlation of the time series. For this estimation, we calculate the scaled, adjusted range:

$$\frac{\max(0, w_1, w_2, \cdots, w_n) - \min(0, w_1, w_2, \cdots, w_n)}{S(n)} \qquad (1)$$

where $s(n)$ is the standard deviation and for each $k = 1, 2, 3, \cdots, n$, w_k is given by: $w_k = (x(1) + x(2) + \cdots + x(k)) - kE[x(n)], k = 1, 2, \cdots, n$ and $E[x(k)]$ is the sample mean.

In order to obtain the Hurst exponent, this computation is repeated for different scales n. And then, the slope of the log-log plot for $\frac{R(n)}{S(n)}$ versus n can be calculated.

For correlated data, a power law relation between $\frac{R(n)}{S(n)}$ and n will be obtained:

$$\frac{R(n)}{S(n)} \sim cn^H \qquad (2)$$

The parameter H, called the Hurst exponent, represents the dependence properties of the data. The values of the Hurst exponent range between 0 and 1. If $H = 0.5$, there is no correlation at all and the time series is simply a white noise; if $H < 0.5$, the data is anti-correlated; and if $H > 0.5$, a persistent long-range power-law correlation is present in the time series.

3 Data and Conditional Time Series

3.1 Sunspot Time Series

Recently, the statistical properties of sun activity have been investigated by previously work [7-9]. In this paper we would like to characterize the complex behavior of sunspot time series through the computation of the conditional correlation which quantifies the correlation exponents H of conditional series. The sunspot data we used is downloaded from SIDC's website. The records collected from January, 1749 to March, 2009 through about 260 years has a sinusoidal trend, with a frequency is equal to the well known cycle of sun activity, approximately 11years, which can be easily seen from the monthly sunspot data.

3.2 Conditional Time Series

Here, we describe the conditional time series exceeding a certain threshold q in detail. For a time series $\{x(k)\}$, a conditional time series $c(j)$ occurs if $x(i-1) > q$, then $c(j) = x(i)$. Fig.1 shows a section of the conditional time series $c(j)$ for a time series for threshold $q = \overline{x}$, where $\overline{x}$ is the mean of the time series. To make a detailed investigation, we consider the conditional time series exceeding threshold $q = 0.15\overline{x}, 0.30\overline{x}, 0.45\overline{x}, 0.60\overline{x}, 0.75\overline{x}$ and $0.90\overline{x}$ respectively.

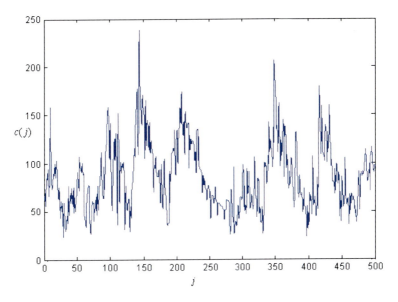

Fig. 1. The conditional time series for threshold $q = \overline{x}$

4 Results and Discussion

4.1 The Correlation Properties of Original Series and the Conditional Time Series

In order to investigate the effect of different threshold q on correlations, we then analyze how threshold q affects the Hurst exponent H of the conditional time series in this Section.

First, we apply the rescaled range method to quantify the correlation properties of sunspot series. The result is shown in Fig.2 where the symbol 'o' denote the original series. Obviously, the fluctuation function $\frac{R(n)}{S(n)}$ can be approximated by a power-law function indicate that the long-range correlation is exist. Next we calculates the scaling exponent using the equation (2) for conditional series for $q = 0.15\bar{x}, 0.30\bar{x}, 0.45\bar{x}, 0.60\bar{x},\ 0.75\bar{x}$ and $0.90\bar{x}$ respectively. The fluctuation function $\frac{R(n)}{S(n)}$ vs. n is also shown in Fig.2 denoted by symbol '*'.

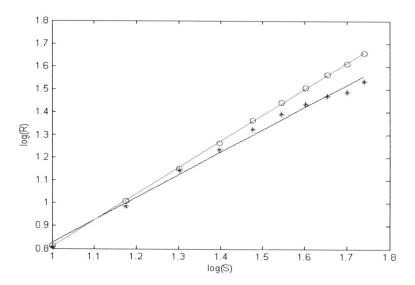

Fig. 2. The plot of $\frac{R(n)}{S(n)}$ versus n

4.2 The Effect of Q on Correlation Properties for the Conditional Time Series

In order to investigate the effect of the different threshold q on the conditional time series, we plot the Hurst exponents for the conditional series vs. threshold q from $0.15\bar{x}$ to $0.90\bar{x}$, where $\bar{x}$ is the mean of the time series $x(t)$. Fig. 3 shows the dependence of correlation properties on different thresholds q for the conditional time series. In details, we can see that with the increasing of q from $0.15\bar{x}$ to $0.90\bar{x}$, the range of the Hurst exponents value H decrease from 1.1264 to 0.9896. Obviously, the Hurst exponent is a decreasing function of the independent q.

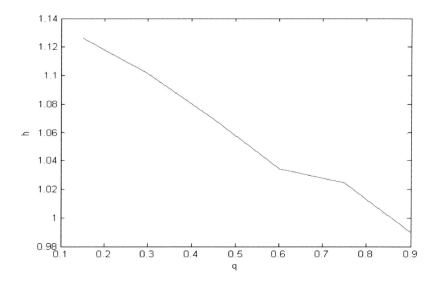

Fig. 3. The dependence of Hurst exponent on threshold q

5 Conclusion

In this paper, we first investigate the correlation properties of original sunspot series and their conditional time series. The conclusion indicate that the correlation exist in between them. For estimating the effect of different threshold q on the dependence properties of the conditional time series, we calculate the Hurst exponent h with the increasing of q from $0.15\overline{x}$ to $0.90\overline{x}$. We find the range of the Hurst exponents value H decrease from 1.1264 to 0.9896, which indicates the correlations become weaker at the large threshold q.

References

1. Hurst, H.E.: Long-term Storage capacity of Reservoirs. Transactions of the American Society of Civil Engineers 116, 770–808 (1951)
2. Hong, Z., Keqiang, D.: Multifractal Analysis of Traffic Flow Time Series. Journal of Hebei University of Engineering 26, 109–112 (2009)
3. Kumar, S., Deo, N.: Multifractal properties of the Indian financial market. Physica A: Statistical Mechanics and its Applications 2009, 1593–1602 (2009)
4. Cajueiro, D.O., Tabak, B.M.: Time-varying long-range dependence in US interest rates. Chaos, Solitons & Fractals 34, 360–367 (2007)
5. Yang, Y.-g., Yuan, J.-f., Chen, S.-z.: R/S Analysis and its Application in the Forecast of Mine Inflows. Journal of China University of Mining and Technology 16, 425–428 (2006)

6. Mandelbrot, B., Wallis, J.: Robustness of the rescaled range R/S in the measurement of noncyclic long-run statistical dependence. Water Resources Research 5, 967–988 (1969)
7. Dewan, E.M., Shapiro, R.: Are sunspot-weather correlations real? Journal of Atmospheric and Terrestrial Physics 53, 171–174 (1991)
8. Cole, T.W.: Periodicities in Solar Activity. Solar physics 30, 103–110 (1973)
9. Hanslmeier, A., Denkmayr, K., Weiss, P.: Longterm prediction of solar activity using the combined method. Solar Phys. 184, 213–218 (1999)

Hurst Exponent Estimation Based on Moving Average Method

Nianpeng Wang[1], Yanheng Li[2], and Hong Zhang[1]

[1] Department of Mathematics, Hebei University of Engineering,
Handan, P.R. of China
[2] Key Laboratory of Resource Exploration Research of Hebei Province,
Hebei University of Engineering, Handan, P.R. of China

Abstract. In this paper, we introduce moving average method to estimate the Hurst exponent of the Hang Seng Index data for the 22-year period, from December 31, 1986, to June 6, 2008 in the Hongkong stock market, a total of 5315 trading days. Further, we present a detailed comparison between the regular rescaled range method and the moving average method. We find that the long-range correlations are present by both the new method and the regular method.

Keywords: Hurst exponent, Moving average method, Time series analysis, Rescaled range, Long-range correlation.

1 Introduction

In recent years the Hurst exponent H introduced by H. E. Hurst to model the fluctuations of Nile river in 1951[1] has been established as an important parameter for the determination of fractal properties and the detection of long-range correlations in noisy signals. It has successfully been applied to diverse fields such as economics time series [2-4], geology [5, 6], and hydrology [7], as well as other fields [8]. Lots of methods are commonly used for measuring accurate estimates of H, such as the rescaled range analysis (R/S) and wavelet packet methods. The rescaled range analysis originated in hydrology where it was used by Hurst to determine the design of an optimal reservoir based on the given record of observed discharges from the lake. Mandelbrot and Wallis further developed the rescaled range statistic and introduced a graphical technique for estimating the so-called "Hurst exponent ", a measure of persistence or long memory, in a time series.

Recently, Many studies have asked the question whether stock market prices are predictable from their past [9-13]. The main focus in early research was on existence of long-range or short-range dependence in stock returns. For this purpose, the dynamic behaviors of stock markets have been studied by various techniques, such as distribution functions, correlation functions, multifractal analysis. Fortunately, stock market indexes around the world have been precisely recorded for many years and therefore represent a rich source of data for quantitative analysis.

In this paper, we will detect the long-range correlations of the Hang Seng Index (HSI), which recorded every day of trading from December 31, 1986, to June 6, 2008 in the Hongkong stock market. Specifically, we estimate the Hurst exponent H under the method of rescaled range analysis and the moving average method, respectively. The result indicates that the new method is valid to estimate the correlation properties of the time series.

The outline of this paper is as follows. In Section 2, we briefly introduce the rescaled range analysis method and the moving average method. Section 3 is about the details of the stock market data we used in this paper. In Section 4, we discuss the result of the moving average method. The conclusions of this paper are presented in Section 5.

2 Methodology

The moving average method consists of two steps. The primary step is essentially identical to the conventional rescale range procedure.

Step 1: Let us suppose that $x(k)$ is a series of length N, and the scaled, adjusted range is given by:

$$\frac{R(n)}{S(n)} = \frac{\max(0, w_1, w_2, \cdots, w_n) - \min(0, w_1, w_2, \cdots, w_n)}{S(n)} \tag{1}$$

where $S(n) = \sqrt{\frac{1}{n}\sum_{i=1}^{n}(x(i) - \overline{x}_i)^2}$, w_k is given by

$$w_k = (x(1) + x(2) + \cdots + x(k)) - \overline{x}_1 - \ldots - \overline{x}_k, k = 1, 2, \cdots, n$$

$$\overline{x}_i = \frac{1}{n}[x(i - \left\lceil\frac{n}{2}\right\rceil) + \ldots + x(i) + \ldots + x(i + \left\lceil\frac{n}{2}\right\rceil)] \tag{2}$$

Step 2: To obtain the Hurst exponent, we plot $\frac{R(n)}{S(n)}$ versus n on a log-log plot and compute the slope.

If we get a power law relation between $\frac{R(n)}{S(n)}$ and n :

$$\frac{R(n)}{S(n)} \sim cn^H \tag{3}$$

3 Data

In this paper, we investigate the price changes of the Hang Seng Index (HSI), which recorded every day of trading from December 31, 1986, to June 6, 2008 in the Hongkong stock market, a total of 5315 trading days. The records are continuous in regular open days for all trading days, and the times when the market is closed have been removed.

We define

$$g(t) = |X(t + \Delta t) - X(t)| \qquad (4)$$

as the change of the HSI index level between t and $t + \Delta t$, where $X(t)$ is the value of the index at time t. We obtain volatilities $g(t)$, and then analyze the correlation of the time series.

4 The Correlation Properties

4.1 The Hurst Exponent by the Rescaled Range Analysis Method

In this section, we investigate the correlation properties of the price changes $g(t)$ obtained by Eq.(4) for the Hang Seng Index. With the Eq.(3), the Hurst parameter H of the time series can be calculated. To get a clearer view of this phenomenon, this paper calculates the scaling exponent and plots the fluctuation $\frac{R(n)}{S(n)}$ versus n, as shown in Fig. 1. It is obviously that the time series exhibits long-range dependence behavior as we can calculate from Eq.(3) by where the Hurst exponent H=0.7331 can be obtained.

4.2 The Hurst Exponent by the Moving Average Method

In this section, we obtain the Hurst exponent H applying the moving average method instead of the rescaled range analysis method. Obviously, we calculate the standard deviation $S(n)$ by subtracting the moving average from original time series, which differ from the rescaled range method. In order to investigate the effect of the new method on time series, we calculate the Hurst exponent H with the moving average method. Fig. 2 shows the dependence of $\frac{R(n)}{S(n)}$ on n. It is quite clear that, for varied scales n, the fluctuation $\frac{R(n)}{S(n)}$ exhibit approximate power-law behavior, and the slope is the Hurst exponent H.

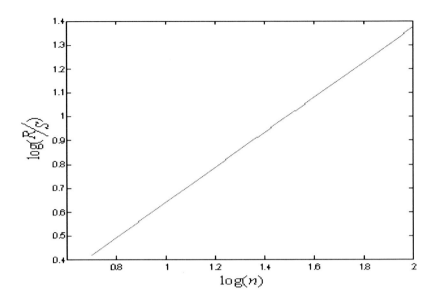

Fig. 1. The plot of $\dfrac{R(n)}{S(n)}$ versus n by conventional rescale range method

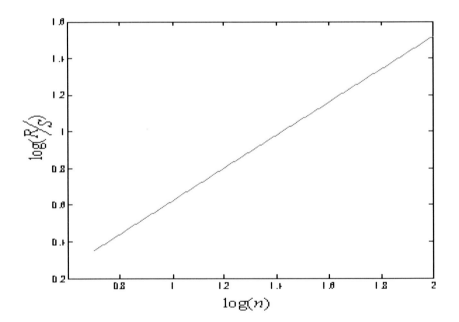

Fig. 2. The plot of $\dfrac{R(n)}{S(n)}$ versus n by moving average method

Comparing Fig.1 with Fig. 2, we can clearly see that the Hurst exponent *H* obtained by both the rescaled range analysis method and the moving average method are arranged in the interval [0,1]. This conclusion proves that the new method is valid to detect the long-range dependence properties in time series.

5 Conclusions

We introduce a moving average method to estimate the Hurst exponent H in this paper. In order to test the validity of this method, we consider the price changes of the Hang Seng Index collected at the Hongkong stock market from December 31, 1986, through June 6, 2008, and study the correlation properties of the change of the HSI index. We show that the Hurst exponent H obtained by the moving average method in the same way as it by the rescaled range analysis method, that is to say, the new method is valid to detect the long-range correlation properties of return intervals.

The above results from the present investigation provide positive evidence regarding the extendence of the rescaled range analysis method. With the develop of study on Hurst exponent , the rescaled range analysis method will take more significant roles in providing foundation theories for stock time series predictions.

References

1. Hurst, H.E.: Long-term Storage capacity of Reservoirs. Transactions of the American Society of Civil Engineers 116, 770–808 (1951)
2. Carbone, A., Castelli, G., Stanley, H.: Time dependent Hurst exponent in financial time series. Physica A 344, 267–271 (2004)
3. Couillard, M., Davison, M.: A comment on measuring the Hurst exponent of financial time series. Physica A 348, 404–418 (2005)
4. Ausloos, M., Vandewalle, N., Boveroux, P.: Applications of statistical physics to economic and Financial topics. Physica A 274, 229–240 (1999)
5. Chen, C.-c., Lee, Y.-T., Chang, Y.-F.: A relationship between Hurst exponents of slip and waiting time data of earthquakes. Physica A: Statistical Mechanics and its Applications 387, 4643–4648 (2008)
6. Yang, Y.-g., Yuan, J.-f., Chen, S.-z.: R/S Analysis and its Application in the Forecast of Mine Inflows. Journal of China University of Mining and Technology 16, 425–428 (2006)
7. Koutsoyiannis, D.: Nonstationarity versus scaling in hydrology. Journal of Hydrology 324, 239–254 (2006)
8. Hong, Z., Keqiang, D.: Multifractal Analysis of Traffic Flow Time Series. Journal of Hebei University of Engineering 2009 26, 109–112 (2009)
9. Yau, H.-Y., Nieh, C.-C.: Testing for cointegration with threshold effect between stock prices and exchange rates in Japan and Taiwan. Japan and the World Economy 21, 292–300 (2009)

10. Mazouz, K., Joseph, N.L., Joulmer, J.: Stock price reaction following large one-day price changes: UK evidence. Journal of Banking & Finance 33, 1481–1493 (2009)
11. Yudong, Z., Lenan, W.: Stock market prediction of S&P 500 via combination of improved BCO approach and BP neural network. Expert Systems with Applications 36, 8849–8854 (2009)
12. Hsu, Y.-T., Liu, M.-C., Yeh, J., Hung, H.-F.: Forecasting the turning time of stock market based on Markov–Fourier grey model. Expert Systems with Applications 36, 8597–8603 (2009)
13. Majhi, R., Panda, G., Sahoo, G.: Development and performance evaluation of FLANN based model for forecasting of stock markets. Expert Systems with Applications 36, 6800–6808 (2009)

Comprehensive Evaluation of Regional Independent Innovation Ability: Based on Chinese Enterprise Level

Zishuo Feng and Shukuan Zhao

School of Management, Jilin University, No. 5988,
Renmin Street, Changchun, P.R. China
fengzishuo@yahoo.com.cn, zsk@jlu.edu.cn

Abstract. In this paper we establish the evaluation index system of independent innovation ability, and then employ unitary principle component analysis method to dynamically analyze the enterprises' independent innovation ability of Jing-jin-ji area, Yangtz river delt area, South coastal area and Northeast area from 2005 to 2008, the results of this research provide important references to recognize enterprise' independent innovation ability of each economic area in China.

Keywords: independent innovation ability; evaluation index system; unitary principle component analysis; enterprise level.

1 Introduction

Since the early 1990s of the last century, China's academic evaluation of regional innovation capability to conduct a more in-depth systematic research, the results of the regional innovation system not only to promote in-depth theoretical study, but also for China's regional innovation system provides guidance on the practice of building[1-2]. To sum up, these studies have the following characteristics: (1)In the evaluation of the object, the main focus on the provincial-level administrative regions or innovation ability evaluation of the big or middle cities, cross-administrative divisions of the large economic is relatively small; (2)In the use of evaluation data, mainly using cross-sectional data on the current status of the static analysis and evaluation of cross-sectional data will be used in conjunction with the historical data, thus the current characteristics and future trends in integrated assessment and judgments of the study also are rate.

In the new century, the science and technology and economic of our country have some new features and trends. First, acceleration of the integration process of regional science and technology, since 2003 has singed the Yangtze River Delta, Pearl River Delta and the Northeast old industrial base in the regional innovation system construction agreement, so that this process has been substantial advance. The result of development of regional science and technology integration is not the type to promote the convergence of different regions or provinces within a region of uniform qualitative development, but in the framework of inter-administrative divisions formed a number of distinctive advantages of full of innovation and dynamic regional, regional science and technology have begun in the pattern of division of labor; The second is regarded in various region to enhance self-innovation capacity-building as a key link in the regional core competencies, however, in the independent innovation capacity-building strategy and tactics of design, make different plans and arrangements; The third, the enterprise independent innovation capacity-building as of independent innovation capacity building of regional core content, cultivate leading enterprises to create innovative cluster of independent innovation capacity-building as a regional point of entry[3].To face of the new situation of China's regional scientific and technological development, the regional innovation ability evaluation studies must also be achieved in the following four areas of innovation and breakthroughs, first in the evaluation object, the more attention to inter-administration divisions of the major economic zones innovation capability assessment. This assessment and analysis of China's greater integration of geographical space, innovation and resources, and thus the formation of innovation has important practical value. Second, in evaluating the content, should be in the general sense of the technological innovation capability assessment, based on independent innovation capacity assessment of the regional general theory and methods, in particular, is to evaluate the ability of independent innovation enterprises into a more prominent position, their evaluation findings for capacity building of regional innovation strategies provide a basis for selection; Third is to make the evaluation results as far as possible reflect the regional characteristics and core competitiveness, thus derive policy recommendations for inter-regional cooperation in the rational division of labor and the depth to provide reference views; Four indicators it to evaluate the formation mechanism of sorting and capacity analysis of organic combined in order to dig more in-depth level of information behind the data so that it can not flow in the evaluation of a simple sort, but can have an impact on regional development and promotion of the practice.

2 Samples and Model Building

2.1 Sample Selection

Based on the above ideas, this study was based on the enterprise level, representative of China's four major economic zones of independent innovation ability evaluation of a trial. The four economic zones are: Beijing and Tianjin, the Yangtze River Delta region, the southern coastal areas and northeast. Beijing and Tianjin area includes Beijing, Tianjin municipalities and Hebei Province. The region has China's total 35% of the national key laboratories, accounting for 34% of the national key disciplines, accounting for 38% of the National Engineering Technology Research Center, and is China's basic applied research center. Yangtze River Delta region includes Jiansu, Zhejiang and Shanghai. In 2006, the Yangtze River Dalta, the country's total high-tech products export areas became China's largest high-tech industrial R&D and manufacturing base. The southern coastal areas include Guangdong, Fujian and Hainan. The region is China's largest high-tech products export base, with a batch of internationally competitive enterprises and products, China's effective participation in international science and technology division of pre-emptive area. Northeast is an important equipment manufacturing and defense industrial base of China, the state sector scientific and technological resources are more abundant compound, by a relatively large concentration of technical innovation has great potential.

2.2 Establishment of Evaluation Index System

Evaluation index system in complying with the general principles established based on the selection of indicators in the evaluation of this study, the first attention to the selection will reflect the "autonomy" indicators of innovation capability; second, to emphasize choice to reflect business innovation indicators. Accordingly, the index system includes items as shown in table 1[4].

Table 1. Evaluation index system of the Independent innovation ability (enterprise)

Indicators	Sub-indicators	unit
Resources Capacity	The number of scientists and engineers	Million
	R&D expenditures	Million
	R&D Full-time equivalent staff	Person years
Environmental capacity	GDP	Billion
	The proportion of non-state economy	%
	Technology Market Turnover	Million
Outcome capacity	The amount of invention patent applications	Parts
	The output value of new products	Billion
Brand capacity	The amount of industrial design patents	Parts
	Well-known Trademark	Parts
	High-tech exports	Million $

(1) Resource capacity

Independent innovation resources include human resources, financial resources and material resources. Resource capacity is the capacity that the main innovation holds or actually operates these resources. Resource capacity is the basis for the formation of independent innovation ability, which reflects the trends and the possible function space of independent innovation ability to a certain extent. Here we choose the number of corporate scientists and engineers, R&D expenditures and R&D staff's working full-time as the evaluation indexes of resource capacity.

(2) Environmental capacity

The environment is the social support system of independent innovation activity, which is the institution, policy and cultural protection of the implement of the resource elements and vector elements. Environmental capacity refers to the support degree and protects level of the specific nation or region to independent innovation. GDP, the proportion of non-state economy and technology market turnover are able to scale the region's economic development level and market-oriented to some extent, which can be the secondary index to evaluate the environmental capacity.

(3) Result capacity

Result is the technological invention or innovative product in the form of patents and new products, the output of previous innovation activities, but also the investment of follow-up innovation activities, which is the direct embodiment material of independent innovation ability. Result capacity is the ability that the main innovation operates the innovative resources to provide scientific discoveries, technological innovation and market brand to the community through the innovative carriers. Here we choose the patent applications amount of business invention and new product output value as the secondary index to evaluate the result capacity.

(4) Brand capacity

The brand is a well-known product and enterprise with a good reputation and enough market shares, which characterized the markets achievement of independent innovation ability. Brand capacity reflects the degree that innovations transfer to productive forces and the extent that brand owners impact and control the value distribution. Here we choose the patent licensing volume of business industrial designs, the well-known trade marks and high-tech product exports as the secondary index to evaluate the brand capacity.

2.3 Principal Component Analysis Model

Since there are 11 indicators contains in the evaluation index system of this paper, which may exist a certain linear relationship between the various indicators, so it is more suitable to use the principal component analysis. On the one hand, it can be overcome the deviation caused by the weights of subjective determination; on the other hand, it can be able to handle indicators and data of large samples. In addition, this paper mainly evaluates the independent innovation ability of Beijing and Tianjin, the Yangtze River Delta, southern coast and northeast area between 2005 and 2008 at the enterprise level. Data interval is the planar data sheet series based on chronological sequence and all the data tables have the same sample points name and the same variable indicator name completely. It has T data sheet series based on chronological sequence, which is the significant difference compared with the plane table, just like a data cartridge, so it is known as time series three-dimensional data table. If we do the principal component analysis to each data table separately, different data tables have complete different main hyper-plane, so we can not guarantee the unity, integrity and comparability of systems analysis. Therefore, doing principal component analysis to this three-dimensional data tables is to find a simplified uniform subspace for all the data tables, getting a unify public factors of principal component, extracting important information from the three-dimensional tables quickly, and then to conduct evaluation and analysis to the sample. The calculation steps of global principal component analysis are as follows:

(1) If we unify n regions and use the same p indicators x1, x2...xp to describe it, there is a data table $X^t = (x_{ij})_{n \times p}$ in t years, where n is the number of sample points, p is the number of variables. One table in a year, T tables in t years. Then arrange the T tables from top to bottom together, which constitute a large nT * p matrix. We defined this matrix as the global data table, denoted as

$$X = (X^1, X^2, ...X^T)'_{nT \times p} = (x_{ij})_{nT \times p} \qquad (1)$$

Each row is a sample in the Matrix, there are nT samples totally. Look at the image, the global data table is expanding the three-dimensional time series data sheet by the time vertically. Then we can implement the classical principal component analysis to the global data table.

(2) To standardize the X in the global data table:

$$x'_{ij} = \frac{x_{ij} - \bar{x}_j}{s_j}, \; i=1, 2, \ldots nT, j=1,2,\ldots p. \tag{2}$$

Where $\bar{x}_j = \frac{1}{n}\sum_{i=1}^{nT} x_{ij}; s_j^2 = \frac{1}{n}\sum_{i=1}^{nT}(x_{ij} - \bar{x}_j)^2$ .For convenience, we still donate the standardized data sheet as X.

(3) Calculate the covariance matrix R of $\dot{x}$, where R is called the global covariance matrix and X have been standardized. Therefore, R is the correlation coefficient matrix of X and R is a positive definite matrix.

(4) Find the eigenvalues $\lambda_1 \geq \lambda_2 \geq \ldots \geq \lambda_p \geq 0$ of R, the corresponding eigenvectors u1, u2, ... up, they are orthonormal and are called the global axis, denoted as u=(u1,u2,…up), $uu' = u'u = I$. Thus we can get the k-th principal component: Fk = Xuk, k = 1,2, ... p, and obtain the variance contribution rate and cumulative variance contribution rate of the principal component $F_1, F_2,\ldots F_p$:

$$a_k = \frac{\lambda_i}{\sum_{i=1}^{p}\lambda_i}, \quad a_1 + a_2 + \ldots + a_m = \frac{\sum_{i=1}^{m}\lambda_i}{\sum_{i=1}^{p}\lambda_i} \tag{3}$$

Select the corresponding principal component $F_1, F_2,\ldots F_p$ of prior m largest eigenvalues, so that the cumulative variance contribution rate can be more than 85%.

(5) Find the correlation coefficient r_{ij} of X_i and F_j, and obtain the correlation coefficient matrix $A = (r_{ij})$, which is also known as the factor loading matrix. r_{ij} means the load of the i-th variables X_i in j-th common factor F_j, which could explain the information on what variables the principal component F_j mainly contains. The principal component F_j is a common factor.

(6) And then we can obtain the factor score function by the regression:
$F_j = \beta_{j1}X_1 + \beta_{j2}X_2 + \ldots + \beta_{jp}X_p$, j=1,2,…m

Therefore, we can calculate the score of each sample that selected m common factors.

(7) Finally, combining the contribution rate a_k of principal components, we can obtain a consolidated score function $F = a_1F_1 + a_2F_2 + \ldots + a_mF_m$. And then calculate the composite score of all samples to evaluate the differences among the samples.

3 Empirical Study

3.1 Data Collection

According to the indicators set previously, we refered to China Statistical Yearbook, China Statistical Yearbook on Science and Technology in relevant years, collected and organized the data to get the original data for evaluating the independent innovation ability in Southern coastal regions, the Yangtze river delta region, Jin-jing-ji region and the Northeast region.

3.2 The Evaluation of Independent Innovation Ability in the Four Major Economic Zones

Using SPSS to calculate the global principal components, we process the original data for standardization firstly, and calculate the Eigen value, the variance contribution rate (table 2), the principal component loading matrix (table 3), and the principal component score coefficient matrix (table 4).

Table 2. The eigenvalue and the variance contribution rate

Principal component	Eigenvalue	Variance contribution rate	Accumulated contribution rate
1	8.059	73.264	73.264
2	1.224	11.131	84.395
3	1.086	9.874	94.269
4	0.372	3.382	97.651
5	0.151	1.372	99.024
6	0.057	0.519	99.542
7	0.028	0.250	99.792
8	0.008	0.075	99.867
9	0.006	0.057	99.924
10	0.005	0.047	99.971
11	0.003	0.029	100

Generally, some principal components whose cumulative contribution rate take more than 90% can represent the most of the information of many original indicators. We can see from table II that the top three principal components can be representative 94.269% information of the original 11 indicators. We can see from table IV that the relation of Z_1, Z_2, Z_3 from $X_1, X_2, ..., X_{11}$ is expressed as follows:

$Z_1=0.0958X_1+0.1175X_2+0.1112X_3+0.1212X_4+0.0966X_5+0.045X_6+0.1158X_7+0.118X_8+0.1115X_9+0.1141X_{10}+0.0955X_{11}$

$Z_2=0.2193X_1-0.1793X_2+0.0047X_3+0.036X_4-0.1817X_5+0.626X_6-0.0346X_7+0.2181X_8-0.229X_9+0.2382X_{10}-0.3903X_{11}$

$Z_3=-0.5133X_1+0.0132X_2-0.3914X_3-0.1087X_4+0.455X_5+0.4429X_6-0.0499X_7-0.0337X_8+0.0788X_9+0.1637X_{10}+0.2286X_{11}$

The scores of the principal component in the four major economic zones are gained ,after putting original data which is processed for standardization into the above equation. then according to Table-3, we consider the variance contribution of Z_1,

Z2, Z3 to be the coeffiency, sum them with weights to get a comprehensive score function on measuring independent innovation capability of every economic region on enterprise level:

P=0.7326Z1+0.1113Z2+0.0967Z3

Finally, we put the scores of the principal component in the for major economic zones into the above equation to get the scores of dynamic evaluation of independent innovation ability in Jin-Jing-Ji region, the Yangtze River Delta region, Southern coastal regions, and the Northeast region as shown in Table 5.

Table 3. Principal component loading matrix

	Z1	Z2	Z3
X1	0.7723	0.2686	-0.5575
X2	0.947	-0.2195	0.0143
X3	0.896	0.0057	-0.4251
X4	0.977	0.044	-0.118
X5	0.7787	-0.2225	0.4941
X6	0.3627	0.7665	0.4811
X7	0.9331	-0.0423	-0.0542
X8	0.9511	0.267	-0.0366
X9	0.8987	-0.2804	0.0855
X10	0.9192	0.2917	0.1778
X11	0.8017	-0.4779	0.2483

Table 4. Principal component score coefficient matrix

	Z1	Z2	Z3
X1	0.0958	0.2194	-0.5133
X2	0.1175	-0.1793	0.0132
X3	0.1112	0.0047	-0.3914
X4	0.1212	0.036	-0.1087
X5	0.0966	-0.1817	0.455
X6	0.045	0.626	0.4429
X7	0.1158	-0.0346	-0.0499
X8	0.118	0.2181	-0.0337
X9	0.1115	-0.229	0.0788
X10	0.1141	0.2382	0.1637
X11	0.0995	-0.3903	0.2286

Table 5. The dynamic comprehensive evaluation scores of the independent innovation ability

Year	Jin-Jing-Ji region scores	sequence number	Yangtze River Delta region scores	sequence number	Southern coastal regions scores	sequence number	Northeast region scores	sequence number
2005	-0.5575	3	0.2546	1	-0.2946	2	-0.9328	4
2006	-0.377	3	0.5753	1	-0.0173	2	-0.8486	4
2007	-0.1278	3	1.0183	1	0.1618	2	-0.7688	4
2008	0.0784	3	1.8544	1	0.5999	2	-0.6183	4

In order to understand the composition and the changes in trend of the independent innovation ability further in every region, we make respective evaluation of the resource capacity, environmental capacity, achievement capability and brand ability, and the results of the dynamic evaluation of sub-indicators shown in table 6.

Table 6. The dynamic comprehensive evaluation scores of the independent innovation ability

Sub-indicators	Year	Jin-Jing-Ji region scores	sequence number	Yangtze River Delta region scores	sequence number	Southern coastal regions scores	sequence number	Northeast region scores	sequence number
Resource capability	2005	-0.8874	4	1.6549	1	-0.1607	2	-0.6068	3
	2006	-1.8215	4	1.8711	1	0.3357	2	-0.3853	3
	2007	-1.6664	4	2.1693	1	0.2225	2	-0.7255	3
	2008	-1.6231	4	2.2680	1	0.1483	2	-0.793	3
Environmental capability	2005	-0.3334	3	1.5007	1	0.4970	2	-1.6644	4
	2006	-0.3368	3	1.5476	1	0.4710	2	-1.6818	4
	2007	-0.3177	3	1.6106	1	0.3877	2	-1.6805	4
	2008	-0.2970	3	1.5868	1	0.4106	2	-1.7069	4
Achievement capability	2005	-1.2598	3	1.2526	1	0.3147	2	-1.4074	4
	2006	-0.7519	3	1.8263	1	0.2114	2	-1.2858	4
	2007	-0.506	3	1.9254	1	-0.0491	2	-1.3704	4
	2008	-0.9078	3	1.867	1	0.2931	2	-1.2522	4
Brand capability	2005	-0.9255	3	1.0394	2	1.5643	1	-1.6782	4
	2006	-0.9199	3	1.1303	2	1.5072	1	-1.7177	4
	2007	-0.8689	3	1.2848	2	1.3668	1	-1.7828	4
	2008	-0.8336	3	1.3154	2	1.3298	1	-1.7217	4

4 Results

(1) According to the overall evaluation results, the independent innovation ability can be divided into three levels in the four major economic zones, and the ranking didn't change between 2005-2008. The Yangtze River Delta region is classified as the first category, Southern coastal regions as the second category, Jin-Jing-Ji region and the Northeast region as the third category, and there is a wide gap of abilities among the three regions. However, we are still unable to make a persuasive explanation for the reason of the difference according to first grade indexes. So we need to analyze the second grade indexes to make a specific interpretation for the characteristics, strengths and weaknesses in a number of regions.

(2) Among the four second grade indexes, the resource capacity, environmental capacity, achievement capability in Yangtze River Delta region ranks first every year, but the brand capacity ranks second every year. It shows that the resource capacity, environmental capacity, achievement capability in Yangtze River Delta region have obvious advantages compared with other regions and deserve the first place, however, the brand capability is slightly weaker than southern coastal areas, but the gap is small. It must be noted that there isn't a paralogous relationship among the brand capability, resource capacity and loading capability. That is, the technology of considerable brands is exogenous, and some brands lack of technological element of application.

From the core competitiveness point of view, the real self-owned brand must base on the premise that has mastered the key (core) technology. Thus, it could be further considered that there exist dislocation phenomenon among brand capability, resource capacity and loading capability in the Yangtze River Delta region. Therefore, it is the major issue of independent innovation enterprises that how to achieve the combination of brand capability, resource capacity and loading capability in this region.

(3) Among the four second grade indexes, the brand capability in Southern coastal region rank first every year, but the resource capacity, environmental capacity, achievement capability rank second every year. It shows that the matching degree of the constituent elements of independent innovation ability is better in the enterprises in the southern coastal area. The characteristic of the southern coastal region is that brand-name products and the well-known enterprises which are supported by independent intellectual property rights is higher than other regions. The international competitiveness of the industry was significantly higher than other regions, which shows that the downstream link of the enterprise independent innovation value chain that is close to the market segments in the region have a greater advantage and potential.

(4) The other three indictors in Jing-Jin-Ji region rank the third except that resource capability ranks the fourth every year, which shows that there exists a interaction relationship among various capacity-factors on the basis of lacking of innovation resources. And independent innovation forms a virtuous circle mechanism among resource aggregation, capability forming and market realization.

(5) Except the resource capacity in the northeast ranks the third, the others rank the fourth. It shows that there are some advantages in independent innovation resource inventory; however, the output of the independent innovation ability in enterprises didn't match with the resource ability, and the environment for innovation and brand capabilities have lower-ranking. It shows that there exists a wide gap compared with other areas in independent innovation; cultivate independent brands and high-tech products in international market development of the enterprises in the old industrial bases in northeast region.

References

1. Sheng-gang, R., Jian-hua, P.: The Evaluation and Comparison of Regional Innovation Capacity Based on Factor Analysis. Systems Engineering 2, 87–92 (2007)
2. Rui, S., Jin-tao, S.: Secondary analyses on regional innovation ability in China using factor and cluster method. Studies in Science of Science 6, 985–990 (2006)
3. Kai-yuan, L.: Research on the Across Provinces Innovation System. Forum on Science and Technology in China 6, 50–54 (2004)
4. Ruo-en, R., Hui-wen, W.: Multivariate Statistical Analysis—Theory, Methods, Case. Nation Defense Industry Pressing, Beijing (1997)

Research of Chord Model Based on Grouping by Property

Hongjun Wei, Jun Yan, and Xiaoxia Li

Information & Electronic Engineering
Hebei University of Engineering
Handan, China
way_hj@163.com

Abstract. Chord is one of the peer-to-peer systems that searching algorithm for resource,simplicity stabilize and scalable. But its reckon without properties of node while mapping out. Every node is equal under the system. This leads to some problems. In this paper, considering instance of application in practice, we propose a scheme that improved Chord Model Grouping by Property(GP-Chord). And results show that its satisfactory through simulation.

Keywords: P2P, Chord, Grouping, Property.

1 Introduction

Peer-to-Peer(P2P) network is the new schema application. It is the equivalent, and self-governing entity consisting of a self-organizing system, the aim is a distributed networked environment, shared resources, to avoid the centralization of services[1]. Server as the core in the traditional computer networks based on Client/Server(C/S) model application. Unlike this, every node is equal in P2P system, sharing resources of computing information and communication by exchange directly. Participative nodes all are both consumer and provider, for direct access to shared resources, without going through an intermediate entity.

According to the relationship of resource position and location of the node, P2P can be divided into two types: (1) Unstructured P2P systems. In these systems, the location of each resource that is the node for sharing, there is no provision of resources to the resource location identifier mapping, the number of resource locating steps can not be determined. Typical representative, such as Napster[2], Gnutella[3] and so on. (2) Structured P2P systems. The system placed resource precisely on each determined node, and provide mapping from resources characteristic to resources location, ensure that number of resource locating steps is limited and accurate, robust, scalable. Therefore it is a hot topic researched current in the front domain of P2P systems. Typical representative such as based on Distributed Hash Table(DHT) in Chord [4], Pastry [5] and so on.

1.1 Chord System

Chord presented by Stoica from the Massachusetts Institute of Technology(MIT) in 2001. Its a DHT-based peer resource discovery and routing protocols. Algorithm is simple and easy to achieve, and its basic operation is the keyword (key) mapped to node. Chord using the SHA-1[6] as a hash function, DHT keyword is one of the m-bit identifier, that is a integer between [0,2^m-1]. Identifier form a ring topology modulus of the 2^m based one-dimensional. Each data item and the node associated with an identifier. Data identifier is a keyword key, the node identifier is an ID value. So, in form, (key, value) pairs (k, v) from the ID holding not less than k nodes, called a successor of the keyword k. According to consistent hashing algorithm, Chord will assigned key to its clockwise direction recent follow-up node on the ring, denoted by successor(k). Chord's contribution is present the distributed searching protocol, specified keyword key mapped to the specifying node. Figure 1 shows an identifier circle with m=3. The circle has three nodes:0,1 and 3.Keyword 1 located at the node 1. 2 locatedd at 3, K6 located at 0.

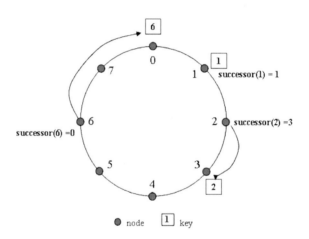

Fig. 1. Chord Space

1.2 Chord Routing

Each node in Chord maintains a routing table called Finger Table, to provide routing information on the node. The table up to m entries, node n, pointing to the table the first i ($1 \leq i \leq m$) entry is the value of n +2^{i-1} follow-up in the Chord ring nodes, if s = successor (n +2^{i-1}) mod 2^m, that called s is the ith finger of node n, denoted by n.finger[i].node, Table 1 shows meaning of the other items in the pointer.

Table 1. Definition of variables node n

Notation	Definition
finger[k].start	$(n+2^{k-1})$ mod 2^m, $1 \cdot k \cdot m$
.interval	(finger[k].start,finger[k+1].start)
.node	First node •n.finger[k].start
successor	The next node on the indentifier circle finger[1].node
predecessor	The previous node on the identifier circle

Figure 2 shows a 6-bit Chord space, with 10 nodes and 7 data items. Keyword K5's successor is N8, K43 successor is N43, K61 located on the N8. The second pointer of the node N8 is N10, the third is N15. The pointer of the first node is always identifier that its direct successor.

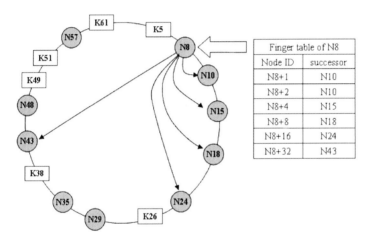

Fig. 2. Finger table of N8

Chord algorithm using information of the finger table stored in each node. When the request reaches a node n, if keyword k is located between n and successor of n, then the node n report their subsequent as a request response. For N nodes, the degree is $O(\log^2 N)$, the average routing delay is $1/2\log^2 N$, dynamic maintenance costs is $O(\log_2^2 N)$.

2 Grouping by Properties Model

2.1 Weakness of Chord

From the above study, we found that all nodes are "equal" in Chord model, reckon without the practical application scenarios. Actually each node has different properties in network. For example, the node's physical location, performance, degree of concern for resources (hot-resources) and so on. Seeing that, we propose a improved Chord model grouping by node properties(GP-Chord). According to different scenarios, taking into account the different properties of the node clustering, and thus the network is divided into several groups. When the node initiate a request, check there first be in the inner group. Small-World[7] theory shows that the sub-group model can improve the positioning performance of P2P system's resources.

2.2 Grouping

Node properties are diverse and can choose a suitable property for scenarios. As an example, we use the node region information to illustrate the implementation process. This division can solve the inconsistencies from physical topology to logical in Chord system. Because the search may only be carried out within the LAN, it will greatly enhance the search efficiency. We made a careful analysis of node's region character after the network initialized, and divided it into regions adjacent nodes in the same group, called "neighbors group", there may be several groups in the network obviously, these groups constituted and called "global group."

2.3 Transformation of the Finger Table

We will still using the Chord routing algorithm for resource location in the neighbors group and global group. The difference is that, each node not only belongs to neighbors group but global group, it is necessary to transform the nodes finger table. Each neighbors group elect a proxy node (called "leader node"), and each ordinary node add its information in order that point to it, while the leader nodes in addition to protecting the inner nodes information, but also increase the global group nodes information.

2.4 Searching Process

Assuming that the node n to initiate a query for keyword key, then the searching process as follows:

Step 1. To determine whether the key was at the successor node in the neighbors group, s = successor(n), if the successor node returned show that resources in the neighbors group, the process succeeded and end. Otherwise go to Step 2.
Step 2. The query forwarded to the leader node, it is responsible global searching, to determine whether the key was at the successor leader node in the global group, if true the successor leader node returned, else query forwarded to the

successor leader node. if resources be found then return node in that group else then the query forwarded to the next leader node to continue.

Step 3. Repeat (1) and (2), until after the end of the success or failure.

A typical campus network topology shows in Figure 3, Figure 4 (a) shows it's Chord model, (b) shows the GP-Chord proposed in this paper.

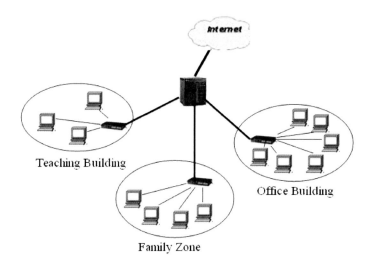

Fig. 3. Campus Network Topology

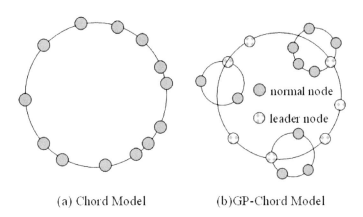

Fig. 4. Chord Model and GP-Chord Model

2.5 Node Joins

Suppose a new node n to be added to the network, you will need the following steps:

Step 1. The node provide geographic information to the system to determine its target group.
Step 2. The node n is added to the global group by Chord algorithm if the target group not found, as leader node, to create a new neighbors group at same time. Otherwise, go to step 3.
Step 3. Suppose n node to be added to the neighbors group called G1, n' one of the G1, then n asked n' check point for its finger table item.
Step 4. Update other nodes finger table call itself function. Of course including leader nodes.
Step 5. Transform keyword n, on successor nodes inner group, to node n.

If the system has M nodes, time complexity is $T_M=O(\log^2 M)$ for finding and updating all nodes before grouping. After is $T_{GN}=O(\log^2 N)$ or $T_{GR}=O(\log^2 R)$, N is the number of leader nodes and R is the number of nodes neighbors group to joined. We take the grouping time complexity $T_G=MAX(T_{GN},T_{GR})$, clearly $N \leq M$, $R \leq M$, so $T_G \leq T_M$. This shows that the grouping is valid.

2.6 Node Leaves

If it is an ordinary node, leave the neighbors group by Chord algorithm and informed the leader node. If it is a leader node, copy its finger table to successor node and then update other nodes table in the neighbors group, and leave the global group finally.

3 Experiment and Analysis

We have compared the searching performance before and after through simulation. The routing hops and the average delay of two parameters were observed. Using the

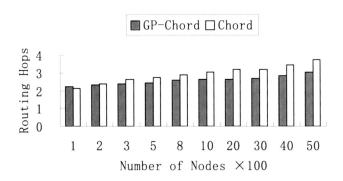

Fig. 5. Routing Hops Comparison

GT-ITM[8] Transit-Stub model to generate network topology. Set the delay time is 50ms between transit domain, 10ms from stub domain to transit domain, 3ms in stub domain, the number of nodes taken 50 to 5000, and generated number of neighbors groups randomly. Routing hops comparison in Figure 5 and time delay in Figure 6 after analysis of experimental data. Results show that the performance of GP-Chord better than the Chord, especially in the case of large number of nodes.

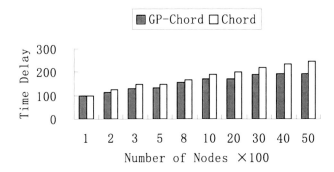

Fig. 6. Average Delay Comparison

4 Conclusion

This paper proposed a Chord model Grouping by Property(GP-Chord), and conduct simulation experiments. According to geographical attributes of the node to grouping. This division can solve the inconsistencies from physical topology to logical. Results show that the improved model to reduce hops and time delay during searching, improved system performance.

The next work include: (a) Node properties are diverse, to determine what the best for performance; (b) What the relationship between each of the properties, and verify the performance of grouping model by multi-properties.

References

1. Steinmetz, R., Wehrle, K.: Peer-to-Peer-Networking & -Computing. Informatik-Spektrum 27(1), 51–54 (2004) (in german)
2. Napster Website, http://www.napster.com
3. Gnutella Website, http://www.Gnutella.com
4. Stoica, I., Morris, R., Karger, D., Kaashoek, F., Balakrishnan, H.: Chord:A Scalable Peer-To-Peer Loopup Service for Internet Applications. In: Proceeding of the 2001 ACM Sigcomm Conference, pp. 149–160. ACM Press, New York (2001)
5. Rowstron, A., Pastry, D.P.: Scalable, Distributed Object Location and Routing for Large-Scale Peer-to-Peer Systems. In: Proc. of Middleware. ACM Press, New York (2001)

6. FIPS 180-1.Secure Hash Standard. U.S. Department of Commerce/NIST, National Technical Information Service, Springfield, VA (April 1995)
7. Watts, D.J., Strogatz, S.H.: Collective dynamics of small-world networks. Nature 393(6684), 440–442 (1998)
8. Zegura, E.W., Calvert, K.L., Bhattacharjee, S.: How to Model an Internetwork. In: Proc. of the INFOCOM 1996, Institute of Electrical and Electronics Engineers, Inc., New York (1996)

Research of Some Autopilot Controller Based on Neural Network PID

Jinxian Yang, Bingfeng Li, and Hui Tao

School of Electrical Engineering & Automation,
Henan Polytechnic University,
454003 Jiaozuo, China
`yangjinxian@hpu.edu.cn`

Abstract. First, established submersible vehicle movement mathematic model; then analyzed disadvantage of PID autopilot effect. Mainly, a multi-controller method with a neural network control and PID control was adopted, and researched the submersible vehicle autopilot control technology, then a neural network identification(NNI) was designed, and to identify the submersible vehicles space mathematical models on line, though the submersible vehicle mathematical models can not fully determine, the submersible vehicles approaching mathematical models also can be identified on line by NNI through the real time input and output of the submersible vehicle under large interference. At the same time, the multi-layer prior neural network as neural PID controller (NNC) was adopted, and it improved accuracy. The simulation results show that neural network PID control autopilot has very good performance and is better than traditional PID autopilot in robustness and practicability.

Keywords: some autopilot; PID control; neural network control; simulation.

1 Introduction

The submersible vehicle controls not only the course but also the depth and ubiquitous-vertical navigation poses etc., and some auto-rudder technology is a very important part. The classic proportion-integral- differential (short for PID) control theory is adopted to operate the submersible vehicle auto-rudder, and the its mechanisms are based on the precise mathematical models, but in fact submersible vehicle 's accurate mathematical models are not easy accessed, what's more submersible vehicle s-this controlled targets which are serious nonlinear, therefore PID control is based on the controlled targets which are linear in control strategy, and so it has a larger error, poor adaptability and robustness. This is mainly due to large changes of the scope of work by the forced coupling among the submersible vehicle at all kinds of freedom role, and linear process is more difficult. Thus the methods have considerable limitations. In complex cases, the effects of the only use of PID control are poorer, and the paper used neural network control methods combined with PID to control the submersible vehicle autopilot, and it has achieved very good results.

2 Submersible Vehicle Space Movement Model

The fixed coordinate system $E\text{-}\xi\eta\zeta$ and the movement coordinate system G-xyz as follows:

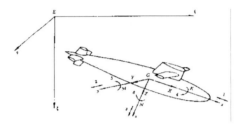

Fig. 1. Movement coordinate system and fixed coordinates system

The submersible vehicle movement features:

(1) The vertical plane submersible vehicle movement against the impact of water movement can ignore.
(2) Submersible vehicle changed course, vertical and horizontal dumping and the dumping of a principal by the first submersible vehicle to drift-angular velocity and horizontal movement caused.
(3) Water power coefficient values to a certain extent reflect the size of the water for the submersible vehicle movement impact levels.

In the light of the foregoing conclusions, we get the following simplified model through careful selection and computer simulation experiments.

1) axes equation:

$$U \approx u \approx U_0 \left(1 - e^{-0.52u / |\dot{w}|L}\right) \qquad (1)$$

2) transverse equation:

$$m[\dot{v} + ur] = \frac{1}{2}\rho L^4 \left[Y'_r \dot{r} + Y'_p \dot{p}\right] + \frac{1}{2}\rho L^3 \left[Y'_{\dot{v}} \dot{v} + Y'_r ur\right] + \frac{1}{2}\rho L^2 \left[Y'_{uv} uv + Y'_{v|v|} v \left(v^2 + w^2\right)^{\frac{1}{2}}\right] + \frac{1}{2}\rho L^2 \left[Y'_{\delta_r} u^2 \delta_r\right] \qquad (2)$$

3) horizontal equation:

$$m[\dot{w} - uq + vp] = \frac{1}{2}\rho L^4 [Z'_q \dot{q}] + \frac{1}{2}\rho L^3 [Z'_w \dot{w} + Z'_q uq + Z'_{vp} vp]$$
$$+ \frac{1}{2}\rho L^2 \left[Z'_0 u^2 + Z'_w uw + Z'_{w|w|} w \left| (v^2 + w^2)^{\frac{1}{2}} \right| + Z'_{vv} v^2 \right] \quad (3)$$
$$+ \frac{1}{2}\rho L^2 [Z'_{\delta_s} u^2 \delta_s + Z'_{\delta_b} u^2 \delta_b]$$

4) movement relations

$$\dot{\xi} = u\cos\psi\cos\theta + v(\cos\psi\sin\theta\sin\phi - \sin\psi\cos\phi) + w(\cos\psi\sin\theta\cos\phi + \sin\psi\sin\phi) \quad (4)$$

$$\dot{\eta} = u\sin\psi\cos\theta + v(\sin\psi\sin\theta\sin\phi + \cos\psi\cos\phi) + w(\cos\phi\sin\psi\sin\theta - \cos\psi\sin\phi) \quad (5)$$

$$\dot{\varsigma} = -u\sin\theta + u\cos\theta\sin\phi + w\cos\theta\cos\phi \quad (6)$$

$$\dot{\phi} = p + q\tan\theta\sin\phi + r\tan\theta\cos\phi \quad (7)$$

$$\dot{\theta} = q\cos\phi - r\sin\phi \quad (8)$$

$$\dot{\psi} = \frac{q\sin\phi}{\cos\theta} + \frac{r\cos\phi}{\cos\theta} \quad (9)$$

Where: u is submersible vehicle speed, m/s; v is horizontal speed, m/s; w is vertical speed, m/s; p is horizontal swimming angle velocity, rad/s; q is obliquitous-vertical angle velocity, rad/s; r is head swimming angle velocity, rad/s; h is submersible vehicle stable highness, m; δ_s is submersible vehicle head elevator angle, rad; δ_b is submersible vehicle back elevator rudder angle, rad; δ_r is submersible vehicle rudder angle, rad; ϕ is obliquitous-horizontal angle, rad; θ is obliquitous-vertical angle, rad; ψ is course angle, rad; L is submersible vehicle length, m; m is mass, kg.

3 Autopilot Controller Structural Design

Though traditional PID autopilot has a simple structure and faster response etc., steered submersible vehicle model is no n-linear, and slow variational characteristic. Thus it needs frequent steer rudder, large rudder angle scope, much power of rudder machine, difficult parameter adjustment etc.. Now analytic and climbing mountain methods are used in controlling optimized parameters in domestic. Despite these fast methods being convergent, they need submersible vehicle precise movement equations which are not easy got in some interference. However single PID control results will appear fluctuation in the case of random interference. Next, we introduce wave interference in the submersible vehicle voyage course, and analyze PID autopilot control effect and system response.

White noise through shaping filter firstly turns out colored noise, it simulates wave interference. The wave simulant signal:

$$S_\zeta(\omega) = \frac{8.1 \times 10^{-3 g^2}}{\omega^5} \exp(-\frac{3.11}{h_{1/3}^2 \times \omega^4}) \quad (10)$$

A rational function of the shaping filter: $G(s) = ks/s^3 + a_2 s^2 + a_1 s + a_0$

Where working out: k=1.217, a_0=0.4802, a_1=0.8483, a_2=1.4396. Thus

$$G(s) = \frac{1.217 s}{s^3 + 1.4396 s^2 + 0.8483 s + 0.4802} \quad (11)$$

After being transformed by the shape, the noise seeing Fig.2.

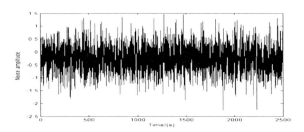

Fig. 2. Noise curve

The simulations results of the course and rudder angle of submersible vehicle PID autopilot control as follows Fig.3 and Fig.4.

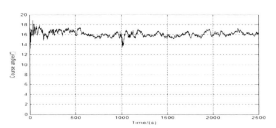

Fig. 3. Course of PID autopilot control with interference

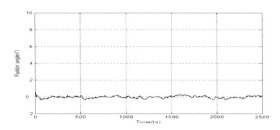

Fig. 4. Rudder angle of PID autopilot control with interference

From the above figures, if PID autopilot control parameters cannot be adjusted, the control effect is very poor in the case of interference. The autopilot controller structure is composed of neural PID controller (NNC) and neural network identification (NNI), its concrete frame seeing Fig.5.

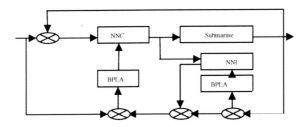

Fig. 5. Neural network PID controller

The controlled target is the single-input/output nonlinear systems, that is, input expected course angle or depth, and output actual course cents or depth. The mathematical model of the input and output system is

$$y(t) = f[y(t-1),\cdots,y(t-n),u(t-1),\cdots,u(t-m)] \quad (12)$$

Where: $y(t), u(t)$ are respectively input and output of the system, and n, m are respectively exponential of $y(t), u(t)$; f is nonlinear function.

Neural network identification (NNI) has three-layer networks, and the BP algorithm is adopted in its identification. The input and output of the network are respectively $u(t)$ and $y(t)$. The identification network input layer has $n_1=n+m+1$ nerve cell, and its structure:

$$x_i(t) = \begin{cases} y(t-i) & 0 \le i \le n-1 \\ u(t-i+m) & n \le i \le n+m \end{cases} \quad (13)$$

The number of the hidden network $n_H \ge n_1$, its I/O relation:

$$\begin{cases} net_i(t) = \sum_{j=1}^{n_H} w_{ij} x_j(t) + \theta_i \\ O_i(t) = g[net_i(t)] \end{cases} \quad (14)$$

Where: w_{ij} is weight coefficient ; θ_i is hreshold ; $g(x)$ is work function, as $g(x) = (1-e^{-x})/(1+e^{-x})$.

The number of the network output layer is 1, and the output:

$$\hat{y}(t+1) = \sum_{i=1}^{n_H} v_i O_i(t) + \gamma \quad (15)$$

Where: $\hat{y}$ is the output of NNI; v_i is weight coefficient;

γ is valve value, the performance index of learning is:

$$J_m = [y(t+1) - \hat{y}(t+1)]^2 / 2 \tag{16}$$

Let J_m minimum, and get adjustment law of the coefficient:

$$\begin{cases} \Delta v_i(t) = \alpha \varepsilon(t+1) O_i(t) + \beta \Delta v_i(t-1) \\ \Delta \gamma(t) = \alpha \varepsilon(t+1) + \beta \Delta \gamma(t-1) \\ \Delta w_{ij}(t) = \alpha \varepsilon(t+1) g'[net_i(t)] v_i(t) x_j(t) + \beta \Delta w_{ij}(t-1) \\ \Delta \theta(t) = \alpha \varepsilon(t+1) g'[net_i(t)] v_i(t) + \beta \Delta \theta(t-1) \\ \varepsilon(t+1) = y(t+1) - \hat{y}(t+1) \\ g'(x) = 0.5[1 - g^2(x)] \end{cases} \tag{17}$$

Where: $\Delta x(t) = x(t) - x(t-1)$; α, β are respectively learning correction coefficient and inertia coefficient, its value $\in (0,1)$.

NNC is two-layer linear network, and its input has three elements, the input is respectively

$$\begin{cases} h_1(t) = e(t), h_2(t) = \sum_{i=0}^{t} e(i) \\ h_3(t) = \Delta e(t) = e(t) - e(t-1) \end{cases} \tag{18}$$

Where: $e(t)$ is system warp, that is $e(t) = r(t) - y(t)$. In the last equation, $r(t)$ is a fixed system value. And know from (16), the note input layer of NNC is respectively system warp and its integral and differential value, firstly pre-process system warp, and then feedback to the relevant note. The output of the network is:

$$u(t) = K_1 h_1(t) + K_2 h_2(t) + K_3 h_3(t) \tag{19}$$

Where : K_i is weighted coefficient, $u(t)$ is weighting sum of input layers signal. And the controller has PID control structure. The performance index

$$J_c = [r(t+1) - \hat{y}(t+1)]^2 / 2 \tag{20}$$

and use J_c minimum to train the controller parameter. Because of after proper learning, $\hat{y}$ approached y, thus

$$J_p = [r(t+1) - y(t+1)]^2 / 2 \tag{21}$$

J_p minimum may replace J_c minimum, and in fact take estimated system: Jacobian information $\partial \hat{y}(t+1) / \partial u(t)$ instead of $\partial y(t+1) / \partial u(t)$.

According to the equation (18), optimize it with grad, and get a correct mathematic formula of the related Neural Network PID controller.

$$\Delta K_i(t) = \lambda [r(t+1) - \hat{y}(t+1)] h_i(t) \frac{\partial \hat{y}(t+1)}{\partial u(t)} \quad (22)$$

Where: $0 < \lambda < 1$, $\partial \hat{y}/\partial u$ is

$$\frac{\partial \hat{y}(t+1)}{\partial u(t)} = \sum_{i=1}^{n_H} \frac{\partial \hat{y}(t+1)}{\partial O_i(t)} \frac{\partial O_i(t)}{\partial net_i(t)} \frac{\partial net_i(t)}{\partial u(t)} = \sum_{i=1}^{n_H} v_i(t) g'[net_i(t)] w_{in}(t)$$

Therefore, we can get adaptive neural network online PID control algorithms:

1) Using (-1,1) random value to initialize weight of neural network.

2) Sampling data, calculate $e(t)$, $\sum e(t)$ and $\Delta e(t)$.

3) $u(t)$ generated by the neural network controller, and meanwhile send $u(t)$ to the target and NNI.

4) Using (17) to correct weighted value of NNI.

5) Using (21), (22) to correct network weighted value of NNC.

6) as $t = t+1$, let $y(t)$, $u(t)$ and $e(t)$ shift process, then return step(2).

4 Realizing Neural Network PID Controller

4.1 The Layers of Network

Because of taken BP algorithm, we only consider design of the BP network. In theory proven: a network of a deviation and at least an s style hidden layer plus one linear output layer, can approach any function. This actually gives us a basic principle of the basic network design. Design principles: proper a number of the adding layer; adding a number of neural cells.

4.2 The Neural Cell Number of Hidden Layer

To enhance network training precision, according to the experimental results: for example, the neural cell number is 2 or 3, while the network cannot learn well and need many training times; If many neural cell numbers, cycle times and training time will increase; When the neural cell number =8 or 9, the effect is best.

4.3 Select Initialized Weighted Value

Because of a nonlinear system, initialized value is a very important relation for whether get local minimum with learning, constringe and training time. In general, the output of each neural cell is expected to zero, and assure maximal adjustment of each weighted value. Thus initialized value $\in (-1,1)$.

4.4 Learning Efficiency

Learning efficiency determines weighted variation of each cycle training. High learning efficiency may cause to instable; while low convergent speed may cause to slow. Thus learning efficiency $\in (0.01, 0.8)$.

4.5 The Controller Performance Simulation

In case of the initialization term: submersible vehicle velocity V=6 knots, $\Psi_0=\varphi_0=\theta_0=\zeta_0=0$, expected course angle $\Psi_0=30^0$, without environment interfere, the simulation result as follows, Fig.6, Fig.7, Fig.8 are respectively change of submersible vehicle course, the output of NNI and helm angle.

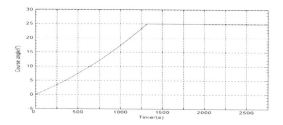

Fig. 6. Change of submersible vehicle course

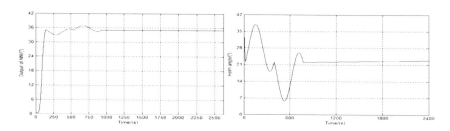

Fig. 7. Output of NNI　　　　　**Fig. 8.** Change of helm angle

The initialization term does not variety, and we get simulation results after adding interfere of wave noise as follows: Figure 9, Figure 10.

In the same initialization terms (considered wave interference), supposed as original depth=0, and expected stable voyage depth is 100m underwater. The simulation result Figure 11, Figure 12, Figure 13 as follows:

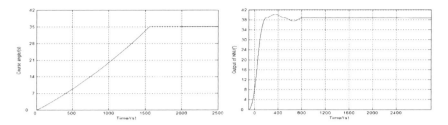

Fig. 9. Chang of submersible vehicle course after adding interference

Fig. 10. Output of NNI after adding interference

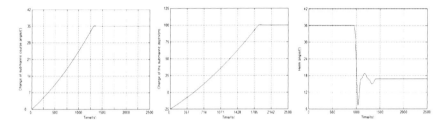

Fig. 11. Change of course after adding interference

Fig. 12. Change of depth (with wave interference)

Fig. 13. Change of helm angle

We can see from the above simulation curves, the neural network PID controller has a very good control in the submersible vehicle course and depth. Especially in random waves interference circumstance, the nerve network PID autopilot has better anti-interfere capability, and its control result is more stable, the fewer number of steering the helm and better robustness.

5 Conclusion

After we combine the neural network with PID technology, the new controller can adapt its parameters itself in real time, so we can expect better performance of the neural network PID controller than the single PID controller. The simulation results show that neural network PID control autopilot has very good performance and is better than traditional PID autopilot in robustness and practicability especially in random interference. We reasoning believe nerve network control is better perspective.

References

1. Lina, X.: The Nerve Network Control. Harbin Industry Press, Harbin (2007)
2. Xiaoxi, Z.: The Design of an Autopilot Control System. Harbin engineering university A Dissertation for the Degree of M. Eng. (2004)

3. Cao, L., Kejun, W.: Progress of the Submersible vehicle Control Methods Study. Shipping Engineering (4), 15–18 (2006)
4. Castro, E.L., Van Der Molen, G.: A Submersible vehicle depth control system design. INT, J. CONTROL, 279–308 (2007)
5. Wang, W.J., Fan, Y.T.: New Output Feedback Design in Variable Structure Systems. Journal of Guildance, Control and Dynamics, 183–190 (2008)

Dependent Failure Reliability Assessment of Electronic System

Wenxue Qian[1], Xiaowei Yin[2], and Liyang Xie[1]

[1] School of Mechanical Engineering & Automation,
Northeastern University, Shenyang, 110004, P.R. China
[2] Department of Mechanical Engineering,
Shenyang Institute of Engineering, Shenyang, 110136, P.R. China
Qwx99@163.com

Abstract. Electronic systems are widely used in many engineer and civil fields, such as aviation, energy, communication, military and automatic etc. Usually the more the components that a system includes the poorer the system reliability is. But the failures of components are usually dependent. In this paper, a reliability model of dependent failure of electronic system is built and the influences of strength decentrality and load decentrality is discussed. The results of two examples prove the validity and superiority of the method in the application of the reliability modeling and assessment of system and the model that considered failure dependent is more reasonable than those conventional reliability models.

Keywords: Reliability assessment; Electronic system; Dependent failure.

1 Introduction

Since the last century the reliability of products attract more and more regard. Usually reliability is defined as the probability of a device performing its intended function for a specified period of time under the specified operating environment. This concept of reliability as a probability, typically quantified by assessing the mean time to failure (MTTF), implies that field failures are inevitable. In today's very competitive electronic products market, a commitment to product reliability is necessary. Estimating system reliability is an important and challenging job. Now the modern machine products, such as large machine tools, aeroengines, ships etc are more and more complex and large. Also as the sophistication and complexity of electronics systems increases, the need for subsystems and components with high reliability levels increases. So it put forward a more and more high request to the modeling and assessment of system reliability and fault diagnosis of system. It is usually impossibly for complex products to carry on a great deal of full-scale system-class reliability experiment because of cost and organization. How to well make use of unit and system of various experiments information to carry out an accurate valuation to the system reliability is a complicated problem, as a result causes many scholars concern [1-7].

Traditional reliability valuation methods such as Fault Tree Analysis(FTA) and Reliability Block Diagram (RBD) are widely used. However these methods can only deal with binary failure system and can't handle functional dependencies between components. So we must do many hypotheses when using these methods and the results may become lack of credibility. It is also difficult to assess the influence to the system when one or several components are failure. So it is necessary to develop some new tools to do this.

Actually a majority of failures of electronic systems are dependent. How to predict and assess this kind of system accurately is regarded by many government regulatory bodies, designers and scholars as essential. Within this field, a number of research issues are unresolved. The first is finally some models are built to assess the reliability of dependent failure system, but many of them do so based on the hypothesis that the failure of each weak site is independent. This hypothesis does not agree with the practice, as the failure of each weak site is dependant on the other sites [8-9].

In this paper, the modeling of multiple weak site components is studied, and the failure properties of multiple weak site components are analyzed.

2 System Reliability Modeling

In conventional reliability assessment of electronic system, only one component is chosen to analyze the reliability and the result is considered to determine the reliability of the component. Usually this proves a too pessimistic result as it has been derived by analyzing only one component. For many electronic systems, which contain more than one component; the failure of any one of the component can cause the failure of the system. In terms of system reliability, it can be regarded as a serial system. Many practical cases show that the reliability obtained through using a conventional system reliability model is markedly less than the actual reliability of component.

However, if only the conventional serial system reliability model is used to calculate the reliability of a electronic system, an invalid result will be obtained. The reason for this is that the failure of each component is not independent but dependent. Conventional serial and parallel system reliability models derive from some special system analysis; they are based on the hypothesis that the failure of each element is independent. For some special systems, under certain conditions this hypothesis can be thought to be approximately correct. However for most systems this hypothesis is almost always incorrect because the failure of an element in a system is dependent and can not be thought of as independent from other elements in the system. Large systems such as aircraft, cars and military tanks usually comprise of thousands of components and the relation of each to the other is serial. When using a conventional, serial system reliability model to calculate the reliability of a system, even if the reliability of each component is very high, the reliability of the system can be very low, because the system reliability is the product of the individual element reliabilities. This is not in accordance with actual situation.

The components of an electronic system usually bear the same load, when the load is uncertain; the failures of the components are dependent on each other. The strength of and the stress imposed on each component are random variables and accord with a certain distribution. Conventionally the strength of each component is considered as independent and identically distributed. Suppose that the probability density function of strength is $g(s)$ and the distribution function is $G(s)$, the probability density function of stress is $f(\delta)$ and the distribution function is $F(\delta)$. Suppose that the system comprise n component, and the observed strength values of the components ranked from smallest to biggest are $s_1, s_2, \ldots, s_n$. From probability and statistics theory it can be shown that $s_1, s_2, \ldots, s_n$ are the order statistics[11]. By using the multinomial distribution function, the probability density function of order statistics of n strength is

$$h(y) = \frac{n!}{(r-1)!(n-r)!} \left[\int_{-\infty}^{y} g(s)ds \right]^{r-1} g(y) \left[\int_{y}^{\infty} g(s)ds \right]^{n-r} \tag{1}$$

For the strengths of components, the most important matter is to determine if the lowest value of strength is less than the stress. When the minimum strength of all components of a system is greater than the stress, then the system does not fail. Thus the reliability of the system is the probability that the minimal order statistic of system strength is greater than the stress in the domain. By using the stress – strength interference theorem, the reliability of a system is

$$R = \int_0^{\infty} f(\delta) \int_{\delta}^{\infty} h(y_1)ds d\delta = \int_0^{\infty} f(\delta) \left(\left[1 - G(s)\right]^n \right) d\delta \tag{2}$$

It can be seen that model (2) is different from conventional system reliability models as it considers the failure dependence of components bearing a common random load.

3 Examples of Reliability Assessment of Electronic Systems

3.1 Series Electronic Systems

There is a series electronic system shown as Fig.1. Suppose the strengths of the components within the system are independent and follows identically distributed normal distribution random variables. The distribution parameters of strength are $\delta(300, 12^2)$, and the distribution parameters of stress are $s(240, 40^2)$.

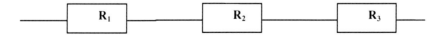

Fig. 1. Reliability model of series electronic system

From Fig.1 it can be seen that the failure of any component can cause the failure of the system.

3.2 Parallel Electronic Systems and Other Complex Electronic Systems

Here the reliability of dependent failure electronic system in relation to the number of component will be studied. Suppose the strengths of the components within a certain system are independent and governed by an identically distributed normal distribution random variable. The distribution parameters are $s(\mu_s, \sigma_s)$, and all the components bear a common load, with the distribution parameter of stress, $\delta(\mu_\delta, \sigma_\delta)$. In order to research the influence of the number of components on the reliability of a dependent failure system, let $\mu_\delta = 300$MPa and $\mu_s = 240$MPa. Fig.2 is a reliability model of parallel system and Fig.3 is a reliability model of series-parallel electronic system and Fig.4 is a reliability model of parallel-series electronic system.

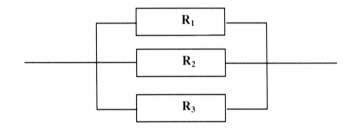

Fig. 2. Reliability model of parallel electronic system

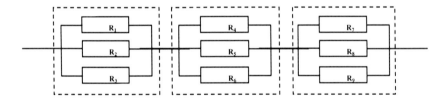

Fig. 3. Reliability model of series-parallel electronic system

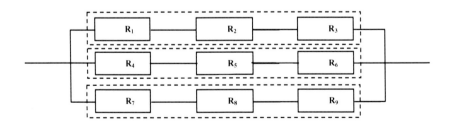

Fig. 4. Reliability model of parallel-series electronic system

For a system with dependent failure components, if only one component is considered, that is to say the reliability of the component is thought of as the reliability of the system, then $R = R_1$, where R is the reliability of system and R_1 is the reliability of any component. When using a conventional serial system reliability model to calculate the reliability of a multiple weak site component, the result is $R = R_1^n$, where n is the number of component. However as this result does not consider the effect of load roughness, the reliability is usually far less than the actual value.

Simple computation can show the relationship of the reliability and the numbers of components for systems under the independent failure hypothesis. It can be seen that the reliability of components decreases very quickly as a result of unreasonable assumptions with the result being too conservative and meaningless in a practical sense.

4 Conclusions

In this paper, a new method is introduced to assess the reliability of dependent failure electronic system. Through this method it can be seen that not only the numbers of components have an important effect on the reliability of dependent failure electronic system. When using conventional serial system reliability models to assess the reliability of dependent failure electronic system, the result is too conservative.

Acknowledgments. This work was partially supported by the Research Fund for the Doctoral Program of Higher Education of China (Grant No. 20070145083), the Hi-Tech Research and Development Program of China (Grant No. 2007AA04Z428), the Scientific Research Fund of Liaoning Provincial Education Department (Grant No. 2009A543) and the National Natural Science Foundation of China (Grant No. 50905031, 50775030, 50805070).

References

1. Kononenko, I.: Machine learning for medical diagnosis: history state of the art and perspective. Artificial Intelligence in Medicine 23(1), 89–109 (2001)
2. Neil, M., Tailor, M., Marquez, D., Fenton, N., Hearty, P.: Modelling dependable systems using hybrid Bayesian networks. Reliability Engineering & System Safety 93(7), 933–939 (2008)
3. Wilson, A.G., Huzurbazar, A.V.: Bayesian networks for multilevel system reliability. Reliability Engineering & System Safety 92(10), 1413–1420 (2007)
4. Wen-Xue, Q., Xiao-Wei, Y., Li-Yang, X., Xue-Hong, H.: Reliability analysis of disk fatigue life based on Monte-Carlo method. Journal of System Simulation 19(2), 254–256 (2007)

5. Slattery Kerry, T., Riveros Guillermo, A.: A parametric template format for solid models of reinforced concrete structures. In: Zhang, T., Horigome, M. (eds.) 17th Analysis and Computation Specialty Conference, p. 15 (2006); Zhang, T., Horigome, M.: Availability and reliability of system with dependent components and time-varying failure and repair rates. IEEE Transact. Reliab. 30, 151–158 (2001)
6. Varghese, J., Dasgupta, A.: An experimental approach to characterize rate - dependent failure envelopes and failure site transitions in surface mount assemblies. Microelectronics and Reliability, 9 (2006)
7. Xie, L., Zhou, J., Hao, C.: System-level load–strength interference based reliability modeling of k-out-of-n system. Reliability Engineering and System Safety 84, 311–317 (2004)
8. Qian, W., Xie, L., Huang, D., Yin, X.: Systems Reliability Analysis and Fault Diagnosis Based on Bayesian Networks. In: 2009 International Workshop on Intelligent Systems and Applications, May 23-24, pp. 1–4 (2009)
9. Wenxue, Q., Liyang, X., Xiaowei, Y.: Reliability Modeling and Assessment of Isomorphic Multiple Weak Site Component. In: 15th Issat International Conference on Reliability and Quality in Design, August 6-8, pp. 143–147 (2009)

Research of RFID Authentication Protocol Based on Hash Function[*]

Li Heng, Gao Fei, Xue Yanming, and Feng Shuo

Information and Electron College
Beijing Institute of Technology
Beijing, China
liheng1109@163.com

Abstract. The introduction of authentication protocols for RFID system provides the security to it, authentication protocol based on hash function is one of the most commonly used authentication protocol, which the hash function they used is a unilateral function, with a relatively high security, and it is easy to implement in the tag of RFID, so it has a wider application in RFID system. As the hash function, the characteristics of its own will have hash table conflicts, and thus would have resulted in security vulnerabilities, aim at the hash table conflict, this paper proposes a specific solution, and simulate the improved authentication protocol.

Keywords: Authentication protocol; hash function; hash table conflict; improve protocol.

1 Introduction

RFID (Radio Frequency Identification) is a non-contact automatic identification technology, its basic principle is the use of radio frequency signal and the spatial coupling transmission characteristics, and achieve automatic Identification of objects. RFID systems are generally composed of three parts: tags, readers and back-end database. The tag storage a variety of information of objects, the reader can read the data in the tag with non-contact methods and write the processed data back into tag. [1]

As an emerging technology, RFID has been widely used in warehouse management, parking management, anti-theft systems, animal management, and so on. With the further development of RFID technology, its applications of fields will become wider, including many areas which have stringent requirement on data security, so people take more and more attention on data security.

The transmission channel of RFID system can be divided into wireless channel between tag and reader and wired channel between reader and back-end database,

---

[*] Fund Item: 863 project" Research of communication test technology in RFID applications" (2006AA04A106).

wired channel is usually as a high-security communications channel to be studied, compared with it, the wireless channel is vulnerable to be attacked and eavesdropped, so the current security studies have focused on the wireless channel.

In order to ensure the security of data communications, develop a comprehensive security authentication protocol is necessary and effectively. Aim at The characteristics of RFID system, people have made a number of security authentication protocol, such as Hash-lock protocol, Distributed RFID inquiry - response authentication protocol, LCAP protocol, based on hash ID change protocol, anti-tracking and cloning of the lightweight protocol, etc. These security authentication protocols provide a certain security guarantees to RFID system. Hash-lock protocol is one of the widely used protocols; People carried out a series of improvements on it, and advance a great degree of security of the data [2].

Because Hash-lock protocol is an important security authentication protocol, this article studies the detailed of it, and simulates the protocol by MATLAB to analyze the important role in RFID security.

2 Hash-Lock Protocol Analysis

2.1 Hash-Lock Protocol Process

Hash-lock protocol is an access control mechanism which was designed by Sarma, and it was base on the unilateral Hash function. As the RFID system have certain requirements to the tag cost and size, it is not possible to use complex encryption algorithm on it. To achieve a Hash function modules only need about 1,000 gates, so it is easy to achieve in the RFID system. At the same time, the Hash function is a unilateral function, anybody who want to steal the Hash value is impossible to restore the characteristic value, so the Hash function has a very high security.

In order to avoid the security threats within sending tag ID clear-text, Hash-lock protocol introduce into a key value. Before RFID system communication, tags stored the following data first: tag ID, key, the back-end database stored the following data: tag ID, key, Hash (key), and the reader in the communication process play the role of receiving and transmitting data. The workflow of Hash-lock protocol is as follows [3]:

First, the reader sends a request to the tags which enter the working scope, for certificating the reader;

When tag receives the request, it will use its own key value to compute Hash (key) and send to the reader;

When reader receives the Hash (key), it will use the Hash (key) stored in reader's own memory to compare, if they are not equal, then refuses to pass the certification; if they are equal, then the certification of reader to tag is pass, it can read their data. At the same time, reader will send the key value stored in its own memory to the tags, ask for tag certification;

When tag receives the key, it will compare with the key stored in tag's own memory, if they are not equal, then refuses to pass the certification; if they are equal, then the certification of tag to reader is pass, tag can send its data to reader and receive the data from reader.

2.2 Matlab Simulation Hash-Lock Protocol

In order to analyze how the introduction of Hash-lock protocol improve RFID system security, we use MATLAB simulate of it, and compare with the system which not authentication protocol. In the simulation process, we have selected 50 tags and 10 readers, of which there are three legitimate tags and a legitimate reader, that is, the key value of this three tags and one reader is legal value, and the rest are fake tags and readers. The simulation purpose is to verify that if legitimate tags and readers can pass the certification, and if the counterfeit will deceive the system authentication protocol.

Communicate these 50 tags with 10 readers one by one, carry through authentication first in accordance with Hash-lock protocol, through the implementation of simulation program, the three tags with valid key value have been certified ,and the remaining does not pass the authentication, thus have been excluded from the RFID system. The reader send the key value of itself after certified legitimate three tags, request the certification of tag. There are only one reader has a valid key value in these 10 readers, after the authentication by Hash-lock Authentication Protocol, the legitimate reader is be certified, so the certified tags and reader constitute an RFID system that can transfer data.

2.3 The Direction of Improvement to Hash-Lock Protocol

The introduction of Hash-lock authentication protocol elevate the system's security greatly, and easy implement in low-cost tags, but after we study it in-depth and found that there still exist some security vulnerability.

First of all, Hash-lock protocol is based on Hash function, and in the course of design, Hash function is deemed to an ideal single function, that means is, each different characteristic value is only corresponding a different Hash function value. In the actual implementation, it is difficult to find such an ideal Hash function, Hash table often exist conflict, that is, two different characteristic values are likely to corresponding to the same Hash function value. If a fake tag uses a different key to gain the legal Hash value with corresponded the legal key value, then it will naturally result in spoofing attacks.

When the Hash-lock protocol accomplish authentication, tags and readers begin to transfer the data to each other. First, the tag transfer its ID to reader clear-text, because the wireless communication channel is an insecure channel, are vulnerable to be eavesdropped, therefore transfer tag ID clear-text is very easy to been subjected to theft, so the attacker possible to forge a legitimate tag.

In every authentication process, the tags are sent with a same Hash (key) to the reader, the attacker can tap into this same Hash (key) to determine the location of tags, which can cause tags location of leaks, it also will cause the system data security risks.

For the security risk issues of transferring ID tag by clear-text and transferring a same Hash (key), There are already some corresponding to the improved method, for example, through add a random number module to improve the fixed Hash (key) issues and transmit in the tag ID through Hash transform, thus preventing the issue of ID leaks [4].

Because the Feature of Hash function is mapped a long series to a short hash table, it will easily lead to the inevitable conflict of Hash table, which are not to be considered in the preliminary design of secure authentication protocol. In order to solve this problem, this paper proposes an improvement of it.

2.4 Hash Table Improvement of the Conflict

Hash-lock protocol, the conflict due to the presence of Hash table of possible error caused by authentication, even though we are careful choice of Hash function, Hash table due to the characteristics of still can not find a completely satisfactory Hash function. To avoid this security risk, we can consider adding another different type of Hash function module in the RFID system, and calculate two different Hash values of one key value, the different types of Hash function is such as a linear function or a power function. After a large number of calculations, the probability of two different Hash function obtained the same eigenvalue is extremely small, in practical application, we can consider that these two Hash values are different. So it is avoid conflict by adding another Hash function module [5].

The specific steps of authentication protocol after improved are as follows:

First, the back-end database stored another data G (key);

Tag receives a request, and then it will computing the Hash (key) and G (key) through the two Hash function modular, and sent Hash (key) and G (key) to the reader;

When the reader receives the Hash (key) and G (key), it will compared them with the reader's own memory of the Hash (key) and G (key), if one of the data is not equal, then consider this is an illegal tag, refuses to pass the certification; if they are equal respectively, then the reader through the certification of reader to tag. At the same time, readers will send the key value stored in its own to the tags, request certification of tag to reader;

Tag compare the key receives from the reader to its own, if the two are not equal, then refuses to pass the certification; if they are equal, then it will pass the authentication of reader to tag, and send its data to reader or receive data from reader.

In order to verify the improved protocol can solve the problem of conflict of Hash table, we choose a Hash table of conflicts arising RFID systems, such as

Research of RFID Authentication Protocol Based on Hash Function   181

shown in Figure 1. A total of 50 tags, of which only four have the legitimate key value, due to unavoidable conflicts Hash table, there is a key value of the illegal tag through authentication due to the Hash conflict, so the RFID system seen it as a legitimate tag, and there are five tag pass the authentication finally. (Red circle stand for the authenticated tags and readers).

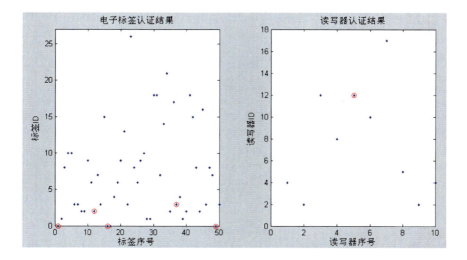

Fig. 1.

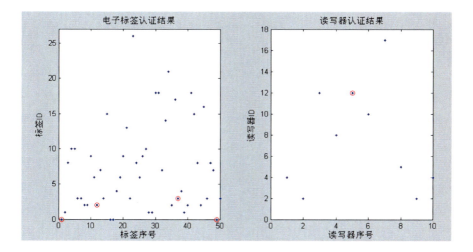

Fig. 2.

Use the improved authentication protocol of two different Hash function modules, we can see through it the illegal key value was verified out, and this tag was locked, not through authentication. using the improved protocol authenticate the above 50 tags and 10 readers again, the result was showed as the Figure 2, the legitimate four tags are pass authentication, and the illegal tag causing by Hash table conflict was denied to authentication. This shows that the improved authentication protocol can solve the Hash table conflict, and the security of RFID system has a good upgrade.

3 Concluding Remark

The applied occasion of RFID system is more and more important, in order to improve the system of data security, people adopt the authentication before data communication, the authentication protocol based on Hash function is one of common protocol. After analyzing and comparing the introduction of Hash-lock authentication protocol in RFID system than a system does not use authentication protocol, the security of system is advanced, not only to stop the illegal tag to communicate, but also to ensure that legitimate tag can participate in communication in system.

Although the Hash-lock authentication protocol for RFID system has brought a large extent of protection, but through people's analysis found that the authentication protocol are still some security bugs. it is mainly reflected in Hash table conflict, the tag ID transmitted by clear-text and fixed position leaks caused by a same value, in order to construct a more secure system, these three issues need to be improved.

References

1. Zhanqing, Y., Sujian, L.: The theory and application of RFID technology. Electronic Industry Press, Beijing (2004)
2. Zhenhua, D., Jintao, L., Bo, F.: Research on RFID Security Authentication Protocol based on Hash function. Computer Research and Development 46(4), 583–592 (2009)
3. Shuguang, Y., Hongyue, D., Shengli, L.: Research on RFID Authentication Protocol based on Hash function. Computer Engineering 34(12), 141–143 (2008)
4. Caixia, Z., Lianglun, C.: Design of RFID Security Protocol based on Hash function. Packaging Engineering 30(1), 115–117 (2009)
5. Yulan, Z.: Data Structures and Algorithms. Tsinghua University Press, Beijing (2008)

Energy Efficient Message Routing in a Small World Wireless Sensor Network

Suvendi Chinnappen-Rimer[1] and Gerhard P. Hancke[2]

[1] Department of Electrical Engineering Science,
University of Johannesburg, South Africa
`suvendic@uj.ac.za`
[2] Department of Computer Engineering,
University of Pretoria, South Africa
`gerhard.hancke@eng.up.ac.za`

Abstract. Routing data messages in a wireless sensor network is different from traditional networks because sensors do not have a unique IP address, knowledge of the network topology or routers and routing tables. Also, the limited power resource places restrictions on the number of messages sent and received within a network. The purpose of this article is to develop an algorithm based on small world networks, to route sensor data to the nearest sink node. A wireless sensor network, with optimally placed sink nodes, does lead to efficient routing of data between nodes while preserving battery lifetime.

Keywords: Wireless sensor network, small world network, routing.

1. Introduction

A Wireless Sensor Network (WSN) is a collection of sensor nodes, deployed within an application area, to monitor a specific event or set of events. As the term *"wireless"* implies, there is no fixed physical connection between sensors to provide continuous energy and an enclosed communication medium. This creates two problems, firstly, the sensor has a finite amount of energy, which once depleted, renders the sensor unusable. Secondly, all transmitted messages will be detected by any listening device within receiving range, which then has to decide whether to accept, forward or ignore the message.

A significant proportion of current research on WSNs is focused on energy preservation, to ensure longer node and hence network lifetime. This need to conserve energy, due to the lack of a continuous energy source, impacts on communication between nodes. It costs energy to transmit and receive messages, and the wireless medium means that all nodes within range receive a message. Hence, a significant amount of research has focused on reducing the number of messages transmitted within a WSN.

Routing data in a wireless sensor network differs from traditional network routing schemes because unique IP type addressing and routing tables are not available. There are no centralized routers with complete knowledge of the network topology. Each node uses only locally available information to route messages. The recipient node in a typical WSN application is usually a sink node which has specialized equipment that transmits the data to an end user [1]. Since most nodes communicate primarily with a sink node, the possibility develops that sink nodes can be placed within the application area at predetermined points. We have shown in [1] that a WSN can be modeled as a small world network, by placing a number of sink nodes in a WSN application area, so that a message from any node within the application area will reach a sink node within a small predetermined number of hops.

The question we will try to answer is can the number of message re-transmissions required to route a message be reduced by placing sink nodes at specific points in the application area. The properties of WSN that indicate similarities with small world networks are [2, 3]:

- The overlap in wireless communication range means that most immediate neighbors of a sensor node are also neighbors of each other.
- Sink nodes communicate with remotely located users (outside WSN area). The remote station coordinates communication between sink nodes. (This long edge provided by the sink nodes reduces the diameter of wireless sensor networks)
- The average transmission distance between two nodes is short.
- There is no central authority to determine optimal routes. Routing is performed at each node with only information about the nodes nearest neighbors.

2. Algorithm Design

2.1 Initialization Algorithm

Sinks are placed at predetermined points in the application area as described in [1]. After the sinks have been placed in the application area, an initialization message (IM) is transmitted by each node in a staggered time format. The time lag can be pre-determined before implementation using a small subset to determine time differentials between receiving multiple IMs at each node. The idea behind the IM is to create a routing table for each node, which can be used to transmit a message from source to destination. As each node only transmits an IM once, the energy cost over the total lifetime of the WSN is low. The routing table is constructed as follows:

- Each node starting from each sink node transmits an IM containing a unique node ID and a list of neighbors. This list is initially empty.
- With the exception of the sink node, a node will wait to receive an IM before transmitting an IM. This ensures that the nodes furthest from the sink node receive valid routing information.
- When a node receives an IM, it will wait a specified time before transmitting its IM. This is because a node may receive two or more IM from its neighbors. By introducing a time lag, we ensure that the node updates its routing table with information from all neighbors.

- If the hop count is less then the specified maximum number of hops required for a message to reach its destination, then the receiving node adds the node's ID, as well as the transmitting node's neighbor list to its neighbor list.
- The node replaces the neighbor list in the IM with its updated neighbor list and re-transmits the message.
- This process continues until the hop count exceeds the maximum hop count.
- All nodes are required to transmit an initialization message.

On initialization N number of IM will be sent within the network, where N is the number of nodes. Each node will only transmit the IM once, thus limiting power usage. Thus all nodes can build up a list of nearest neighbors, and one or more routes to the sink. Although each node sends only one IM, a node will receive multiple IMs depending on the number of neighbors in range. Each received IM route and its hop count will be added to the receiving nodes route table.

Sometimes conditions within the application area raise concerns about the reliability of the wireless medium. This can result in a node not receiving an IM. If a node does not receive an IM, it must send a query to its nearest neighbors. When a neighbor receives a query message, it must respond with its neighbor list. This ensures that all nodes will receive a message containing the path to take to the sink in the required number of hops.

The algorithm assumes that the sink nodes are static. In a mobile application, before a sink node changes location, it should send a message to all neighboring nodes informing them of its change of location, and to allow each node to delete the sink node from its routing table. When a sink node reaches its new location, an IM has to be re-sent so that all neighboring nodes can update its routing table. This obviously incurs an energy cost that would not be required in a once-off stationary sink placement application.

Figure 1 shows a few of the possible routes a message can take to reach the sink node. Routes a, b and c are not all possible routes. Depending on when an IM reaches a node, additional routes such as 1, 2, 7, 12, 0 are possible. Having multiple routes also provides redundancy against node failure. In real-life, sensors will not be optimally placed as in Figure 1, and the routing table will also store the number of hops to the sink, for instances where a real time response is important.

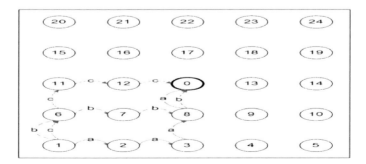

Fig. 1. Possible routes from sensor to sink

2.2 Routing Algorithm

The nature of most applications in a WSN is to detect an event within the network's application area and only then to transmit data. Depending on the number of nodes placed in a particular area, multiple nodes may have events triggered at the same time, and attempt to transmit the data. Current WSN applications generally use some form of data aggregation to reduce the number of messages transmitted in the network. We assume that the data has been aggregated by the nodes, and a single node transmits the data to a sink recipient node. If the sink node is not in the immediate wireless range of the transmitting node, one or more other nodes re-transmit the message until the message reaches the intended destination.

After initialization, each sensor will have one or more routes to one or more sinks in the application area. When an event occurs and the node needs to send a message, it will choose the route at the top of the table. The next node in the route (i.e. one of its nearest neighbor's) will be the intermediate destination and the sink node will be the actual destination. Nodes which are not the intermediate destination or the actual destination and receive the message will update the route and topology information but will not re-transmit the message. The chosen route will be moved to the bottom of the table. This ensures that the same nodes are not used all the time to send a message to the sink. When a node receives a message destined to the sink, it updates its route and topology data. This allows a node to build a reverse direction view of the network topology.

For applications that place many nodes within close range of each other, the following rules will apply:

- nodes that are too close to the transmitting node ignore the message
- a node will use received signal strength to determine if it is at least $\mu * R$ from the transmitting node, where $0.5 \leq \mu \leq 1$ and μ is a variable whose value is set so that the received signal can be accurately decoded.

These checks are done to ensure that nodes which are too close to each other do not re-transmit the message, which results in the hop count reaching its maximum without reaching the actual destination node. The small world routing scheme proposed is a combination of multi-hop routing, where a routing table is built and used to calculate the shortest path algorithm to determine the next neighbor node to forward a message to, and AODV concepts during the initialization stage, when each node sends an IM to its immediate neighbors.

3 Related Work

There are various categories of WSN routing protocols including cluster-based, data-centric, hierarchical, location based, quality of service, network flow or data aggregation protocols [4]. Although our solution can be considered to share certain similarities with clustering, in clustering, each sensor has one cluster head that performs data aggregation to improve energy efficiency. In the small world model, a sensor node is not assigned to one specific sink (cluster head). Instead, a routing table will be created with routes to any sink that is located within the specified number of hops from the sensor node.

Heinzelman has proposed a data-centric routing protocol (SPIN) [5] that initially advertises the data, waits for an interested sensor to request the data and then only sends the data to the requesting node; as well as a combination hierarchical and cluster based scheme that groups sensors and appoints a cluster head to transmit messages to the sink, thus saving the surrounding nodes energy (LEACH) [6].

Niezen et. al [7] compare flooding, multi-hop routing, LEACH and ad-hoc on demand distance vector (AODV). In their results, flooding proved to be worse than multi-hop routing and the LEACH protocol in terms of time for node(s) to fail, while AODV sends even more messages in the network than the flooding protocol and was not initially designed for a WSN low power environment.

Krishnamachari [8] et. al. argues that aggregating similar data from multiple sources to a single destination conserves a sensor's power, and that the real-time lag caused by the data aggregation is not sufficient to significantly impact on the responsiveness of the system. Helmy [9] showed that it is possible to model a WSN as a small world network by adding a small number of shortcuts. He shows that the path length is reduced through the introduction of shortcuts without any significant impact on the structure of the network. Sharma et. al [10] suggest that by adding a few wires to a WSN, the average energy expenditure per sensor node, as well as the non-uniformity in the energy expenditure across the sensor nodes is reduced. The position of a single sink node is arbitrarily chosen.

Guidoni et. al [11] discuss creating a heterogeneous sensor network based on small world concepts in order to optimize communication between sensors and the sink node. They consider two methods to model a small world network, namely directed angulation towards the sink (DAS) and sink node as source destination (SSD). In DAS, node location awareness is crucial to determine the routing path to the sink node. In our solution, nodes do not need to be aware of the exact location of the sink; they utilize an available route path discovered during the initialization phase. In SSD, short-cuts are added directly to the sink node, to create a small world network and reduce the number of hops. The added short-cuts are dependent on the range of the short-cut sensor and will be located close to a sink node. Our model uses multiple sinks so each sensor is at most the maximum specified hops from a sink node.

4 Experimental Simulation

A program to simulate calculation of number and placement of the sinks was developed. This program was used to create the node topology (scenario file) for use in the Network Simulator (NS-2). NS-2 was used to compare the routing capabilities of the small world inspired routing algorithm against the routing capabilities of flooding and gossiping. Afterwards, an indirect comparison of our results against the results presented by [5] using flooding and gossiping as the common base is discussed.

To ensure consistency in comparison with [5], the WSN network shown in Figure 1 was set-up in NS-2. The network assumes nodes are placed in a 40x40 2-dimensional grid. The network consists of 25 nodes, one sink node placed in the centre of the grid and 24 sensor nodes placed 3 or less hops from the sink. We assume all nodes will eventually receive an IM. Similar values were used as those described in [5]. In summary, each node has an initial energy of 1.6J, an accurate message range of 10m between neighbors, the power used to send a message is 600mW and the power used to receive a message is 200mW. Each message size is 500 bytes.

The experiment was run for flooding, gossiping and routing using our algorithm. Two types of experimental topologies are used:

1) Sink placed in the centre of the application area.
2) Sink placed in the top right hand corner of the application area.

In the first scenario, messages were sent from nodes 1, 5, 20 and 24 to the destination sink, node 0 (Figure 1). These nodes were chosen because the nodes are the furthest from the sink node in the given topology. Four messages were sent from nodes 1, 5, 20 and 24 to destination node 0. In the second scenario, messages were sent from node 1 to the destination sink node 24, (for consistency with [5]). In gossiping, a pseudo-random function chose the next node to send the message to. In flooding the message is sent to all neighbors within range. In small-world routing, a once-off initialization message is sent from the sink node, to create a routing table at each node. The small-world algorithm uses this routing table to decide which node is the next recipient node to route a message to the destination node.

5 Results and Discussion

Sink nodes were placed as described in [1]. A message was sent four times from node 1, 5, 20 and 24 (Figure 1). Our solution requires a once-off IM that requires each node to send one message. As the IM builds a routing table in each node, this allows a node to send a direct message along a specific path to the destination. Therefore, each transmission by a node uses the specified number of hops or less to reach the destination node.

As gossiping is dependent on a pseudo-random function that chooses the next node to send the message to, the destination is not always reached. The gossiping results discussed here are best-case scenarios (destination node actually reached). The best-case scenario in gossiping occurred when the pseudo-random function did not loop back to previously used nodes. When there was loop back, the destination node was not reached. In flooding, messages are broadcast to all neighboring nodes, even after a message reaches its destination. In gossiping and small world routing, the message re-transmissions stop after they reach the specified destination.

5.1 Scenario 1

Figure 2 shows the number of total messages sent and received from nodes 1, 5, 20 and 24 to reach the destination node (i.e., node 0) over time elapsed. From Figure 2, there is no time delay in routing a message for small-world routing without the IM. In small world routing with the IM, there is a time lag at initialization to send the IM and calculate the route table. For this particular scenario, a total of 70 messages were required at initialization to create a routing table at each node. The number of messages sent and received to reach the destination node in small world routing with and without the added cost of the IM, shows a significant improvement over both flooding and best-case gossiping. Figure 3, shows the energy used to route a message from nodes 1, 5, 20 and 24 to reach the destination node (i.e., node 0). From the reduced number of messages transmitted in the small world scheme, there is less energy usage which implies longer network lifetimes can be achieved.

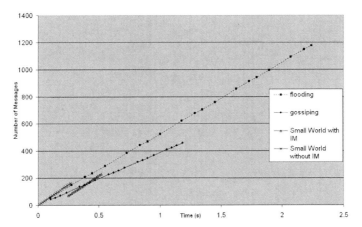

Fig. 2. Total messages from nodes 1, 5, 20 and 24 to node 0

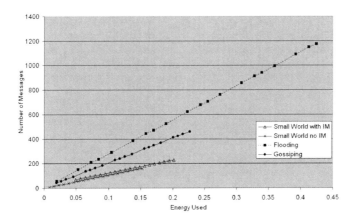

Fig. 3. Energy used to send a message from nodes 1, 5, 20 and 24 to node 0

5.2 Scenario 2

When a message was sent from node 1 to node 24, i.e. the furthest destination from each other, small world routing still outperforms flooding. The direct small world (without IM) send was best as the maximum number of sends and receives are limited to the number of node hops plus the sending and receiving nodes. As shown in Figure 4, the number of messages increases during the IM phase and then levels off as only send-direct messages are routed. Flooding was the worst performer, even though we stopped counting messages, once the destination node has received a message. When the gossiping algorithm performed optimally, gossiping performed slightly better than small world routing with the IM included. However, its larger gradient indicates that it would eventually perform worse then small world routing as the number of messages from the transmitting node increases. When the IM is not included, then the small world routing performs better then gossiping as shown in Figure 5.

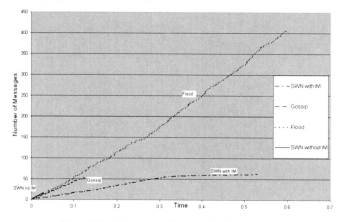

Fig. 4. Total messages from node 1 to node 24

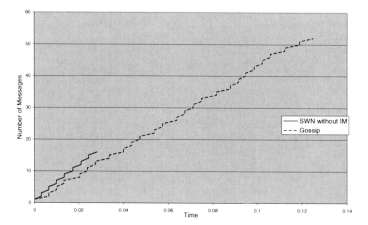

Fig. 5. SWN without IM performs better then gossiping

Figure 6 shows the energy used to send an IM from node 24 to node 1, and thereafter, a normal message from node 1 to node 24. It also shows the energy used to send a message from node 1 to node 24 using flooding and gossiping. As can be seen, there is initial large energy consumption that tracks flooding until all nodes have sent an IM. Thereafter, the energy consumption levels off significantly as nodes used a specific path to send messages. When the pseudo-random function chose the optimum route in gossiping; gossiping outperformed small world routing. However when the IM is discounted as a once-off cost in small world routing, then small world routing outperforms gossiping as shown in Figure 5.

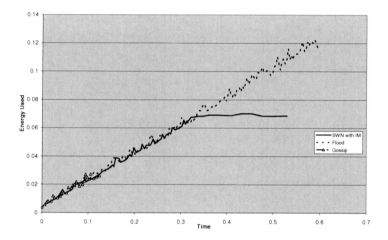

Fig. 6. Energy usage to send message from node 1 to node 24

From the figures, we can see that the IM carries a cost. If only one message is transmitted, then gossiping can be more effective. However as the number of transmitted messages increase, the cost of our algorithm increases at most by the number of hops plus one, (for e.g. 4 sent messages equates to sending node plus 3 hop sends).

We could not perform a direct comparison against SPIN [5] because we did not have access to the NS-2 agents developed for use in SPIN. However, both SPIN and our algorithm appear to perform better than flooding. We achieve better results with gossiping then [5], but this is dependent on the pseudo-random implementation used.

6 Conclusion

We have presented a solution to the routing problem in WSN based on small world network concepts. Small world based routing offers significant advantages over other forms of routing, such as flooding and gossiping, provided more than one message is sent. We have shown that the number of messages required to route a message is restricted to the number of hops from a sink. Our proposed solution imposes a once-off cost at initialization. It is hoped that these costs could be included with other initialization messages when the network is set-up to reduce the energy usage. The solution assumes a two-dimensional grid application area where nodes are stationary and not randomly distributed. In the event of random scattering, our solution may not work as well and as future work we will study implementing small world routing in random topologies.

References

1. Chinnappen-Rimer, S., Hancke, G.P.: Modelling a wireless sensor network as a small world network. In: Proceedings of the International Conference on Wireless Networks and Information Systems, pp. 7–10 (2009)
2. Newman, M.E.J.: The structure and function of complex networks. SIAM Review 45, 167–256 (2003)
3. Kleinberg, J.: The small-world phenomenon: an algorithm perspective. In: Proceedings of the 32nd annual ACM symposium on Theory of computing, pp. 163–170 (2000)
4. Akkaya, K., Younis, M.: A survey on routing protocols for wireless sensor networks. Elsevier Journal of Ad Hoc Networks 3(3), 325–349 (2005)
5. Heinzelman, W., Kulik, J., Balakrishnan, H.: Adaptive Protocols for Information Dissemination in Wireless Sensor Networks. In: Proceedings of 5th ACM/IEEE Mobicom, Seattle, pp. 174–185 (1999)
6. Heinzelman, W., Chandrakasan, A., Balakrishnan, H.: Energy-efficient communication protocol for wireless sensor networks. In: Proceedings of the Hawaii International Conference on System Sciences, Hawaii (2000)
7. Niezen, G., Hancke, G.P., Rudas, I.J., Horváth, L.: Comparing wireless sensor network routing protocols. In: Proceedings of IEEE AFRICON, pp. 1–7 (2007)
8. Krishnamachari, B., Estrin, D., Wicker, S.: Modeling Data-Centric Routing in Wireless Sensor Networks. In: Proceedings of IEEE INFOCOM, New York (June 2002)
9. Helmy, A.: Small worlds in wireless networks. IEEE Communications Letters 7(10) (2003)
10. Sharma, G., Mazumdar, R.: Hybrid Sensor Networks: A Small World. In: Proceedings of the 6th ACM International Symposium on Mobile Ad Hoc Networking and Computing archive, pp. 366–377 (2005)
11. Guidoni, D.L., Mini, R.A.F., Loureiro, A.A.F.: On the design of heterogeneous sensor networks based on small world concepts. In: International Workshop on Modeling Analysis and Simulation of Wireless and Mobile Systems, Proceedings of the 11th international symposium on Modeling, analysis and simulation of wireless and mobile systems, Vancouver, British Columbia, Canada, pp. 309–314 (2008)

Based on the Integration of Information Technology and Industrialization to Propel and Accelerate Industrialization of HeBei Province

Aifen Sun[1] and Jinyu Wei[2]

[1] Department of Management
Tianjin University of Technology
Tianjin, China
sunaifen@yahoo.cn
[2] Department of Management
Tianjin University of Technology
Tianjin, China
weijinyu@tjut.edu.cn

Abstract. China is on the important accelerating step to industrialization country, taking a new road of industrialization of integrating information technology application with industrialization is a very crucial job for our economy. The traditional industrialization that relies on resources and high capital input accelerated the rapid economic growth and at the same time, inevitably, caused a number of problems. For example, the overall efficiency of the national economy was dragged down, all sectors of society slackened the drive to promote technological innovation and efficiency, the development of service industry was suppressed and the shortage of basic resources like water, land, coil, electricity and oil was intensified and the ecological environment was worsened. All of these make the traditional way difficult to carry on and call for a new type one. In this paper, take HeBei Province for an example, to discuss how to increase the industrialization from the angle of integration of information technology and industrialization.

Keywords: Industrialization, Information technology, Integration, Strategy.

1 Introduction

The 17th Party Congress associated "information technology application" with industrialization for the first time to take a new road of industrialization through promoting the integration of information technology application and industrialization so as to push our large industries to grow stronger. Under the "super-ministry system" reform Technology was established, symbolizing that our country starts to take a new road of industrialization of integrating information technology application with industrialization. It also means that China's economy will face new opportunities for development in the process of using information technology to propel industrialization. The integration of information technology and industrialization can not only promote China's industrialization and Information round development, but also help China to speed up the formation of the modern industrial system, promote the industry from large to strong, make the country to strengthen the industry easily .Vigorously push forward the integration of information technology and industrialization, it is the current response to the World Gold Financial crisis, the implementation of the central economic work conference, "to maintain growth, expanding domestic demand, adjust the structure, "the practical requirements, but also implement the scientific concept of development in practice, Acceleration optimization and upgrading of industrial structure in HeBei Province's development needs.

2 The Connotation of Integration of Information Technology and Industrialization

The so-called integration of information technology and industrialization, that is, in the industrial R & D, production, circulation, business areas such as extensive use of information equipment, information products, information technology, to promote research and development of digital design, intelligent manufacturing equipment, Production process automation and Management Network, continuously improve production efficiency, improve the production work Arts, optimizing industrial structure, promote industrial information technology is generally improved over process. From the inherent relationship, Mutual integration of industrialization and information technology is inseparable in mutual integration mutual promotion.

First, industrialization is the source and foundation of information technology, industry develop to a certain stage will directly lead generation of information technology generation, and create materials, energy, capital, talent, market and so on basic condition for the development of information technology.

Second, information is the industrialization of engine and power, through the guidance of industrial development, enhance industrial development speed and improve industrialization Level, make the modern industry towards high value-added direction development, and make a vast space for further development needs.

Third, mutual integration of information technology and industrialization, not only provide a solid material foundation for information technology, but also promoted the industrialization to in-depth development, and more importantly, in the integration process of development, breed a batch of new industries and new trend of marketing , promote pattern change of economic development and the whole society economic transformation.

2.1 The Integration in Micro-enterprise Level

Integration of information technology and industrialization is first reflected in the business strategy. In order to survival and rapid growth in the increasingly competitive market, the companies or through the production and marketing intensive, the pursuit of economies of scale; or through internal growth or external expansion, make an inventory of owned factories and marketing, management of the capital sources, saving the unit cost, increase economic efficiency and effectiveness. Can not be ignored, regardless of what kind of cost effectiveness is needed to pursue the technical conditions. The corporate pursuit of economic effectiveness, the technical pre-conditions is in the existing industry or new industry technology leader, with independent intellectual property rights. Therefore, information technology as the representative of the high-tech, constantly replace traditional techniques, to enhance core competitiveness, professional way to go. The company's foreign expansion must be kept able to absorb information technology as the premises, the most striking is the information industry. It is precisely because the miniaturization of integrated circuits, high-capacity, computer-based and small to large-scale of co-existence of trends and high-capacity technology, running high-speed, optical communications, satellite communications, the rapid development of communications technology has created integration of related fields or relevant areas, the pursuit of larger, more cost-effective super-large enterprises among each other. The integration of information technology and industrialization made the production, operation, management service implementation information, the core business digital, network, auto-intelligent.

2.2 The Integration in Meso-industry Level

The integration of Information technology and industrialization in micro-enterprise level inevitable spread to industry levels of integration and the formation of industrial clusters. Within the cluster to form a new relationship between competition and cooperation, both overcome the inertia and rigidity, but also speed up the competing elements of innovation, to focus on the science and technology investment, information, infrastructure and human resources development on the edge, bringing economic benefits for the entire industry and the region greater. The integration of Information technology and industrialization upgrade the industry structure to form a high new technology industries as the leader, basic industries and traditional industries as the support, to promote economic growth mode from extensive to intensive type of changes, push the industrial economy to information economy transition.

2.3 The Integration in Macro-social Level

To promote information technology and industrialization in the macro-social level integration, can promote the information technology and traditional production technology integration, greatly contributed to the liberation of productive, forces to enhance social production efficiency; can promote the information technology and traditional life style integration, resulting in the new life style, and thus effectively enhance people's quality of life; can facilitate information the greatest degree of spread of civilization, prompting people to change the original production and living ideas and mode of thinking, and promote social harmony and stability; to make the social economic foundation, structure, productivity and the relations of production from an industrial society to information society, the transition to ensure the realization of social economic information.

3 The Qualifications Required of HeBei Province to Promote Integration of Information Technology and Industrialization

3.1 The External Conditions

3.1.1 Driven by the Wave of the World's Information Technology

At present, the world's developed countries striving to develop and improve information technical level, vigorously develop the information industry, as soon as possible to improve national capacity, thereby enhance the country's comprehensive, in the leading position in the 21st century competition. Information industry has become developed countries, national economy and powerful driving force. Therefore, the information technology capacity has become a measure of national strength and international competitiveness, an important Logo. At the regional economy and accelerating the pace of global economic integration today, trade protection become more and more difficult to implement, International harmonization of economic and trade rules to abide by the guidelines have become compliance standards, must face the increasingly fierce global competition in the market. In the world division of labor finding our location, give full play to its comparative advantages to develop core competitiveness.

3.1.2 Construction of the National Importance of Information

Since the Sixteenth Congress, the CPC Central Committee continued deep understanding of information technology. Emphasis on information technology has been increasing. General Secretary Jintao Hu in his report to the fifth part of the 17 "to promote national economic good and fast development "put forward:" the development of modern industrial system, and vigorously promote the integration of information technology and industrialization. This fully reflects from the using information to promote industrialization, to promote information based industrialization, to promote the integration of information technology and industrialization the CPC Central Committee better grasp the actual of China's information technology and industrialization at the current stage, seizing the nature relationship between information technology and industrialization. According to the national "Eleventh Five-Year" Outline the general requirements of the State Council, Plan put forward to 2010, China overall goal of the development of information technology as well as in the applications of information technology, infrastructure, information industry, technological innovation and national information in five areas. The successive policies and regulations are a reflection of our Government's pay an import attention to information technology and information re-construct work.

3.1.3 Driven by Other Economic Growth Pole

Since the reform and opening up, knowledge growth and technological progress significantly speed up. The eastern coastal areas, especially the Yangtze River Delta and Pearl River Delta Economic Zone is transformation form industrial to knowledge economy, information technology and information industry. The three regions are experiencing the change in the pattern of economic growth, promote the industrialization of HeBei Province and make challenges for the new mode of growth.

3.2 Internal Conditions

To speed up the "the integration of information technology and industrialization" for HeBei take the lead in the transformation of economic development mode and enhance the capability of independent innovation, promote the optimization and upgrading of industrial structure have a great significance. Thirty years of reform and opening up, HeBei step into the fast economic and social development Lane, urban economic strength has increased notably, industries have been expanding, information-based rapid development, these has laid a solid foundation for the integration of information technology and industrialization.

Industrialization is the basis of information, industrialization provides the material basis and capital accumulation for the information technology, and to promote society demand for information technology? expand the market capacity for information technology.

Information dominates the development direction of the new era of industrialization, so that it development towards high value-added. Information technology can bring the industrial investment and consumption demand ,added a large number of employment opportunities for industrial ; information technology expand the scope of industrialization resources, information resources has increasingly become important resources in the industrialization development process; Information technology to improve the quality of human resources of industrialization, increased the overall quality and overall competitiveness.

The effectiveness of industrialization

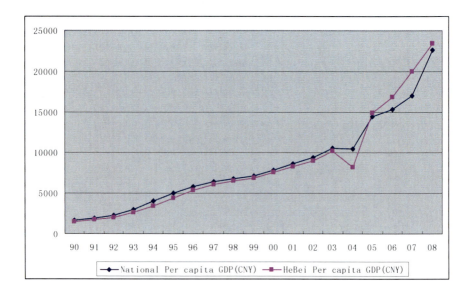

4 Measures for the Industrialization Status of HeBei

The process of industrialization is the evolution of economic structure, during which, the proportion of agricultural earnings and population is dropping respectively in national income and the total population, while non-agricultural sectors like industry are playing ever more significant role. At the present time, the new-type industrialization refers to blazing the new trail to industrialization featuring high scientific and technological content, good economic returns, low resources consumption, little environmental pollution and a full display of advantages in human resources. As a result, we may make breakthroughs in modern information technology, modern information infrastructure and information technology application to transforming traditional industries in using information technology to propel industrialization.

As information infrastructure is the material foundation for information technology application, the efforts should be continuously made on strengthening information infrastructure and creating ubiquitous network environment. For information network security system that is a part of information infrastructure, we must reinforce legal system of information technology application and comprehensive management to ensure the security and reliability of information and network. The high-tech development can be guaranteed reliably and effectively, only when the strategy is upgraded to be national laws. In the future, the focus of scientific and technological legal system will be put on developing high-tech and its industrialization laws in connection with information and biological technology development, high-tech enterprises licensing and policy support and planning and management of hi-tech industry development zone as well as risk financing mechanism so as to bring high-tech licensing, development, industrialization, policy support and protection into the line of ruling by law.

The core of the integration of information technology and industrialization is innovation, should pay more attention and to stimulate the application of innovation. Innovation is the soul of the integration of information technology and industrialization, involving technology, products, markets and even in policy, organization, system and so on aspects. Innovation is not equal to inventions, intellectual property rights, innovation more need is to create new value and applications, so as to meet the people, daily needs, to meet the customers, market demand. Thus, should pay more attention and stimulate the application of innovation, and is recognized by society and the market, driving economic growth, to create new value.

Finally must be emphasized that the key to promote the integration of information technology and industrialization is talent, in particular, a group of both professional knowledge and practice experience of the compound talents. In a sense, the integration of information technology and industrialization in essence are "mechanization, electrification, automation" and "digital, intelligent network" integration, involving software development, info-communications, products design, equipment manufacturing, automatic control and other technologies, is a large number of research and integration of technological innovation, the integration of information technology and industrialization is not only a technical issues, but also a rich connotation of the socio-economic problems. Thus, need for cross-cutting, both information technology expertise, but also understanding the economic and industrial management industry, innovative, versatile talents co-operation.

References

[1] Kang, X., Li, L., Aiting, T.: Integration of Informatization and Industrialization, Technical Efficiency and Convergency. Management Review 21, 3–6 (2009)
[2] Boyong, X., Keping, Z.: The discuss about connotation of the fusion of industrialization and informatization. Manufacturing Automation 31, 34–37 (2009)
[3] Hao, C.: Micro-Research on Foundation Theory of the Integration of Industrialization and Informatization. Reformation & Strategy 25, 56–58 (2009)
[4] Juan, L.: The Syncretization of Informationization with Industrialization for Boosting the Revitalization of Old Industrial Bases in Northeast China. Journal of Harbin University of Commerce(Social Science Edition) 4, 98–101 (2009)
[5] Shulian, Z.: Emphasizing Informatization and Energetically Promoting the Integration of Informatization with Industrialization. Journal of China Executive Leadership Academy Jinggangshan 13, 90–93 (2008)
[6] Jinjie, W., Dong, J.: The accomplished way and strategy of the fusion of industrialization and informatization. Journal of Xi an University of Post and Telecommunications 1, 5–8 (2008)

Pivoting Algorithm for Mean-CVaR Portfolio Optimization Model

Yanwu Liu and Zhongzhen Zhang

School of Management, Wuhan University of Technology, Wuhan, China 430070
liuyanwu8818@126.com

Abstract. The volatility of financial asset return is getting more and more drastic in recent years. This situation makes regulators and investors pay more attention to high loss risk of portfolio. Conditional value at risk (CVaR) is an effective tool to measure the risk of high loss. In order to control the large loss risk of portfolio, the paper established mean-CVaR portfolio optimization model, and presented pivoting algorithm to solve the model. Based on the real trade data of composition stocks of SZ50 Index, we calculated the efficient frontier of the mean-CVaR model. The computational results showed that the pivoting algorithm has high calculation efficiency which can satisfy the computing demand of investment practitioner.

Keywords: pivoting algorithm, mean-CVaR model, portfolio optimization, risk of high loss.

1 Introduction

Traditional tools for measuring portfolio risk assume that the portfolio return is normally distributed which is often not the fact[1, 2]. The normal distribution ofen underestimate the high loss risk of portfolio since the volatility of financial assets is getting more and more drastic in last years. Under this situation, practitioners and regulators pay more attention to manage downside loss risk, especially large loss risk[3]. With Value-at-risk (VaR) being written into industry regulations, it becomes one of the most important tool to manage the risk of high loss. Although VaR is a very popular measure of risk, it has undesirable mathematical characters such as a lack of subadditivity an convexity[4]. Because of these shortcomings of VaR, researchers have proposed using CVaR rather than VaR. CVaR is a coherent measure of risk, and is convenient to compute[4, 5]. With aid of an auxiliary function, mean-CVaR model can be equivalently converted into a convex programming model which is easy to solve. In cases where the uncertainty is modeled by scenarios and a finite family of scenarios is selected as an approximation, the problem to be solved can even reduce to linear programming[5]. When the number of scenarios is large, the linear programming model is a large scale model which requires high efficient algorithms. Pivoting algorithm for linear programming can deal with variances with upper and lower bounds, equation constraints, and superfluous constraints conveniently[6]. During the course of calculation, pivoting algorithm need not add any auxiliary variables. Therefore, the algorithm can solve the mean-CVaR model with a number of scenarios.

This paper is organized as follow. We establish mean-CVaR optimization model and its equivalent linear programming model in Section 2. Section 3 presents the steps of pivoting algorithm for the linear programming model. Section 4 demonstrates how to apply the pivoting algorithm to computing the efficient frontier of mean-CVaR model based on real trade data of composition stocks of SZ50 Index. The computational results show that the pivoting algorithm has high calculating efficiency. Section 5 summarizes the paper.

2 Basic Model

Suppose that the return RP on the portfolio over some forthcoming period is a weighted sum of the n security returns $R = [R_1, R_2, ..., R_n]^T$.

$$R_P = R^T x$$

where the weights $x = [x_1, x_2, ..., x_n]^T$ are chosen by investors. Assuming that R_i is random variable with finite mean,

$$r_p = r^T x$$

where r_p is the expected return of the portfolio, $r = [r_1, r_2, ..., r_n]^T$ are the expected returns on the n securities. Suppose that $CVaR_\beta(x)$ represents CVaR of portfolio x for a given time horizon and confidence level β. The mean-CVaR model can be formulated as follows:

$$\begin{aligned} \min \ & CVaR_\beta(x) \\ s.t. \ & r_p = r^T x \geq r_0, \\ & x_1 + x_2 + ... + x_n = 1, \\ & x \geq 0 \end{aligned} \quad (1)$$

where r_0 represents the required return of portfolio.

It is difficult to handle CVaR because of the VaR function involved in its definition, unless we have an analytical representation for VaR. A much simpler function $F_\beta(x, v)$ can be used instead of CVaR, where $v \in R$. It is proven that we can use $F_\beta(x, v)$ for the optimization of CVaR, i.e.

$$\min_{x} CVaR_\beta(x) = \min_{x,v} F_\beta(x, v)$$

Therefore, model (1) can be converted equivalently into the following model

$$\begin{aligned} \min \ & F_\beta(x, v) \\ s.t. \ & r_p = r^T x \geq r_0, \\ & x_1 + x_2 + ... + x_n = 1, \\ & x \geq 0, v \in R \end{aligned} \quad (2)$$

By introducing auxiliary variables z_i, model (2) can be converted into the following LP model

$$\min \ v + (1/(J(1-\beta)))\sum_{j=1}^{J} z_j \quad (3)$$

$$\text{s.t. } r_p = r^T x \geq r_0,$$
$$x_1 + x_2 + \ldots + x_n = 1,$$
$$z_j \geq -x_1 R_{1j} - x_2 R_{2j} - \ldots - x_n R_{nj} - v, j = 1, 2, \ldots, J,$$
$$z_j \geq 0, j = 1, 2, \ldots, J,$$
$$x \geq 0, v \in R$$

where R_{ij} represents the return of security i under scenario j.

3 Algorithmic Steps

The number of scenarios in model (3) is usually large, so model (3) is a large scale linear programming model. The pivoting algorithm can be used to solve large scale LP because of its many advantages. The algorithmic steps are as follows:

Step 1 Initialization. Let $c = (0, \ldots, 0, 1/(J(1-\beta)), \ldots, 1/(J(1-\beta)), 1)$, let $X = (x_1, \ldots, x_n, x_{n+1}, \ldots, x_{n+J}, x_{n+J+1})$ where $x_{n+i} = z_i$, $i = 1, 2, \ldots, J$, $x_{n+J+1} = v$; let e_j be the jth row of the identity matrix of order $(n+J+1)$ and M be a number large enough.

Take the system of $x_1 \geq 0, \ldots, x_n \geq 0, z_1 \geq 0, \ldots, z_J \geq 0, v \geq -M$ as the initial basic system whose coefficient vectors are $e_1, \ldots, e_n, e_{n+1}, \ldots, e_{n+J}, e_{n+J+1}$ respectively and whose initial basic solution $x^0 = (0, \ldots, 0, 0, \ldots, 0, -M)^T$. Thus the initial table is constructed as shown by Table 1.

Table 1. Initial table

	e_1	...	e_n	e_{n+1}	...	e_{n+J}	e_{n+J+1}	σ_i
c	0	...	0	$1/(J(1-\beta))$	...	$1/(J(1-\beta))$	1	
a_1	r_1	...	rn	0	...	0	0	$-r_0$
a_2	1	...	1	0	...	0	0	-1
a_3	R_{11}	...	R_{n1}	1	...	0	1	M
...	...	...	...	...	...	...	...	...
a_{J+2}	R_{1J}	...	R_{nJ}	0	...	1	1	M

where $a_1, a_2, a_3, \ldots,$ and a_{J+2} are the coefficient vectors of $r^T x - r_0 \geq 0$, $x_1 + x_2 + \ldots + x_n - 1 = 0$, $z_1 + x_1 R_{11} + x_2 R_{21} + \ldots + x_n R_{n1} + v \geq 0, \ldots,$ and $z_J + x_1 R_{1J} + x_2 R_{2J} + \ldots + x_n R_{nJ} + v \geq 0$ respectively. The suitable value of M can be set according to the innate need of real model.

Step 2 Preprocessing. Put non-basic equalities into the basic system as many as possible.

Let I0, I1, I2 and I3 be the index sets for basic equalities, basic inequalities, non-basic inequalities and non-basic equalities respectively. Suppose

$$c = \sum_{j \in I0 \cup I1} w_{0j} a_j ,$$

$$a_i = \sum_{j \in I0 \cup I1} w_{ij} a_i , \quad i \in I3 \cup I2.$$

(1) If $I3 = \emptyset$, go to step 3. Otherwise for an $r \in I3$, when the deviation of a_r is negative, positive or zero, go to (2), (2) and (4) respectively.

(2) (a) If $w_{rj} \leq 0$ for any $j \in I1$, the linear programming has no feasible solution, stop. Otherwise

(b) Select a basic inequality $a_s x \geq b_s$ such that

$$w_{0s} / w_{rs} = \min\{w_{0j} / w_{rj} : w_{rj} > 0, j \in I1\}.$$

Carry out a pivoting on w_{rs}, let $I3 := I3 \backslash \{r\}$, $I0 := I0 \cup \{r\}$, $I1 := I1 \backslash \{s\}$, $I2 := I2 \cup \{s\}$, then return to (1).

(3) (a) If $w_{rj} \geq 0$ for any $j \in I1$, the linear programming has no feasible solution, stop. Otherwise

(b) Select a basic inequality $a_s x \geq b_s$ such that

$$w_{0s} / w_{rs} = \max\{w_{0j} / w_{rj} : w_{rj} < 0, j \in I1\}.$$

Carry out a pivoting on w_{rs}, let $I3 := I3 \backslash \{r\}$, $I0 := I0 \cup \{r\}$, $I1 := I1 \backslash \{s\}$, $I2 := I2 \cup \{s\}$, then return to (1).

(4) If there is $w_{rj} > 0$ for an $j \in I1$, go to (2) (b); otherwise if there is $w_{rj} < 0$ for an $j \in I1$, go to (3) (b); otherwise let $I3 := I3 \backslash \{r\}$ and return to (1).

Step 3 Main iterations. Interchange between non-basic inequalities and basic inequalities.

(1) If all the deviations of non-basic vectors are nonnegative, stop. Otherwise return to (2)

(2) Select a non-basic vector a_r ($r \in I2$) with a negative deviation to enter the basis. If $w_{rj} \leq 0$ for any $j \in I1$, there is no feasible solutions, stop. Otherwise let a basic inequality $a_s x \geq b_s$ leave the basis that satisfies

$$w_{0s} / w_{rs} = \min\{w_{0j} / w_{rj} : w_{rj} > 0, j \in I1\}.$$

Carry out a pivoting on w_{rs}, let $I2 := I2 \backslash \{r\} \cup \{s\}$ and $I1 := I1 \backslash \{s\} \cup \{r\}$, return to (1).

4 Numerical Experiments

4.1 Scope and Dataset Description

SZ50 Stock Index is one of the most important stock market indexes in China. We choose the composition stock of SZ50 Stock Index as the delegation of all the stocks traded in Shanghai Security Exchange Market. Chinese stock market experiences a large up and down from June 1, 2006 to October 15, 2008. This period provides a good chance to research how to manage the risk under the worst case. In order to be consistent to the latest composition stocks of SZ50 Stock Index, the paper is based on the composition stocks released by Shanghai Security Exchange Market on January 5, 2009.

The returns of the 21 stocks that belong to SZ50 Stock Index throughout the period from August 21, 2006 to October 27, 2008 were considered. Other remaining 29 stocks lack at least 5% trade data during the period, so they are taken away from the sample stocks. The dataset of the 21 stocks are drawn from Dazhihui Security Information Harbor. The dataset consists of daily returns and has 531 time periods, considered as equal probable scenarios (n = 21, J = 531). We use CVaR at 0.05 confidence level and set M equal to 1000. We write the program of pivoting algorithm for linear programming in MATLAB 7.0 optimization toolbox. M is set to be equal to 1100 in this example.

4.2 Characteristic of Return Distribution

The traditional mean-variance model supposes that the return distribution of security is subjected to normal distribution. In fact, the real return distribution of security usually underestimates the large loss of portfolio because the return distribution of portfolio is not subjected to normal distribution in practice. The bars in Fig. 1 show the real daily return distribution of SZ 50 Stock Index during the period between August 21, 2006 and October 27, 2008 while the curve shows the normal distribution of daily return during the same period whose mean and variance are equal to mean and variance of the real daily return data during the same period. Figure 1 shows clear that the real large loss is higher than that implied by normal distribution obviously. Therefore, the mean-CVaR model based on the real return distribution can control large loss risk better.

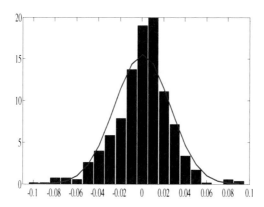

Fig. 1. The comparison between real distribution and normal distribution

4.3 The Computational Results

We solve model (3) based on the dataset mentioned in section 4.1. Table 2 shows the minimum values of CVaR under different returns. Fig. 2 illustrates the mean-CVaR efficient frontier.

Table 2. The minimum CVaR under different daily returns

Expected return	Minimum CVaR	Expected return	Minimum CVaR
0.000869	0.047783	0.001550	0.055005
0.000983	0.048024	0.001664	0.059492
0.001096	0.048394	0.001777	0.067712
0.001210	0.049014	0.001891	0.077069
0.001323	0.049884	0.002004	0.086776
0.001437	0.051942		

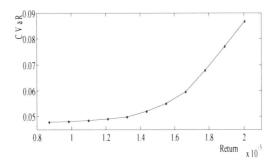

Fig. 2. Mean-CVaR efficient frontier

4.4 Discussion

Figure 1 illustrates that the return distribution of SZ50 Stock Index is not subjected to normal distribution. The return fluctuation of SZ50 Stock Index is more stable than that of its individual composition, so the return fluctuation of individual stock is more drastic and therefore its distribution deviates from normal distribution more likely. Under this situation, traditional mean-variance model cannot reflect the characteristic of portfolio risk, especially the large loss risk. Mean-CVaR model concentrates on the large loss risk; therefore it has obvious advantages for investor to manage financial risk particularly when the financial market is up and down drastically. Furthermore, computing mean-CVaR efficient frontier by model (3) is based on the real distribution of portfolio return which can better reflect large loss risk in contrast with hypothesis of normal distribution.

Although model (3) is a large scale linear programming problem with a number of scenarios, the pivoting can solve the model efficiently. The computational results are shown in Table 2 and Figure 2. Numerical experiments demonstrate that the pivoting algorithm for model (3) has high computing efficiency which can satisfy the computational demand of investors in practice.

5 Summary

Real return distribution of financial asset is always not subjected to normal distribution and shows the characteristic of fat left tail. Traditional mean-variance portfolio model based on hypothesis of norm distribution of return usually underestimates the large loss risk. When the fluctuation of security market is small, mean-variance model is an efficient tool to characterize the risk of portfolio approximately. Under the condition of drastic fluctuation of return, we must apply mean-CVaR model to manage the large loss risk instead of mean-variance. We establish the mean-CVaR portfolio model. Furthermore, we convert the model into linear programming model by linearization technology. Since a number of scenarios make the linear model be a large scale problem, we proposed the pivoting algorithm for linear programming to solve the problem. The pivoting algorithm can deal with equality constrain, free variables, and upper and lower bounds of variables efficiently. In particular, the pivoting algorithm need not add any auxiliary variables during the process of computation. Therefore the pivoting algorithm has high computing efficiency and can be used to solve linear programming of large scale. The numerical experiments based on the real return data of composition stocks of SZ50 Stock Index verify the efficiency of the pivoting algorithm for linear programming.

References

1. Steinbach, M.C., Markowitz, H.: Revisited: Mean-Variance Models in Financial Portfolio Analysis. SIAM Review 1, 31–85 (2001)
2. Alexander, G.J., Baptista, A.M.: A Comparison of VaR and CVaR Constraints with the Mean-Variance model. Management Science 9, 1261–1273 (2004)
3. Jarrow, R., Zhao, F.: Downside Loss Aversion and Portfolio Management. Management Science 4, 558–566 (2006)
4. Rockafella, R.T., Uryasev, S.: Conditional Risk-at-Value for General Loss Distribution. Journal of Bank & Finance 26, 1443–1471 (2002)
5. Rockafellar, R.T., Uryasev, S.: Optimization of Conditional Value-at-Risk. Journal of Risk 3, 21–41 (2000)
6. Zhang, Z.Z.: Convex Programming, pp. 66–98. Wuhan University Publish Press, Wuhan (2004)

The Ontology-Cored Emotional Semantic Search Model*

Juan-juan Zhao, Jun-jie Chen, and Yan Qiang

College of Computer and Software,
Taiyuan University of Technology,
Taiyuan, Shanxi, China
zh_juanjuan@126.com

Abstract. An ontology-cored emotional semantic retrieval model was proposed and constructed in order to overcome the current situation that it is difficult to implement emotional semantic search in the course of image searching. A method Combines Mpeg-7, theory of concept lattices and ontology construction together to construct the kernel ontology library was put forward. The difficulty of this method lies on how to integrate the Mpeg-7 standard descriptor with image emotional ontology properties, and how to auto-generate new concept results. Related experiments were carried out of which the results validated the feasibility of this model on image emotional semantic search.

Keywords: image emotional semantic, retrieval model, ontology, Mpeg-7, theory of concept lattices.

1 Introduction

Image embodies rich emotional semantic meanings. For example, images features including color, texture and shape can reflect rich emotions. Among traditional information retrieval models, there is a relatively bigger semantic difference between logical views of the users' information demand and that of files, especially in image emotional semantic retrieval, such difference may produce "semantic gap".

Ontology is originated from the philosophical concept, and it is the formal specification of the domain's shared conceptual model in the field, represents a common understanding of the domain's knowledge and defines the relationship between domain' concepts. It describes conceptual semantic information based on the relationship between concepts. As a kind of standardized description of the domain's knowledge, ontology is conducive to machine readable and semantic retrieval. Its research results are of important theoretical meaning to research and

---

* Supported by National Natural Science Foundation of China (60970059), Natural Scientific Foundation of Shanxi (2009021017-3).

development of knowledge management, information retrieval, human-computer interaction and and semantic Web.

Researchers have made some achievements in using ontology to represent image features.

2   Image Emotional Semantic Retrieval Model with the Focus on Ontology

Image emotional semantic retrieval model with the focus on ontology is shown in Figure 1, consisting of semantic analysis, image preprocessing and image emotional semantic inference model.

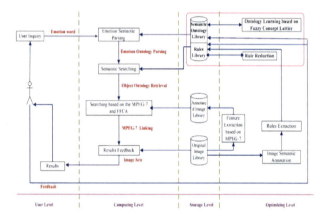

Fig. 1. Image emotional semantic retrieval model

2.1   Semantic Analysis Module

Information demands input by users are varied, including key words for exactly expressing the users' intention; picture information representing users' emotions; fuzzy semantic information. Thus, it becomes necessary to conduct semantic analysis of users' demand information before starting the retrieval. We usually adopt NLP technology, including word extraction, word class tagging, concept identification and mapping of notional word and ontology library.

2.2   Image Preprocessing

Image preprocessing is the standardization process of image preprocessing. Semantic tagging of original image library is realized through unifying pixel, image denoising, image segmentation, image feature extraction and image classification and based on the ontology structure, namely expressing the original image information as formal semantic information that can be understood and processed by the computer.

Image feature extraction. In completing image feature extraction, we studied the developed extraction algorithm related to low-level visual features such as image color, texture and shape in terms of image processing. Principal component analysis (PCA) is applied to extract color features; Gabor filter used for processing texture features; pixel analysis method and Canny operator edge detection are combined to extract shape features of image, and to develop image feature library by integrating such features.

Image classification. Image is classified into numerous categories according to image low-level features. The currently matured machine learning techniques are used for classification, including SVM, Decision Tree, Neural Networks and FSVM. Decision-tree classification is applied in the system.

Image semantic tagging. Owing to many ontology-based semantic tagging tools, we can choose proper tagging tools in accordance with tagging object characteristics, tagging mode or tagging elements. Gate tool taking ontology as the tagging element is applied in the text.

2.3 Emotional Semantic Inference Model

Such inference model is the core of semantic retrieval. Such text selects jena as the reasoning machine. Jena is the development tool kits of Java used for applications development in Semantic Web. Reasoning function is a subsystem of jena, while image emotional semantic ontology library serves as the key issue and the premise to realizing semantic reasoning.

Jena-based emotional reasoning model is described in the following figure.

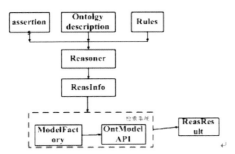

Fig. 2. Emotional semantic inference model

Various models are created based on ModelFactory, and users visit models via OntModel API. Assertions indicates ontology's examples, Ontology description refers to ontology's concept and relationship, Rules represent rule base for ontology acting on the reasoning machine. Their binding with Reasoner produces the reasoning results.

3 Creation of Core Ontology Library

3.1 Relevant Concepts

MPEG-7 is a set of international standards for describing descriptors of diversified multimedia information. Standard Set of Descriptors defined based on MPEG-7 is used for describing various kinds of multimedia data, Description Schemes are used for standardizing generation of multimedia-described subsets and connections among different subsets.

Concept lattice is a mathematization expression based on concept and its levels. It has the following definitions.

Definition 1. formal context: C=(G, M, I) is called as the formal context, including two-tuples consisting of two sets (G, M) and a relation set I of the two-tuples. Of which, G means object, M refers to attribute, I indicates relation, and I$\subseteq$G×M. we usually use (g, m)$\in$I to represent I between G and M.

Definition 2. Define a object set G as G'= { m $\in$ M | gIm, namely g $\in$ G (meaning common attributes set in G). Accordingly, define an attribute set as M'= { g $\in$ G| g Im, namely m $\in$M} (including a collection of attribute objects in all M). Of which, G' means intent, M' indicates extent.

Definition3:if$(G^1,M^1)(G^2,M^2) \in$ F(C)and$(G^1,M^1)(G2,M^2) \in$ F(C),$(G^1,M^1)$is regarded as the sub-concept of (G^2,M^2), and we accordingly call (G^2,M^2)as the super-concept of(G^1,M^1), and mark it as$(G^1,M^1)\leq(G^2,M^2)$. Lattice consisting of partially ordered set(F(C), $\leq$ is regarded as the concept lattice of the formal context, marked as L(C).

3.2 Method for Building the Image Emotional Semantic Ontology Library

Semi-automatic building of image emotional domain ontology framework is realized by combining Mpeg-7(Multimedia Content Description Interface)and Theory of Concept Lattices. The Mpeg-7's ontology structural framework is as follows (see in Figure 3).

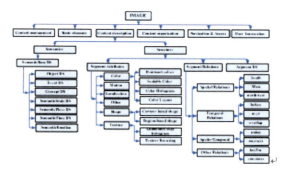

Fig. 3. Framework of image top-level ontology

3.3 Automatic Building of Image Emotional Domain Application Ontology

Hasse chart for creating the concept lattices. Scene image listed in the text contains information related to emotional ontology framework, image color, scene and emotion, Hasse built based on the formal concept background data is shown in Figure 4.

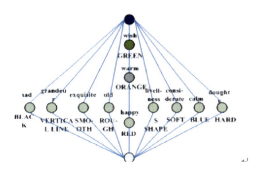

Fig. 4. Hasse Chart for emotion-color's space

3.4 Image Emotional Ontology Framework

Image emotional ontology framework is shown in Figure 6, The corresponding relationship between image low-level features, diversified levels of semantic and emotional semantic features is reflected by combining Mpeg-7 ontology framework and image semantic description model. Since image low-level features can be automatically extracted by use of machine learning techniques, image emotional domain application ontology can be built with the help of such low-level features.

4 System Implementation

Eclipse development platform is applied in the system to build B/S structural emotional semantic retrieval system. We select 300 from 1000 scene images with typical emotional colors as machine learning samples during image preprocessing. We adopt the ontology-building techniques mentioned by in the paper to create the emotional semantic ontology of scene image, and use Jena reasoning technology to realize image emotional semantic retrieval.

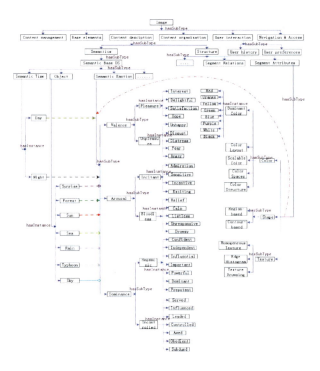

Fig. 5. Framework of Image's emotional ontology

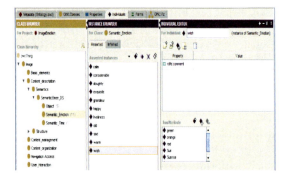

Fig. 6. Structure of image emotional semantic ontology

The Interface of Structure of image emotional semantic ontology emotional semantic search is shown in Figure 6.

The Results of image 's emotional semantic (wish) search is show in Figure 7.

Fig. 7. Results of image emotional semantic (wish) search

5 Conclusions

Based on system verification, we find that image emotional semantic retrieval model with the focus on ontology is an effective means to resolving the current image emotional semantic retrieval. Such model can retrieve corresponding images based onhuman emotional information, thereby effectively realizing image's emotional semantic retrieval. In the future, we will combine text-based ontology automatic-building techniques, and use statistical method to extract conceptual words in image emotional domain, so as to build the fuller image emotional domain ontology by using of the building methods mentioned in the text and to lay a foundation for furthering to realize ontology-based image emotional tagging and retrieval.

References

1. Xiaoyong, D., Man, L., Wang: A Survey on Ontology learning Research. Journal of Software 17(9), 1837–1847 (2006)
2. ISO/IEC JTC1/SC29/WG11N6828, MPEG-7 Overview (version 10)
3. Xianming, C., Xiaoming, W.: The MPEG-7 Video Semantic Description Model Based on Ontology. Journal of South China Normal University(Natural Science Edition) (2), 51–56 (2007)
4. Marek, O., Vaclav, S., Jan, S.: Designing ontologies using formal concept analysis. In: Proceedings of the International Conference on Communications in Computing, pp. 302–308 (2004)
5. Suk-Hyung, H., Hong-Gee, K., Hae-Sool, Y.: A FCA-based ontology construction for the design of class hierarchy. Computational Science and Its Applications 3482, 827–835 (2005)
6. Meili, H., Zongtian, L.: Research on Domain Ontology Building Methods Based on Formal Concept Analysis. Computer Science 33(1), 210–212 (2006)
7. Weining, W., Yinglin, Y.: A Survey of Image Emotional Semantic Research. Journal of Circuits and Systems 8(5), 101–109 (2003)

Multi-agent Task Allocation Method Based on Auction

Xue-li Tao and Yan-bin Zheng

Colleage of Computer and Information Technology, Henan Normal University, 453002,
XinXiang, China
`xueli_xl@126.com, zybcgf@163.com`

Abstract. Task allocation is the core problem of MAS. Capability is the embodiment of agent's own value, and each task requires certain capabilities to complish it. Based on capability, this article present an auction-based task allocation algorithm. Each agent gives a reasonable tender price to select perfect task, according its capability and its preferences for different tasks, and mission requirements. Finally, we make a simulation experiment for transporting problem, and the results show that the algorithm is reasonable and effective, Superior NeA-MRTA[6] algorithm.

Keywords: task allocation, auction, capability, preference.

1 Introduction

Task allocation is one of the problems which must be solved in multi-agent system (MAS). The optimal allocation is an NP problem. In recent years, the auction method based on market mechanism has aroused the interests of many experts and scholars [1-7]. The most prominent characteristic of the auction is that the price is determined by the competition, rather than by the seller. Liu[6] employs SPSB(Second Price Sealed Bid Auction), which is based on the new contract net task allocation algorithm(NeA-MRTA) and task re-allocation algorithm (ReA-MRTA). Hu[7] puts forward to a multi-agent task allocation algorithm which is based on English auction method, with using an improved simulated annealing algorithm with memory. These methods are all based on the homogeneous agent assumption, which is obviously inconsistent with the characteristics of MAS, regardless of the fact that agents have different preferences for different tasks. In this paper, the author puts forward a task allocation algorithm based on the capacity and preference of agent. Experiments show that this algorithm is more reasonable and effective for the result of task allocation.

2 Formal Description

Assume : A is a set of agent to satisfy : $A = \{a_i, 1 \leq i \leq M\}$, T is the collection of task in current system to satisfy : $T = \{t_j, 1 \leq j \leq N\}$.

Definition 1. Utility value signifies the satisfaction of agent for task allocation results. If agent a_i wins the task t_j, then its utility value is u_{ij}, showing the task assigned agent of satisfaction: $u_{ij} > 0$.

Definition 2. The total utility of system for task allocation is marked U.

$$U = \sum_{i=1}^{M} \sum_{j=1}^{N} u_{ij} \tag{1}$$

The optimum allocation problem of multi-agent task allocation can be formed into the linear programming problem:

$$U_{\max} = \max \left(\sum_{i=1}^{M} \sum_{j=1}^{N} \eta_{ij} u_{ij} \right) \tag{2}$$

making:

$$\sum_{i=1}^{M} \eta_{ij} = 1, 1 \le j \le N, \sum_{j=1}^{N} \eta_{ij} = 1, 1 \le i \le M$$

wherein, $\eta_{ij} \in \{0,1\}, 1 \le i \le M, 1 \le j \le N$.

Definition 3. The cost for agent to accomplish the task is the map: $\cos t : resource \to R^+ \cup \{0\}$. Remember the cost of agent a_i to complete the task as : $\cos t(a_i, t_j)$.

Definition 4. The expectancy degree for the task signifies the agent "preference". Remember the preference of agent a_i for the task $t_j : E_i^j \in R^+ \cup \{0\}$.

For a specific application environment, there are various types of tasks and agent executable task is limited. Agent has the different expectation degrees to different tasks. $E_i^j = 0$ this means that agent a_i is not interested in the task t_j, not choosing to undertake the task. $E_i^j = 1$ signifies that agent is very interested in the task. As long as it has such tasks, agent will not choose other tasks.

Capacity is an element employed to evaluate agent[6], which possesses different capacities. Each task also requires a different capacity to accomplish it. Suppose that L consists of a set of of single atom capabilities $c_k : C = \{c_k\}, 1 \le k \le L$, there are M heterogeneous agents in system.

Definition 5. Each agent a_i in the system has kinds of atom abilities, and the set of atom abilities can be difined as an agent's capacity vector C_i^a which is as follows:

$$C_i^a = diag(\alpha_{i1}, \alpha_{i2}, \ldots, \alpha_{il})(c_1, c_2, \ldots, c_l)^T$$

The above formula will be recorded as follows: $C_i^a = [\alpha_{i1}c_1, \alpha_{i2}c_2, \dots, \alpha_{il}c_l]^T$, α_{ik} corresponds with the strength of the ability c_k level of agent a_i, even $\alpha_{ik} \geq 0$. If agent a_i does not have the ability c_k, then its corresponding $\alpha_{ik} = 0$.

Definitions 6. Each task t_j in the system also requires a different capacity to accomplish it, so each task has a corresponding capacity vector C_j^t which can be defined as:

$$C_i^a = diag(\beta_{j1}, \beta_{j2}, \dots, \beta_{jl})(c_1, c_2, \dots, c_l)^T$$

It is recorded as: $C_j^t = [\beta_{j1}c_1, \beta_{j2}c_2, \dots, \beta_{jl}c_l]^T$, β_{jk} corresponds with the strength of the ability c_k level of task t_j, even $\beta_{jk} \geq 0$. If it has no use for capability c_k to complete task t_j, then its corresponding $\beta_{jk} = 0$.

Definition 7. The capability conditions to complete the task: if agent a_i have the capability to complete task t_j, then there are $\alpha_{ik} \geq \beta_{jk}, \forall k, 1 \leq k \leq l$.

3 Auction-Based Task Allocation Arithmetic

In Auction-based method, there are two roles: sellers and buyers. The agent with a task is the seller, and the other agents are buyers. Assuming that the seller be assigned tasks to each have an expectation of the minimum bid MIN_Bid_Cost, and each agent have an assessed value of the task MAX_Exp, When the task is assigned, there is a transaction price ACT_Bid_Cost. If the seller a_i, wins the task t_j via auction, then the satisfaction degree of the agent to the task allocation results will be :

$$\frac{MAX\_Exp}{ACT\_Bid\_Cost} \times E_i^j \quad (3)$$

If the value of Equation (3) is larger, agent to the satisfaction degree for the results will be higher. Seller agent's utility is:

$$\frac{MIN\_Bid\_Cost}{ACT\_Bid\_Cost} \quad (4)$$

If the value of Equation (4) is smaller, the seller agent to the satisfaction for the results will be higher, and its utility will be higher.

3.1 Buyer Algorithm

Assumption: MIN_Bid_Cost is the lowest bidding of the buyer. MIN_Exp is the smallest expectation of agent. $Cost$ is the needed cost to complete the task.

MAX_Exp is the assessed value of agent to the task. Bid_Cost is the consume for one bidding. $Task\_Exp$ is the expectation degree of agent for the task. ACT_Bid is agent's actual bidding. The algorithm is as follows:

```
(1)Seller agent releases all of the agents information
task, including the mission series number, mission
requirements, minimum bids, etc.;
(2)When Buyer agent receives the information task, it
will decide whether to complete the task or not in
light of its capacity; if it can not complete the task,
goto(8);
```
(3) If $MAX\_Exp \leq MIN\_Bid\_Cost$, then go to(8); or else in accordance with the formulated "**bidding strategy**" to bid for the auction task;
```
(4)To submit bid to the seller, waiting for reply;
(5) If the task information arrives in a certain period
of time , that means the acceptance of bid, goto(8);
(6)If the task allocation results have been revealed,
but there is no acceptance of bid , then goto (8);
(7)If there is no task allocation result, indicating
that there is conflict, then proceed to the next round
of the auction. Choose a price "markup strategy" for
those bidding participants; work out the new bid, then
goto(4);
(8)End.
```

3.2 Seller Algorithm

Assume: MIN_Exp_Bid is the lowest expected knocked down price of the seller. CNT for the Auction times, n for the number of buyer agent who give the same highest highest bidder to win the task in an auction. The algorithm is as follows:

```
(1)Issue the task information to all agents;
(2)Accept all the agent's bids, if there are bidding
information, then goto(4);
```
(3) Adjust the value of MIN_Bid_Cost, then goto(1);
(4) $CNT = CNT + 1$;
(5) Select the best bidder from all bidders, assuming there are n, the highest bidding is HP;
(6) If $HP < MIN\_Bid\_Cost$, then goto(3);
(7) If $HP < MIN\_Exp\_Cost$ and even $CNT < 2$, then goto(3);
(8) If $n = 1$, then select the agent, and goto(10);
(9) If $n > 1$, even $CNT \geq 2$, assign a highest bidder randomly, goto(10), if $CNT < 2$, then goto(2);
```
(10)Assign the task to the highest bidder, the trans-
action price is the second high price;
(11)Reveal the distribution result to all the bidders;
```

3.3 Policy Design

3.3.1 Bidding Strategy

If the buyer agent wants to win the task, its bidding ACT_Bid should be higher than the seller's required minimum price, i.e. $ACT\_Bid = MIN\_Bid\_Cost + \Delta p$. At the same time, the buyer agent not only wants to win the task below MAX_Exp, but also hopes to obtain some surplus except the cost of the execution cost, i. e. $MAX\_Exp - (MIN\_Bid\_Cost + \Delta p) - Cost > 0$.

Therefore, $\Delta p = (MAX\_Exp - MIN\_Bid\_Cost - Cost) \times Task\_Exp \times \kappa$. Assuming $Task\_Exp \in [0,1]$, then assuming $\kappa \in [0.1]$, then the value of κ can be chosen according the actual situation.

The bidding strategy of buyer agent:

(1) If the bidding time $CNT = 0$, then $ACT\_Bid = MIN\_Bid\_Cost + \Delta p$.

(2) If $CNT \geq 1$, then $ACT\_Bid = MIN\_Bid\_Cost + \Delta p'$, $\Delta p'$ can be calculated with the same method as Δp, if $\Delta p' < 0$, then the auction is be abandoned.

3.3.2 Markup Price Bidding Strategy

If there are prices conflicts after the first bidding competition, the buyer agent needs to re-auction and $\Delta p'$ should be added on the basis of first bidding. Assuming a Bid_Cost for each consumption, then the condition for agent to re-bid is: $MAX\_Exp - (ACT\_Bid + \Delta p') - Cost - Bid\_Cost > 0$, then the agent can increase the bidding within the range of:

$$0 < \Delta p' < (MAX\_Exp - Cost - Bid\_Cost - MIN\_Bid\_Cost).$$

If multiple bids, on behalf of agent-based CNT number of bids, then: $\Delta p' = (MAX\_Exp - Cost - CNT*Bid\_Cost - MIN\_Bid\_Cost) \times Task\_Exp \times \kappa$.

Buyer agent's markup price bidding strategy:

(1) If $ACT\_Bid = MAX\_Exp$, the auction is abandoned.

(2) If $(MAX\_Exp - ACT\_Bid) < MIN\_Exp$, the auction is abandoned.

(3) Bidding once more, $ACT\_Bid = ACT\_Bid + \Delta p'$.
If $ACT\_Bid > MAX\_Exp$, then $ACT\_Bid = MAX\_Exp$.

4 Simulation

4.1 Experimental Design

In the process of the munitions transportation, each task can be accomplished by an agent, who can estimate the consumption who he reaches the destination. Senior agent is responsible for distributing tasks to other staff agents, who are equal among members of the agents. Communication is reliable between the Senior agent and members of the agents. We use JAVA threading to simulate the algorithm validation. The hardware and software to run experiments are as follows:

- Hardware Conditions: Genuine Intel(R) CPU T1300 @1.66Hz; 1G Memory.
- Software Conditions: Windows XP Professional 2002 Service Pack 3; JDK1.7.

Agent number ascends from 3 to 15. For each simulation, agent and the initial position of the tasks is randomly generated. Each of the combination is simulated for 30 times, the result of which are to be averaged to form the final result in the group.

4.2 Result Analysis

In the strategy design, the buyer's offer and markup price, are based on his expectations of the task given. The algorithm is to be formulated as $Task\_Exp \times \kappa$, wherein κ is an empirical value, and its value affects algorithm implementation.

4.2.1 κ Value on the Satisfaction of Agent

Equation (3) is to evaluate the buyer's satisfaction. The lower the closing price of the task ACT_Bid, the higher agent expectations degree for the $Task\_Exp$ and the greater the value of equation (3), then the buyer satisfaction with the results of task allocation is higher, that is, the higher the effectiveness of the buyer. Equation (4) is used to measure the seller's satisfaction. The greater the task of closing prices of ACT_Bid, the smaller the value of equation (4), the greater the effectiveness. The results are shown in Fig. 1.

4.2.2 κ Value on the Algorithm Execution Time

As can be seen from Fig. 2, under different values of κ, the algorithm running time is basically the same trends. In the case of a single task, with the increase in the number agent, the overall time is on the rise. When $\kappa = 0.1$ and $\kappa = 0.9$, the overall time is relatively low, but the success rate is rather low now. That is because when $\kappa = 0.1$, the buyer's offer often does not meet the minimum expectations of agent transaction price. While $\kappa = 0.9$, the buyer's offers a higher bid and often has no ability to re-bid in the conflict, because the algorithms are defined as: when $ACT\_Bid = MAX\_Exp$ and $(MAX\_Exp - ACT\_Bid) < MIN\_Exp$, the bidding is to be given up. Therefore, in both cases, the task allocation failure rate is the highest, amounting over 80%. Similarly, when $\kappa = 0.3$ and when $\kappa = 0.7$, the task allocation failure rate is also about 50%. But when $\kappa = 0.5$, task allocation failure rate is just 1% or so. In short, this algorithm is feasible when $\kappa = 0.5$, and the algorithm execution can reach the optimal in the case.

4.2.3 Comparison with Similar Algorithms

Under the same experimental conditions and the environment, NeA-MRTA [6] algorithm experiment was also conducted. With an equal number of agents, the two algorithms are used to compare bidding times when the buyer needs to re-bid. The results are shown in Fig.3.

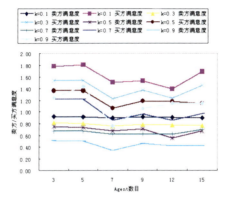

Fig. 1. κ value on the satisfaction of agent

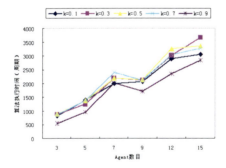

Fig. 2. κ value on the algorithm execution time

As far as algorithm in this paper is concerned, when there is a conflicting bidding, the seller will do nothing but to wait, for the buyer's re-bidding. In this paper, if there are still conflicts after twice bidding, for the same task, then the seller can specify a buyer. Therefore, the speed is faster. NeA-MRTA algorithm designs a tripartite handshake protocol to ensure the distribution of tasks, though it is carried out at the cost of increasing the auction times when the number of agents is

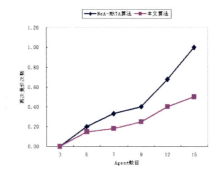

Fig. 3. Bidding times comparison between this algorithm in this paper and NeA-MRTA algorithm when buyers need to re-bid

increased, as can be seen from Fig.3. Under the same experimental background and the environment, the auction times is less in the algorithm in this paper than that in NeA-MRTA algorithm, therefore, as far as the execution time is concerned, the algorithm in this paper is superior to NeA-MRTA algorithm.

5 Conclusion

A rational allocation of tasks can improve the overall performance of the system and vice versa will result in system performance degradation, and even paralysis. In this paper, agent capabilities and mission requirements are described by using capacity-vector, at the same time agent preferences for different tasks are taken into consideration. Thus, not only the buyer's willing is respected to elicit a rational offer, but also the seller's interests are guaranteed. Conflicts can be effectively reduced in the bid. Experiments show that the method is reasonable and efficient, better than NeA-MRTA algorithm in the same experimental conditions and the environment.

References

1. Bernardine Dias, M., Zlot, R., Kalra, N., Stentez, A.: Market-Based Multirobot Coordination: A Survey and Analysis. Proceedings of the IEEE 94(7), 1257–1270 (2006)
2. Gerkey, B.P., Mataric, M.J.: Sold: aution methods for multirobot coordination. IEEE Transactions on Robotics and Automation 18(5), 758–768 (2002)
3. Goldberg, D., Cicirello, V., Dias, M.B., et al.: Task allocation using a distributed market-based planning mechanism. In: Proceedings of the International Conference on Autonomous Agents, pp. 996–997. Association for Computing Machinery, Melbourne (2003)

4. Nanjanath, M., Gini, M.: Dynamic Task Alloction for Robots via Auctions. In: Proceedings of the 2006 IEEE International Conference on Robotics and Autiomation, Orlando, Florida, May 2006, pp. 2781–2786 (2006)
5. Bai, H.: Research of Multi-Unit Combinatorial Auctions Based on Agent. Southwest China Normal University (2005)
6. Liu, L.: Research on Multi-robot System Task Allocation and Formation Control. National University of Defense Technology, Changsha (2006)
7. Hu, J.-j., Cao, Y.-d., Hu, J.: Task assignment of multi-Agent based on English auction protocol. Computer Interated Manufacturing Systems 12(5), 795–799 (2006)

Error Bound for the Generalized Complementarity Problem in Engineering and Economic Equilibrium Modeling

Hongchun Sun[*]

Department of Mathematics, Linyi Normal University,
Linyi, Shandong, 276005, P.R. China
sunhc68@126.com

Abstract. In this paper, the global error bound estimation for the generalized complementarity problem in engineering and economic equilibrium modeling(GCP) is established. The results obtained in this paper can be viewed as extensions of previously known results.

Keywords: GCP; Reformulation; Error bound, Engineering modeling, Economic modeling.

1 Introduction

Let mappings $F, G : R^n \to R^m$, the generalized complementarity problem over a polyhedral cone, abbreviated as GCP, is to find vector $x^* \in R^n$ such that
$$F(x^*) \geq 0, \quad G(x^*) \geq 0, \quad F(x^*)^\top G(x^*) = 0,$$
where F and G are polynomial functions from R^n to R^m, respectively. We denote the solution set of the GCP by X^* and assume that it is nonempty throughout this paper.

The GCP is a direct generalization of the classical linear complementarity problem and a special case of the generalized nonlinear complementarity problem which finds applications in engineering, economics, finance, and robust optimization operations research (Refs.[1,2,3,4]). For example, the balance of supply and demand is central to all economic systems; mathematically, this fundamental equation in economics is often described by a complementarity relation between two sets of decision variables. Furthermore, the classical Walrasian law of competitive equilibria of exchange economies can be formulated as a generalized nonlinear complementarity problem in the price and excess demand variables ([2]), and be also found applications in contact mechanics

---

[*] This work was supported by the Natural Science Foundation China (Grant No. 10771120)and Shandong provincial the Natural Science Foundation (Y2008A27).

problems(such as a dynamic rigid-body model, a discretized large displacement frictional contact problem), structural mechanics problems, obstacle problems mathematical physics, Elastohydrodynamic lubrication problems, traffic equilibrium problems(such as a path-based formulation problem, a multicommodity formulation problem, network design problems), etc ([1]). Up to now, the issues of numerical methods and existence of the solution for the problem were discussed in the literature (e.g., Refs. [5]).

Among all the useful tools for theoretical and numerical treatment to variational inequalities, nonlinear complementarity problems and other related optimization problems, the global error bound, i.e., an upper bound estimation of the distance from a given point in R^n to the solution set of the problem in terms of some residual functions, is an important one due to the following reasons: First, the global error bound can not only give us a help in designing solution methods for it, e.g., providing an effective termination criteria, but also be used to analyze the convergence rate; second, it can be used in the sensitivity analysis of the problems when their data is subject to perturbation (Refs.[6,7,8,9,10]). The error bound estimation for the classical linear complementarity problems (LCP) was fully analyzed (e.g.,Refs. [11,12,13,14,15,16]). Since the GCP is an extension of the LCP, the following two questions are posed naturally: How about the error bound estimation for the GCP? Can the existing error bound estimation for the LCP be extended to the GCP? These constitute the main topics of this paper.

The main contribution of this paper is to establish a global error bound for the GCP via an easily computable residual function, which can be taken as an extension of those for linear complementarity problems ([11,12,14]) or the generalized complementarity problem([17,18]).

We end this section with some notations used in this paper. Vectors considered in this paper are all taken in Euclidean space equipped with the standard inner product. The Euclidean norm of vector in the space is denoted by $\|\cdot\|$. We use R_+^n to denote the nonnegative orthant in R^n, use x_+ and x_- to denote the vectors composed by elements $(x_+)_i := \max\{x_i, 0\}$, $(x_-)_i := \max\{-x_i, 0\}$, $1 \leq i \leq n$, respectively. We also use $x \geq 0$ to denote a nonnegative vector $x \in R^n$ if there is no confusion.

2 Error Bound for GCP

In this section, we would give error bound for GCP, which can be viewed as extensions of previously known results. First, we give the following definition.

Definition 1. *The mapping $f : R^n \to R^m$ is said to be monotone with respect to g if*
$$\langle f(x) - f(y), g(x) - g(y) \rangle \geq 0, \forall x, y \in R^n.$$

To establish error bound for GCP, we can give the needed assumptions for our analysis.

Assumption. *(A1) For The mapping F, G involved in the GCP, we assume that the mapping F is monotone with respect to G,*
(A2) there exists an interior point $\hat{x}$, i.e., $\hat{x} \in R^n, F(\hat{x}) > 0, G(\hat{x}) > 0$.
(A3) Suppose that the set Ω_1 or Ω_2 is bounded, where

$$\Omega_1 = \{x \in R^n \mid |F(x)_i| \leq \mu_1, i = 1, 2, \cdots, m, \text{ for some a constant } \mu_1 > 0\},$$

$$\Omega_2 = \{x \in R^n \mid |G(x)_i| \leq \mu_2, i = 1, 2, \cdots, m, \text{ for some a constant } \mu_2 > 0\}.$$

In the following, we discuss the boundedness of the level set associated with GCP defined by

$$L(\epsilon) \triangleq \{(x) \in \Omega \mid \varphi(x) \leq \epsilon, \epsilon \geq 0\}, \tag{1}$$

where $\varphi(x) = \|(-F(x))_+\|_1 + \|(-G(x))_+\|_1 + (F(x)^\top G(x))_+$.

Theorem 1. *Suppose that Assumption (A1), (A2) and (A3) hold, then $L(\epsilon)$ defined in (1) is bounded for any $\epsilon \geq 0$.*
Proof: Let x be any vector in $L(\epsilon)$, $\|(-F(x))_+\|_1 \leq \epsilon$ and $\|(-G(x))_+\| \leq \epsilon$, we have

$$\begin{aligned}(F(x))_i, (G(x))_i &\geq -\epsilon, i = 1, 2, \cdots, n, \\ (F(x)^\top G(x))_+ &\leq \epsilon,\end{aligned} \tag{2}$$

By Assumption (A1), for any $x \in R^n$, we have

$$\begin{aligned}0 &\leq \langle F(x) - F(\hat{x}), G(x) - G(\hat{x})\rangle \\ &= F(x)^\top G(x) - F(x)^\top G(\hat{x}) - F(\hat{x})^\top G(x) + F(\hat{x})^\top G(\hat{x}).\end{aligned} \tag{3}$$

By (3), we obtain

$$F(x)^\top G(\hat{x}) + F(\hat{x})^\top G(x) \leq F(x)^\top G(x) + F(\hat{x})^\top G(\hat{x}),$$

and

$$\begin{aligned}&\sum_{G(x)_i > 0} F(\hat{x})_i G(x)_i + \sum_{F(x)_i > 0} G(\hat{x})_i F(x)_i \\ &\leq F(x)^\top G(x) + F(\hat{x})^\top G(\hat{x}) \\ &\quad - [\sum_{G(x)_i < 0} F(\hat{x})_i G(x)_i + \sum_{F(x)_i < 0} G(\hat{x})_i F(x)_i] \\ &\leq (F(x)^\top G(x))_+ + F(\hat{x})^\top G(\hat{x}) + \epsilon[\|F(\hat{x})\|_1 + \|G(\hat{x})\|_1] \\ &\leq F(\hat{x})^\top G(\hat{x}) + \epsilon[1 + \|F(\hat{x})\|_1 + \|G(\hat{x})\|_1],\end{aligned} \tag{4}$$

where the second and third inequality is by (2). If $F(x)_i < 0$, by (2), we have $-\epsilon < F(x)_i < 0$. If $F(x)_i > 0$, by (3), we have

$$\begin{aligned}0 < F(x)_i &= G(\hat{x})_i^{-1} G(\hat{x})_i F(x)_i \\ &< G(\hat{x})_i^{-1} (\sum_{G(x)_i > 0} F(\hat{x})_i G(x)_i + \sum_{F(x)_i > 0} G(\hat{x})_i F(x)_i) \\ &< G(\hat{x})_i^{-1} (F(\hat{x})^\top G(\hat{x}) + \epsilon[1 + \|F(\hat{x})\|_1 + \|G(\hat{x})\|_1]),\end{aligned}$$

we show that $F(x)_i$ is bounded. By assumption (A3), $L(\epsilon)$ is bounded.

Similarly, we can also prove that $G(x)_i$ is bounded. By assumption (A3), $L(\epsilon)$ is bounded. □

In the following, we give a conclusion which is easy to deduce.

Lemma 1. *For any $b \in R$, we have $\|(-a)_+\| \le \|\min\{a,b\}\|$.*

Lemma 2. *Suppose that Assumption (A1), (A2) and (A3) hold, then there exists a constant $\eta > 0$, such that*

$$\varphi(x) \le \eta r(x), \forall x \in L(\epsilon),$$

where $r(x) = \|\min\{F(x), G(x)\}\|$.

Proof: Using Lemma 1, there exist a constant $c_1 > 0$, such that

$$\|(-F(x))_+\|_1 + \|(-G(x))_+\|_1 \le 2c_1 \|\min\{F(x), G(x)\}\|. \tag{5}$$

By Theorem 1, for any $x \in L(\epsilon)$, then there is a constant $c_2 > 0$ such that

$$\|\max\{F(x), G(x)\}\| \le c_2.$$

Therefore, By the definition of $\varphi(x)$, we have

$$\varphi(x) \le 2c_1 \|\min\{F(x), G(x)\}\| + [(F(x))^\top (G(x))]_+$$
$$\le 2c_1 \|\min\{F(x), G(x)\}\| + \|\min\{F(x), G(x)\}\| \cdot \|\max\{F(x), G(x)\}\|$$
$$\le \eta(\|\min\{F(x), G(x)\}\|).$$

where $\eta = \max\{2c_1, c_2\}$, the first inequality is by (5), and follows from the fact that $(cd)_+ \le \|\min\{c,d\}\| \|\max\{c,d\}\|, \forall c, d \in R$. □

Lemma 3. *Let $X_\varepsilon \triangleq \{x \in X | r(x) \le \varepsilon, \varepsilon > 0\}$, then X_ε is bounded.*

Proof: By Lemma 2, we have $X_\varepsilon \subseteq L(\eta\varepsilon)$. Combining this with Theorem 1, we know that X_ε is bounded. □

Lemma 4. *Suppose that Assumption (A1), (A2) and (A3) hold, then there exists a constant $\eta_1 > 0$ such that*

$$dist(x, X^*) \le \eta_1 r(x)^{\frac{1}{m}}, \quad \forall x \in X_\varepsilon, \tag{6}$$

where $m = \max\{m_1, m_2\}$, $F(x), G(x)$ are polynomial functions with powers m_1 and m_2, respectively.

Proof: Assume that the theorem is false. Then there exist $\varepsilon_0 > 0$, $X_0 = \{x \in X | r(x) \le \varepsilon_0\}$, for any integer k, there exists $x^k \in X_0$, such that $dist(x^k, X^*) > kr(x^k)^{\frac{1}{m}} \ge 0$, i.e.,

$$\frac{r(x^k)^{\frac{1}{m}}}{dist(x^k, X^*)} \to 0, k \to \infty. \tag{7}$$

Since X_0 is bounded set, and $r(x)$ is continuous, combining (7), we have $r(x^k) \to 0(k \to \infty)$, and there exists a subsequence $\{x^{k_i}\}$ of $\{x^k\}$ such that $\lim_{k_i \to \infty} x^{k_i} = \bar{x} \in X_0$, where $\bar{x} \in X^*$. Since $F(x), G(x)$ are polynomial functions with powers m_1 and m_2, respectively, there exists a constant $\beta > 0$, for all sufficiently large k_i, we have

$$\frac{r(x^{k_i})^{\frac{1}{m}}}{\|x^{k_i} - \bar{x}\|} \geq \beta > 0. \tag{8}$$

On the other hand, by (7), we obtain

$$\lim_{k_i \to \infty} \frac{r(x^{k_i})^{\frac{1}{m}}}{\|x^{k_i} - \bar{x}\|} \leq \lim_{k_i \to \infty} \frac{r(x^{k_i})^{\frac{1}{m}}}{dist(x^{k_i}, X^*)} = 0,$$

this contradicts (8), thus, we have that (6) holds. □

Based on Lemma 4, we have the following conclusion.

Theorem 2. *Suppose that Assumption (A1), (A2) and (A3) hold, then there exist a constant $\eta_2 > 0$, for any $\forall x \in R^n$, we have*

$$dist(x, X^*) \leq \eta_2(r(x) + r(x)^{\frac{1}{m}}).$$

Proof: Assume that the theorem is false. Then for any integer k, there exist $x^k \in \Omega$ and $\bar{x} \in \Omega^*$, such that

$$\|x^k - \bar{x}\| > k(r(x^k) + r(x^k)^{\frac{1}{m}}). \tag{9}$$

It follows that there exist $k_0 > 0, \varepsilon_0 > 0, \forall k > k_0$, we have

$$r(x^k) + r(x^k)^{\frac{1}{m}} > \varepsilon_0. \tag{10}$$

In fact, otherwise, for $\forall k > 0, \forall 0 < \varepsilon < 1$, there exists $\bar{k} > k$, such that

$$r(x^{\bar{k}}) + r(x^{\bar{k}})^{\frac{1}{m}} \leq \varepsilon,$$

we let $\Theta_\varepsilon = \{x \in \Omega | r(x^{\bar{k}}) + r(x^{\bar{k}})^{\frac{1}{m}} \leq \varepsilon\}$. Using the similar arguments to that in Lemma 3, for any $x \in \Theta_\varepsilon$, we have

$$r(x) \leq r(x) + r(x)^{\frac{1}{m}} \leq \varepsilon,$$

i.e., $\Theta_\varepsilon \subseteq L(\eta\varepsilon)$, combining this with Theorem 1, we know that Θ_ε is bounded. Combining this with Lemma 4, for $x^{\bar{k}} \in \Theta_\varepsilon$, there exist $\bar{x}(x^{\bar{k}}) \in X^*$ and constant $\eta_1 > 0$, such that

$$\|x^{\bar{k}} - \bar{x}(x^{\bar{k}})\| \leq \eta_1 r(x^{\bar{k}})^{\frac{1}{m}},$$

combining this with $\|x^{\bar{k}} - \bar{x}(x^{\bar{k}})\| < 1$, we have

$$r(x^{\bar{k}}) + r(x^{\bar{k}})^{\frac{1}{m}} \geq \frac{1}{\eta_1}\|x^{\bar{k}} - \bar{x}(x^{\bar{k}})\|$$
$$+ \frac{1}{\eta_1^m}\|x^{\bar{k}} - \bar{x}(x^{\bar{k}})\|^m$$
$$\geq (\frac{1}{\eta_1} + \frac{1}{\eta_1^m})\|x^{\bar{k}} - \bar{x}(x^{\bar{k}})\|,$$

combining this with (9), for $x^{\bar{k}}$, and $\bar{x}(x^{\bar{k}}) \in X^*$, we have

$$\frac{\eta_1^m}{\bar{k}(\eta_1^{m-1}+1)}\|x^{\bar{k}} - \bar{x}(x^{\bar{k}})\| > \frac{\eta_1^m}{(\eta_1^{m-1}+1)}(r(x^{\bar{k}}) + r(x^{\bar{k}})^{\frac{1}{m}}) \geq \|x^{\bar{k}} - \bar{x}(x^{\bar{k}})\|,$$

i.e. $\frac{\eta_1}{\bar{k}} > 1$. Let $\bar{k} \to \infty$, then we have $\frac{\eta_1}{\bar{k}} < 1$, this is contradiction, we have that (10) holds.

By (9) and (10), we have $\|x^k\| > k\varepsilon_0 - \|\bar{x}\|$, i.e., $\|x^k\| \to \infty (k \to \infty)$.

Let $y^k = \frac{x^k}{\|x^k\|}$, then there exist a subsequence y^{k_i} of $\{y^k\}$, such that $y^{k_i} \to \bar{y}(k_i \to \infty)$, note that $\|\bar{y}\| = 1$. Divide both sides of (9) by $\|x^{k_i}\|$, and let k_i go to infinity, we obtain

$$1 = \lim_{i \to \infty}\frac{\|x_{k_i} - \bar{x}\|}{\|x^{k_i}\|} > \lim_{i \to \infty}\frac{k_i(r(x^{k_i}) + r(x^{k_i})^{\frac{1}{m}})}{\|x^{k_i}\|} \to \infty,$$

this is contradiction, then the desired result is followed. □

Remark. The error bound in the above Theorem 2 is extensions of Theorem 2.6 in [14], Theorem 2.1 in [12], Theorem 4.3 in [11] for linear complementarity problem. and is also extensions of Theorem 4.2 in [17], Theorem 4.1 in [18] for the GCP.

3 Conclusions

In this paper, we established global error bound on the generalized complementarity problems in engineering and economic equilibrium modeling which is the extensions of those for the classical linear complementarity problems and the generalized complementarity problems.

References

1. Ferris, M.C., Pang, J.S.: Engineering and economic applications of complementarity problems. Society for industrial and applied mathematics 39(4), 669–713 (1997)
2. Walras, L.: Elements of Pure Economics. Allen and Unwin, London (1954)
3. Nagurney, A., Dong, J., Zhang, D.: A supply chain network equilibrium model. Transportation Research. Part E 38, 281 (2002)
4. Zhang, L.P.: A nonlinear complementarity model for supply chain network equilibrium. Journal of Industrial and Managment Optimization 3(4), 727–737 (2007)
5. Facchinei, F., Pang, J.S.: Finite-Dimensional Variational Inequality and Complementarity Problems. Springer, New York (2003)

6. Pang, J.S.: Error bounds in mathematical programming. Math. Programming 79, 299–332 (1997)
7. Pang, J.S.: Inexact Newton methods for the nonlinear complementarity problem. Math. Programming 36, 54–71 (1986)
8. Pang, J.S.: A posterriori error bound for the linearly-constrained variational inequality problem. Mathematics of Operations Research 12, 474–484 (1987)
9. Izmailov, A.F., Solodov, M.V.: Error bounds for 2-regular mappings with Lipschitzian derivatives and Applications. Math. Programming. Ser. A 89, 413–435 (2001)
10. Solodov, M.V.: Convergence rate analysis of iteractive algorithms for solving variational inequality problems. Math. Programming. Ser. A 96, 513–528 (2003)
11. Luo, Z.Q., Mangasarian, O.L., Ren, J., Solodov, M.V.: New error bound for the linear complementarity problem. Mathematics of Operations Research 19, 880–892 (1994)
12. Mangasarian, O.L., Ren, J.: New improved error bound for the linear complementtarity problem. Math. Programming 66, 241–255 (1994)
13. Mangasarian, O.L., Shiau, T.H.: Error bounds for monotone linear complementarity problems. Math. Programming 36(1), 81–89 (1986)
14. Mangasarian, O.L.: Error bounds for nondegenerate monotone linear complementarity problems. Math. Programming 48, 437–445 (1990)
15. Mathias, R., Pang, J.S.: Error bound for the linear complementarity problem with a P-matrix. linear Algebra and Applications 132, 123–136 (1990)
16. Ferris, M.C., Mangasarian, O.L.: Error bound and Strong upper semicontinuity for monotone Affine Variational Inequalities. Technical Report 1056, Computer Sciences Department Universityof wisconsin (1992)
17. Sun, H.C., Wang, Y.J., Qi, L.Q.: Global Error Bound for the Generalized Linear Complementarity Problem over a Polyhedral Cone. J. Optim. Theory Appl. 142, 417–429 (2009)
18. Sun, H.C., Wang, Y.J.: Error Bound for Generalized Variational Inequalities and Generalized Nonlinear Complementarity Problem over a convex closed polyhedral. Chinese Journal of engineering mathematics 4, 691–695 (2007) (In Chinese)

Research of Tag Anti-collision Technology in RFID System

Zhitao Guo[1], Jinli Yuan[1], Junhua Gu[2], and Zhikai Liu[2]

[1] School of Information Engineering,
Hebei University of Technology,
Tianjin, China
naory@sina.com, jinli_yuan@hebut.edu.cn,
jhgu@hebut.edu.cn, liu333077@163.com

Abstract. In this paper, the RFID system and its related technologies are introduced, and the inevitable tags collision problem in RFID technology is analyzed. At the basis of an in-depth study on the popular ALOHA algorithm and binary search algorithm to solve the tags collision problem, the basic principles and a variety of improvements of the two algorithms is described in detail, and the emergence of new research trends of Tag anti-collision technology is summarized. Finally, anti-collision technology development tendency is forecasted.

Keywords: RFID; tag collision; anti-collision algorithm; ALOHA algorithm; binary search algorithm.

1 Introduction

Radio Frequency Identification (RFID) system is a non-contact automatic identification technology. It identifies the targets and exchanges the data by means of radio frequency. Compared with the traditional identification technology, RFID technology can automatically identify and manage all kinds of items without having direct contact with them. It can be widely used in materials management, tracking, logistics, location and other fields[1].

RFID system mainly consists of two core components which are the electronic tag and reader. In general, the same RFID tags work in the same frequency band. When there are multiple tags within the working scope of the reader, there may be multiple tags sending information to the reader at the same time. Then mutual interference between the tags will appear, So that the reader can not correctly identify the tags, resulting in tag collision problem.

In order to eliminate or reduce the tags collision problem in RFID system, many researchers conducted a lot of work, and proposed a variety of different anti-collision algorithms. The new algorithms are still emerging. To sum up, the main anti-collision algorithms are ALOHA algorithm and binary search algorithm in the field of RFID. These algorithms have their own advantages and disadvantages and the applications are also different[2].

Through synthesizing various types of literature and conducting in-depth analysis and comparison to the basic principles of these two algorithms and a variety of improved algorithms, the research trend of tag anti-collision algorithm is summarized, as well as the development trend of anti-collision algorithm is prospected in this paper.

2 Tag Anti-collision Algorithm and Its Improvement

Currently the tag anti-collision algorithms are basically based on Time Division Multiple Access (TDMA) access mechanism. According to the different frequency bands of tags, anti-collision algorithms are divided into two broad categories which are ALOHA algorithm and binary search algorithm. In the high-frequency (HF) band, ALOHA algorithm is generally used. In the ultra-high frequency (UHF) band, binary search algorithm is used primarily to solve the tags collision problem. Currently the research of the anti-collision algorithms based on TDMA mechanism mainly focuses on the basic ALOHA algorithm and binary search algorithm and the improvements of them.

2.1 Basic Principle of ALOHA Algorithm

Basic ALOHA algorithm is a random avoidance algorithm. The essence lies in that the tags send information to the reader by selecting over a random period of time when the tags collision happen in order to avoid the collision[3].

The transmission time of each data frame only occupies a small part of the repeated transmission time, so there is a certain probability to make the data frames of two tags avoid collision. In basic ALOHA algorithm, the formula for calculating the throughput S and the average number of data frames G which is sent by communication channel in unit time T is as follow:

$$G = \sum_{1}^{n} \frac{\tau}{T} r_n \qquad (1)$$

$$S = G \cdot e^{(-2G)} \qquad (2)$$

Thereinto: n is the number of the tags in the system. r_n is the data frame sent by the tag n in the time T. According to the relationship between throughput S and the number of data frames G. We can know that the maximum value of S is 18.4% and the best utilization of the channel is relatively low when $G = 0.5$.

2.2 Improvement of ALOHA Algorithm

For the shortcomings that the basic ALOHA algorithm is simple and channel throughput is low, the researchers proposed slot-ALOHA algorithm[4] (SA). This improved algorithm based on analyzing the mathematical model of ALOHA algorithm puts forward to dividing a number of read and write time into the same size as discrete slots. The tags can only send data at the boundaries of the time slot. Then the tags are only sent successfully or completely conflict, two kinds of specific state. Reduce the time interval of collision which is made in basic ALOHA algorithm from $T = 2\tau$ to $T = \tau$. According to the formula $S = G \cdot e^{(-G)}$, we can know that in the improved ALOHA algorithm channel throughput increases from 18.4% to 36.8% when $G = 1$.

In reference [5], frame slot ALOHA (FSA) algorithm[5] proposed is the improvement of SA algorithm. The idea is that a number of time slots are packaged into a frame and the tags must choose a time slot in the frame to transfer data. Moreover the tags can only send data once in a frame. So that it helps to reduce the repetition rate of a single tag in each frame and further reduce conflict and also improve the system efficiency.

After FSA algorithm is proposed, the efficiency of RFID Anti-collision system is effectively improved. However, when the number of tags changes, the algorithm which a fixed frame length is adopted in will make the inevitable waste of time slot resources if the number of tags is a little. And if the number of tags is a lot, time slot resources will be saturation and conflicts will be intensified. According to the number of tags, how to dynamic adjust the number of time slots in a frame in order to make the system achieve maximum efficiency is currently the hot spot of ALOHA algorithm research based on random delaying strategy.

For achieving the dynamic adjustment of the number of frame slots, the key is that we have the ability to estimate the number of tags correctly which is in the working scope of the reader and reasonably adjust the number of time slots in accordance with certain rules. In reference [6], a method based on probability distribution to estimate the number of tags is adopted. After a read cycle, we can safely obtain the information, it includes: idle slot (S_0), successful identification slot (S_1), collision slot (S_k) and total frame slot (S). We can gain further information, it includes: system efficiency ($S_1/S \times 100\%$), system impact probability ($S_k/S \times 100\%$) and the probability which is calculated by the probability distribution of having r tags in a frame within a time slot at the same time (S_r). In reference [7], because RFID tag collision always take place between at least two tags, an idea through a simple formula $T = 2 \times S_r$ to estimate the number of tags is proposed. In reference [8], when RFID system achieves maximum system efficiency, the number of tags collision in a time slot is 2.392. So the idea through the formula $T = 2.392 \times S_r$ to estimate the number of tags is proposed. Considering that the number of slots should be approximately equal to the number of tags and it will facilitate computer processing, The slot number of the next round should be indices of 2 that is the nearest to T.

In reference [9], a dynamic frame slots algorithm based on a multiplying factor is proposed. When the number of tags is much larger than the number of slots, system efficiency is very low. The standard estimation method is no longer applicable. Based on a large number of statistical data, and the relationship among the system efficiency, the number of tags and the number of time slots is analyzed, a proportional relationship between the actual number of tags in the system efficiency within a certain range and the number of time slots currently in use is pointed out in [9] (multiplying factor). So after obtaining the system efficiency, the number of time slots can be quickly adjusted by multiplying factor and basically achieve consistent with the actual number of tags.

In summary, dynamic frame slots ALOHA algorithm is optimal improving method in random delaying ALOHA algorithm. It belongs to the hot spot for research. Major study focuses on a reasonable estimate of the number of tags and dynamic adjustment rules of the number of slots. However there are still a few questions to need to be solved further, such as probability calculation is complex in the process of tag estimation, slot adjustment method is a single and there are large deviations when the difference between the number of tags and time slots is large, and so on.

3 Principle of Binary Search Algorithm and Its Improvement

3.1 Basic Principle of Binary Search Algorithm

Binary tree search algorithm is also called reader control algorithm[10]. The premise is to identify the exact location of the data collision in the reader. Therefore, the selection of the appropriate bit encoding method is very important. Currently, RFID tags communication uses more Manchester encoding. The rising and falling edges of "0" and "1" encoded in Manchester encoding can be superimposed to offset. Once the tags collision happens, the reader receives only the carrier signal. It is convenient for the reader to find the exact collision bit. The basic idea of binary search algorithm is to divide the tags which are in conflict into two subsets of 0 and 1. First search the subset 0. If there is no conflict, the tag will be correctly identified. If there is another conflict, it will divide the subset 0 into two subsets of 00 and 01. Turn and so on. Until it identifies all tags in the subset 0 it follows the steps to search the subset 1.

3.2 Improvement of Binary Search Algorithm

The improvement research of binary tree search algorithm mainly focuses on the improvement based on binary search algorithm of ISO/IEC 18000-6B. The essence is to improve the search efficiency of binary search algorithm by a certain strategy. Thus the average delay and power consumption will be able to be reduced in the process of identifying each tag.

The basic binary algorithm has the following two deficiencies: ①Algorithm strategy is mechanical and simple. It can not take full advantage of tag information so that the recognition efficiency is low. After a tag is identified new identification from scratch is made in the basic binary search algorithm. It doesn't take advantage of tags information that is obtained from the identification process before. ②There is a large number of redundant information in both reader and tag communication in recognition process. It makes the average delay and power consumption longer. All bits behind the maximum conflict-bit are simply set to "1" in the reader request command in basic binary search algorithm. This part of the whole "1" encoded can not provide useful information for tag identification. They belong to redundant information. And all the information before the maximum conflict-bit (including the conflict bit) which the tags return is known to the reader, so the information is also redundant. A large number of redundant information in the reader and tag communication process will inevitably lead to the delay and power consumption in tag identification increasing.

The improvement proposed by domestic and international researchers for the basic binary algorithm focuses on these two aspects. In reference [11], the binary tree search algorithm based on return type proposed by Songsen Yu et al is the improvement to remedy the first defect. In the algorithm when the reader has recognized a tag, it doesn't take the whole "1" from scratch as the parameters of the next request command but obtains from the last request command. For example, the maximum conflict-bit in the last request command is "0". This time, the maximum conflict-bit in the request command will be set to "1".Then the next tag will be identified at once and the recognition process will be greatly shortened[11].

Dynamic binary algorithm[12] is the improvement to remedy the second defect. There is a large number of redundant information in reader commands and tag responses in the basic binary algorithm. Suppose that the length of tag ID is N, the maximum conflict-bit detected currently is M. According that the whole "1" sequences of N-M-1 bits are all redundant in request commands in basic binary algorithm, the data of M+1 bits only needs to be sent. However the former M bits are known in tag responses, the data of N-M+1 bits behind only needs to be sent. Thus when the redundant information is removed, the sum of the length of request command and response data is N in communication process at a time. It full reduced by 50% compared with data volume 2N in communication process in the basic binary algorithm at a time.

Dynamic binary algorithm effectively reduces the redundant information and improves the recognition speed. However, there are still inadequacies in the algorithm. It is mainly manifested that all tags which are active are involved in command comparison every time. This is obviously a waste of the conflict information obtained before. And the interference within the system is increased for the tags which should be excluded from the scope of request. So a multi-state binary search algorithm[13] is proposed by Lifen Jiang et al in the literature 13. In order to take full advantage of the conflict information obtained a sleep counting method is introduced and the tags are set to have three states: standby state, dormant state and inactive states. Making use of the sleep counter the tags which wait to be identified are divided into standby state and dormant state. The count of counter reflects the degree of tags conflict. Only when the count of counter is zero, the tag which is standby state is able to send the response data. The scope of tags is narrowed and the amount of data transferred in identification process is decreased.

In order to reduce the amount of data in communication, save power consumption and improve the communication security, the pretreatment mechanism is introduced in binary search algorithm in reference [14]. After the reader sends request command at the first time, all tags will send their own IDs back. Then the reader will detect all conflict bits at a time and record them. Subsequently the reader will send the conflict data of N bits to the tags. The tags will extract the corresponding conflict-bit data as their new IDs and communicate with the reader again. We can see that if the tags aren't completely conflict, the amount of data sent by the reader and tags will be decreased for any algorithms[14].

In summary, the improvement of binary search algorithm mainly focuses on two major issues which are how to increase the efficiency of the system identification and how to reduce the amount of information redundancy in communication.

3 New Research Trends of Anti-collision Algorithm

4.1 Improvement from the Algorithm Itself to the Specific Mode

With the in-depth research of tag anti-collision algorithm, the anti-collision algorithm based on Time Division Multiple Access (TDMA) has reached a certain degree. This algorithm improves the identification efficiency by improving the algorithm itself. It is difficult to achieve substantial increase in identification efficiency that we improve simply the algorithm from the perspective of its mathematical model. Such as dynamic adjustment of frame slot, improvement of binary tree search strategies, etc.

Many scholars begin to focus from improving the theory for the algorithm itself to the improvement for the specific application mode of the tags. Now some scholars have already improved the algorithm in specific service-oriented application. That is for the priori knowledge such as the specific moving law of tags or the basic information of tags, etc. The RFID identification model that is consistent with the scene feature is established by studying the tag characteristics of different scenes. Then the basic TDMA algorithm could be improved correspondingly.

For example, in reference [15], a "first come first served" (FCFS) algorithm[15] is proposed by Xi Zhao et al considered that traditional anti-collision algorithms tend to have higher leakage rate in specific RFID system which tag movement is regular in. The system blocking is avoided by the rational design of slot limit. When tag movement is regular, this method can effectively reduce the leakage rate.

Research of Tag Anti-collision Technology in RFID System 241

In reference [16], dynamic adjustment algorithm of the binary tree search and polling algorithm[16] are proposed by Zhenhua Xie et al based on the reading and writing features of tags whose ID numbers are two cases of disorder and successive encountered in the application environment. The algorithm improves the channel utilization. The essence of the algorithm is to adjust the search strategy to handle tag anti-collision according to certain priori knowledge.

4.2 Improvement of Algorithm Mechanism

The strategy is shifted from the past mechanism TDMA to CDMA by referring to the idea of spread spectrum communication in communication system. Transmission bandwidth is extended by using "modulation code". Then "tag conflict" is effectively avoided. Each tag modulates the data transferred using different "modulation code". Using the CDMA mode the data transferred by the different Tags could be demodulated by making use of the auto-correlation characteristics of codes in the reader. Then the anti-collision purpose is achieved.

Solving collision is a reason for using this method. And another reason is that the requirements for anti-jamming and security feature in wireless communication link in RFID system are getting higher and higher according that more and more information is carried by the tags In RFID applications, the amount of data exchange is increasing and the environment of RFID system also becomes increasingly complex. And in CDMA mode because spread spectrum communication transmission in the space occupied a relatively wide bandwidth and the receiver demodulated using correlation detection method, the power in unit band is small, the signal and noise ratio is high, the anti-interference feature is strong, the security feature is good and the strongest useful signal can be extracted and separated from the multi-path signals in the receiver. The throughput and channel utilization can be improved greatly in anti-multipath interference mode theoretically.

5 Conclusion

With the increasingly wide application of RFID system how to solve the problem of tags collision effectively will be great significance. Currently, ALOHA algorithm and binary search algorithm which are based on TDMA technology are recommended methods in the international standard. Major study in anti-collision fields focuses on how to make useful improvement to the algorithm and improve the system efficiency. Its specific content includes that how to achieve a reasonable estimate of the number of tags, how to optimize dynamic adjustment strategy for frame slots, how to make full use of the search information in the binary search process, how to minimize the redundant information in communication, and so many aspects. We can see that there is still much room for both Algorithms.

Recently the targeted improvement according to the special application mode of tags for anti-collision algorithm appears. This is a necessary requirement with the increasing popularity of RFID applications. It can significantly improve the efficiency of RFID system that we solve the collision problem by obtaining the priori knowledge of tags and using a targeted approach in a relatively fixed application mode. It is possible for in-depth study in the moving characteristics of tags in special scene, the mathematical model in the change of the number of tags, and so on.

With the increasing amount of data exchange in the RFID system and RFID communication security considered the tag anti-collision algorithm based on CDMA mechanism is also a new research direction in the field. The key is able to find a kind of suitable orthogonal spread spectrum codes to modulate more tags to solve multi-tag collision problem when the encoding length is shorter. It has just begun, not reached ripe and not done a large number of practical tests. However it is an important trend to solve tags collision problem by using CDMA mechanism.

References

1. Finkenzeller, K.: Compiled by Dacai Chen, Radio Frequency Identification Technology, 2nd edn. Publishing House of Electronics Industry, Beijing (2001)
2. Lian, G.: Research on Anti-Collision Algorithm for RFID Systems. Computer Technology And Development 19(1), 36–42 (2009)
3. Finkenzeller, K.: RFID Handbook: Fundamentals and Applications in Contact less Smart Card and Identification, 2nd edn. John Wiley & Sons Ltd., Chichester (2002)
4. Roberts, L.G.: Extensions of Packet Communication Technology to a Hand Held Personal Terminal. In: AFIPS Conf. Proc. Spring Joint Computer Conf., pp. 295–298 (1972)
5. Wieselthier, J.E., Ephremides, A., Michaels, L.A.: An Exact Analysis and Performance Evaluation of Framed ALOHA with Capture. IEEE Transactions on Communications 7, 125–137 (1989)
6. Joe, I., Lee, J.: A Novel Anti-Collision Algorithm with Optimal Frame Size for RFID system. In: IEEE Fifth International Conference on Software Engineering Research, Management and Applications, pp. 424–428 (2007)
7. Vogth: Multiple object identification with passive RFID tags. In: IEEE International Conference on Systems, Man and Cybernetics, pp. 651–656 (2002)
8. Cha, J.R., Kim, J.H.: Novel anti-collision algorithms for fast object Identification in RFID system. In: Proceedings of the, 11th International Conference on Parallel and Distributed Systems (ICPADS), pp. 63–67 (2005)
9. Cheng, L., Lin, W.: Steady dynamic framed slot ALOHA algorithm with high system efficiency. Application Research of Computers 26(1), 86–88 (2009)
10. Wu, Y., Gu, D., Fan, Z., Du, M.: Comparison and analysis of anti-collision in RFID system and improved algorithm. Computer Engineering and Applications 45(3), 210–213 (2009)
11. Yu, S., Zhan, Y., Peng, W., Zhao, Z.: An Anti-collision Algorithm Based on Binary-tree Searching of Regressive Index and its Practice. Computer Engineering and Applications 40(16), 26–28 (2004)
12. Ju, W., Yu, C.: An Anti-Collision RFID Algorithm Based on the Dynamic Binary. Journal of Fudan University (Natural Science) 44(1), 46–50 (2005)

13. Jiang, L., Lu, G., Xin, Y.: Research on anti-collision algorithm in Radio Frequency Identification system. Computer Engineering and Applications 43(15), 29–32 (2007)
14. Choi, J.H., Lee, D., Youn, Y.: Scanning-based pre-processing for enhanced tag anti-collision protocols. In: International Symposium on Communications and Information Techno- logies, ISCIT 2006, pp. 1207–1210 (2006)
15. Zhao, X., Zhang, Y.: Novel anti-collision algorithm in RFID system. Journal of Beijing University of Aeronautics and Astronautics 34(3), 276–279 (2008)
16. Xie, Z., Lai, S., Chen, P.: Design of Tag Anti-collision Algorithm. Computer Engineering 34(6), 46–50 (2008)

An Algorithm Based on Ad-Hoc Energy Conservation in Wireless Sensor Network

Jianguo Liu, Huojin Wan, and Hailin Hu

Department of Automation, Information Engineering College,
Nanchang University, Nanchang, China
ncu_ljg@163.com

Abstract. An improved medium access control algorithm based on energy conservation is proposed in this paper because the energy is constrained in wireless sensor network. In this algorithm, each code can adjust its sleeping probability by its remaining energy, balance energy consumption of net codes, so the life time of whole network is enlarging .The simulating result shows that comparing with traditional algorithm, the max life time can be increased from 1.48×104 time-slot to 3.6×104 time-slot. At sometime, the data groups also can be improved more in life time with the same energy of each code.

Keywords: Wireless sensor network, Medium access control, Sleeping probability, Energy conservation.

1 Introduction

WSN (Wireless Sensor Network) is a kind of network of network nodes which can monitor collect and perceive all kinds of real-time information of Perceptual object(such as Light, temperature, humidity, noise, and the concentration of harmful gases and so on) the observers in node deployment area take an interest in, and process these information, send out them in a wireless way to observers through wireless networks. In military reconnaissance, environmental monitoring, medical care, intelligent household, industrial production control and business fields show broad application prospects.

In wireless sensor networks, the sensor node has terminal and routing functions[1] which, in the one hand, realize data acquisition and processing, and on the other hand realize data fusion and routing, that is, the sensor node gather the data itself collect and the data the other node send to it ,then forward route to the gateway node. The number of a gateway node often is limited, and it use a variety of way (such as Internet, satellite or mobile communication network, etc) communicate with the outside, and often can get energy. While the number of sensor node is very large, and it use wireless communication, dynamic network, so the node energy and processing capacity is very limited, and usually uses the battery without supplement to provide energy, once the node energy be exhausted, the nodes will cannot exercise data acquisition and routing functions, will directly influence the stability and the security of the whole sensor network, and the life cycle. so, wireless sensor networks energy problem is very important, also gradually become one hotspot of wireless sensor network research. Because wireless sensor networks and Ad Hoc networks have larger similarities, MAC(Medium Access Control) protocol originally designed for the Ad Hoc network also be considered applied in wireless sensor networks, But because the Ad Hoc networks only consider network fairness, network real-timing and network throughput and give the node energy problem little consideration, many scholars research on this problem, and also propose T-MAC protocol such as competition, based on the synchronous DEANA,TRAMA, and D-MAC protocols[2-5], etc, while through these protocols consider the energy problem, network throughput is still preferred in its design, commonly according to the network flow, adjustments of the adaptation and the sleep SHCH to achieve the goal of energy saving. In these algorithms, little consideration is given to their residual energy of the nodes which causes that some nodes have been as routing node in the network undertaking the task of forwarding data, so that they die because their energy is exhausted prematurely, and then causes that the survival time of the network reduced. In this article, an improved algorithm is proposed which is based on SEEDEX-MAC algorithm of the Ad-Hoc network. this algorithm ensure network throughput, and in the meanwhile, it introduce the residual energy of nodes into consideration and according to the location of the residual energy in the whole network, determine the dormancy probability, which balance the energy consumption of the network, and maximize the network survival time.

2 Seedex-Mac Algo

For traditional SEEDEX-MAC algorithm, time axis are divided into some timeslots of equal length, whose length mean the time length after that the length of each wireless data frame is divided into equal parts, generally is decided by the carrier bandwidth and carrier rate, and unit is ms or s. A node produce the scheduling list of this node according to their respective pseudorandom seed, that is, in every timeslot the node decide to be in sending state or in accepting state. Node cycle interact with pseudorandom seed who spread two jump, that is, every node know the state of this node ,the neighbor node of this node and the neighbor node of the neighbor node in each timeslot. So, at the beginning of every timeslot the node in the sending state can decide whether to get access to channel or not according to the status of the destination node and the status of other neighbor nodes of the destination node. Through spreading the pseudorandom seed which is used to produce a scheduling list, this algorithm can inform covert terminal and exposed terminal of the state of the node which can effectively avoid conflict, improve the transmission efficiency, and guarantee the network throughput. But if it is applied to WSN mechanically and ponderously, there will be obvious flaw for the node residual energy and network survival time, so based on this, we must fully consider network energy in order to prolong the network life.

3 Improvd Algorithm

Because of the importance of the network energy and the survival time in the WSN, so the first consideration should be network energy, while for the SEEDEX—MAC algorithm, the node is always in the sending state or in the receiving state, that is, it is always on the energy consumption, which will greatly influence the survival time of the network. So first of all, let us introduce another state-dormancy state in which the node has minimal energy consumption. Secondly, residual energy of every node should be an important basis for the production of the node scheduling list, according to its residual node energy, the node can adjust the dormant probability adaptively, which thus will balance energy consumption of each node in the network, and maximize network survival time.

3.1 Description for Algorithm

In this algorithm, time axis is divided into cycle, and each cycle is divided into control and data transmission parts, time axis is divided into timeslots of equal length for the data transmission part, as shown in fig.1, and the algorithm is shown in fig.2.

Fig. 1. Single cycle time-slot

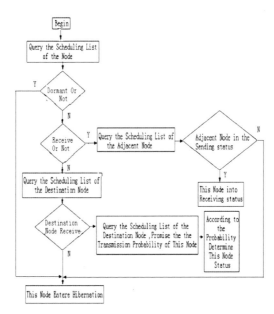

Fig. 2. Flow diagram of data transmit

An example of timeslot in the data transmission part for this algorithm is shown in fig.3. In the present moment, a in sending state send data to c. while c is also in sending state, a choose to be dormant in this timeslot. b will send data to d, d is in receiving state, but b knows that d has another neighbor node and c is also in sending state in the present moment, then b will choose the right sended probability to get access to channel. e will send data to f, f is in the receiving state, f is another neighbor node g is in dormant state, d is in the receiving state, then e transmit data with probability l in the present moment.

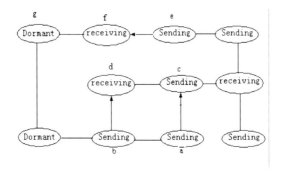

Fig. 3. Data transmit of network

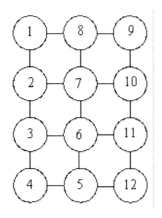

Fig. 4. Network topology

3.2 How to Determine Dormant State and Sending State

In the algorithm process, determination of dormancy probability and sending probability is very important. When the probability that the node is dormant is big, the corresponding energy consumption will be small, so when the residual energy of the node is less, the smaller the probability node is activated, the better it will be. Here, we assume that residual energy of the ith node at the end of the kth cycle is Ei, then the residual energy of this node as in (1).

$$\overline{E_k} = \frac{1}{N}\sum_{i=0}^{n} E_{i,k} \qquad (1)$$

We also assume that dormancy probability of the ith node at the end of the (k+1)th cycle as in(2).

$$P_s(i, k+1) = f(E_{i,k} / \overline{E_k}) \qquad (2)$$

In order to achieve the purpose that when P_s is bigger energy consumption will be smaller, then $f(x)$ must be monotonic decreasing function, $f(x) \in [0,1]$, the relationship between the residual energy and P_s as in (3).

$$E_i(k+1) = g(P_s(k)) \qquad (3)$$

In (3) $g(x)$ is monotonic increasing function. The value method of P_s is as in (4).

$$P_s = \begin{cases} 0, \dfrac{E_i}{\overline{E}} \geq 1 \\ 1 - \dfrac{E_i}{\overline{E}}, \dfrac{E_i}{\overline{E}} < 1 \end{cases} \qquad (4)$$

Finally, mutual restriction between the residual energy and scheduling list of the node, when the node energy is lower, protect the node automatically and prolong the network lifetime.

In an uneven distribution network, the sending probability P_T want to get an optimal value is very difficult, so only approximation method [6] of SEEDX−MAC algorithm can be adopted. That is, Destination node has neighbor nodes in the sending state, we assume that every node sends data with the same P_T, then the probability the data is received by destination node successfully as in (5).

$$(n+1)P_T(1-P_T)^n \qquad (5)$$

When $P_T = \dfrac{A}{n+1}$ and $A=1$, P_T can obtain maximum. But in the actual situation, a node don't send data with the same probability PT, and not all the nodes in the sending state have data to send, so the value of A can be adjusted according to the actual network circumstance. By the experimental results of reference [6] P_T, we can know that when the load of the network is low, the ideal value of A is 2.5, and when the load of the network is high, the ideal value of A is 1.5. P_T is always smaller than 1,so we introduce an operation for taking little, as in (6)

$$P_T = \min\{\frac{A}{n+1}, 1\} \qquad (6)$$

In which n is the number of neighbor nodes in sending state of the destination node, when the load of the network is large, A=1.5, when the load of the network is small, A=1.5.

In this algorithm process, the production method of the scheduling list of a node in every cycle is as follows. When dormancy probability P_T is determined, we take the same value for the probability of a node in the sending state and in receiving state which is all $0.5(1-P_T)$. Thus, valuing method of determining the state of timeslot is: There randomly generates a vsalue x between 0-1 evenly distributed, if $\chi \in (0, P_T)$, then node in this timeslot is in dormant state.

$$x \in (P_T, \frac{1+P_T}{2}) \qquad (7)$$

If the (7) is ture, node sending state in this timeslot; otherwise, node receiving state in this timeslot. Similarly, when dormancy probability P_T is determined, there also randomly generates a value between 0-1 evenly distributed, if the value is smaller than P_T, this node begins to send data, otherwise, it will enter dormant state.

4 Simulation Experiment

In order to analyze the feasibility of the algorithm, we make the following simulation experiment. We select the network with node number 12, and make it in the saturated state, that is, when nodes are in sending state, and they have data to send, then here A can be 1.5; 12 nodes select randomly neighbor nodes as destination node, and the initial energy of these nodes is evenly distributed from 0—15000 units of energy(unit: J/b, that is, energy consumption transmission of every bit information need); We ignore energy consumption in the control part, a node needs a unit of energy consumption when sending data or receiving data at a timeslot; There is 100 timeslots in a cycle of data transmission part. In this experiment, we simulate nine times and compare with the traditional SEEDEX MAC algorithm, the survival time of the network and the number of groups of data transmission is as shown in figure 5 and figure 6. As can be seen from the figure, the maximum survival time of a node can be from 1.48×104 timeslots of traditional algorithm to 3.2×104 timeslots, and under the condition of the same node energy, in network survival time the number of groups of the transfered data raises greatly. This is because of the introduction of the consideration about the nodes energy and the appropriate energy saving measures in the new algorithm, which make the node survival time greatly increase, and the network survival time greatly increase, so when the node energy is the same, the new algorithm makes network can submit more groups of data, which is the goal of the energy saving measures in sensor network.

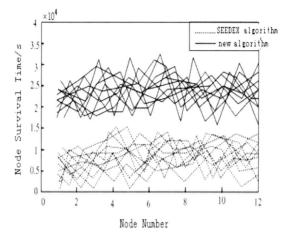

Fig. 5. Each node survival time

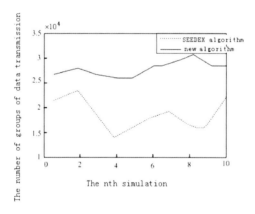

Fig. 6. The number of data transmission in network survival time

5 Conclusion

This paper proposes an improved algorithm based on SEEDEX-MAC algorithm of Ad-Hoc network, which ensure network throughput and determine the probability of dormancy through the introduction of the consideration of the residual energy of nodes and the position of the node residual energy in the whole network, which can balance the energy consumption of the network, maximize the network survival time and there is certain significance for improving the efficiency of wireless sensor networks and prolonging the network survival time.

References

1. Sun, L., Li, J., Chen, Y., Zhu, H.: Wireless sensor network, pp. 197–200. Tsinghua University Press, Beijing (2005)
2. Van Dam, T., Langendoen, K.: An Adaptive energy—efficient MAC protocol for wireless sensor networks. In: Proc. 1st. Int'l. Conf. on Embedded Networked Sensor Systems (SenSys), pp. 317–320 (2003)
3. Bao, L., Garcia, J.: A new approach to channel access scheduling for ad hoc networks. In: Annual Int'l. Conf. on Mobile Computing and Networking (MobiCom 2001), Rome, pp. 207–210 (2003)
4. Rajendran, V., Obraczka, K.: Energy—efficient, collision—free medium access control for wireless sensor networks. In: Proc. 1st Int'l. Conf. on Embedded Networked sensor Systems, SenSys 2003, pp. 181–192 (2003)
5. Lu, G., Krishnarnachari, B., Raghavendra, C.: An adaptive energy — efficient and low—latency MAC for data gathering in wireless sensor networks. In: Proc. 18th Int'l. Parrallel and Distributed Processing Symp., IPDPS 2004, April 2004, pp. 224–230 (2004)
6. Rozovsky, R., Kumar, P.R.: A MAC protocol for ad hoc networks. In: Proceedings of the ACM Symposium on Mobile Ad Hoc Networking computing. MobiHoc 2001, pp. 67–75 (2001)

The Design of the Hardware and the Data Processing Technology of Interbus

Jianguo Liu, Huojin Wan, and Hailin Hu

Department of Automation, Information Engineering College,
Nanchang University, Nanchang, China
`ncu_ljg@163.com`

Abstract. The paper design the hardware of interbus, and introduce a new method of data transmission of the interbus control system. In which the data is processed based on the PCAuto3.6. The interbus control systems, which are in the local area network of a factory, connect with others through Ethernet technology. The interbus can accomplish the issuing of network and transmission of the real-time data efficiently, and it provides a powerful method that can improve the productivity and the management level of the factory markedly.

Keywords: Interbus, Network Transmission, Network Issuing, Ethernet Technology.

1 Introduction

The interbus field bus is mainly applied in the automobile profession, tobacco profession, governs golden profession, process automation profession and so on [1]~[4], which is devised by the Phoenix Contact company of Germany in 1990. The interbus field bus is one of the international standard buses and adopts the data link communication which assures the synchronism and periodicity of data transmission. The communication way of full-duplex and the consistent speed of 2M assure the real-time of data communication. Therefore, the interbus field bus is an international acknowledged high-speed bus; its difference signal transmission and its special ring circuit check assure the formidable anti-interfere.

    The interbus control system has been set up in each plant of the automobile production. In order to display the effectiveness of the interbus control system fully and make the gathering data of each control system to transmit promptly to the higher authority management office, each interbus control system is interlinked together by the Ethernet technology. Through the data sharing, the management level of company and working efficiency of company staff's may be enhanced.

2 Hardware Designs of Interbus Control System

2.1 System Construction of Interbus Controller

The system construction of one interbus controller is divided into three parts of monitoring level, controlling level and equipment level. The monitoring level is a host computer (industrial PC), which exchanges the data with the controlling level by the configuration software of PCAuto3.6. The controlling level is composed by bus control board of the IBS ISA FC/I-T or the RFC 430 ETH-IB of the PHOENIX CONTACT company, which can realize the data acquisition and the procedure control of equipment level and the data transmission with the monitoring level. The equipment level is composed by the call-reply system and the display system. The call-reply system is composed by the bus coupler IBS IL 24 BK-T/U, the digital input module IB IL 24 DO 16, the digital output module IB IL 24 DI 16, the serial communication module IB IL RS232, the scene module FLS IB M12 DIO 8/8 M12 and the buttons. The display system is composed by the LED display monitor and the lamp box. The structure of overall system is shown in the Figure 1. By the structure of interbus system, besides the diagnosis display on the bus control board, each module which disperses in the system also has the display of corresponding system running state. Through diagnosing display content, the operator can obtain the state of interbus system without any other tools. Therefore, the entire interbus system has good system maintenance.

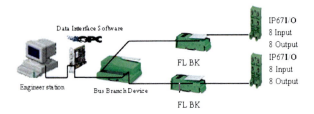

Fig. 1. The system structure drawing of one interbus controller

2.2 System Composing of Networking of Interbus Control System

The each department of automobile manufacture including automobile body factory, the stamping factory and the assembling factory has set up one INTERBUS control system successively and the structure of each interbus control system is similar. In order to make the most use of the data resources of each interbus control system, each interbus controller system is interlinked with the local area network of the factory by the Ethernet technology, the system topology is shown in the Figure 2. When each INTERBUS controller system is networked, management level may realize the real-time management, statistics and inquiry of production data.

3 Data Transmission and Processes of Interbus Control System

3.1 Data Transmission and System Control of Interbus Control System

Interbus control system is a data link structure. The bus adaptive control board is the central equipment of the data link control, which exchanges the data of transmission in the data link in serial with the high level computer system and low level interbus equipment. The data exchange is carried on between two sides synchronously and periodically. The data link has a distributional shift register structure. Each interbus equipment connects the process periphery equipment of numeral or simulation to the interbus system through his data register.

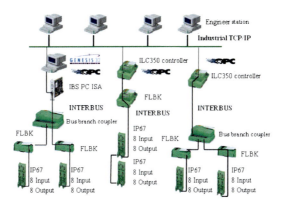

Fig. 2. The topology of interbus controller system networking

Each equipment of the interbus system has a ID register, which includes module type and the length, condition and error message of register in the data link. The interbus has two periodicities- identification periodicity and data periodicity. The running of identification periodicity initializes the interbus system. In the identification periodicity, the adaptive control board of the bus reads ID register of the equipment and uses this information to generate the process of reflection. Generally, the identification periodicity performs when system start, so it can distinguish the configuration of connection and compare with the configuration saved in the bus adaptive control board. After identification periodicity carries out successfully, the interbus equipments switch to data register in the interior and only carry out the data periodicity. The data periodicity is responsible for the data transmission, and in the data periodicity, the adaptive control board of the bus simultaneously updates all input and output data of the interbus equipment.

The program of system control is programmed based on the PCWORX software of PHOENIX CONTACT company including the hardware configuration and control programs. The control program adopts the structure of level and the programming thought of modular. The lower level module is transferred by the upper level module. By designing the basic function module of each level, the function of task control, serial communication and online diagnosis are realized in the topmost level. The input data of all the interbus equipment transmitted by the adaptive control board of the bus is taken as the input data of the control program. After processing by the control program, the input data and output data of all the interbus equipment are updated simultaneously by the adaptive control board of the bus.

3.2 Data Processing of Upper Computer in Interbus Controller System

OPC is based on the technology of the component object model and the distributed object model, adopts client/server mode and defines a set of COM object along with its interface standard. OPC server is the provider of data, provides the needed data for the OPC client. OPC client is the user of the data, which deals with the data provided by the OPC server. The interbus OPC server is an special OPC server developed by Phoenix company for the INTERBUS system and the configuration software of the PCAuto3.6 is worked as its OPC client, which communicates with the interbus device conveniently through the interbus OPC server. The monitoring and data processing of the whole system can be realized through the secondary development in the upper computer by the configuration software of the PCAuto3.6.

3.2.1 Create the History Database and Save the History Data

In the manufacture process of the automobile, the data is taken as the criterion to evaluate the working performance of the staff, included the requirement for the materials of every station, the running status of every device and the response information of the maintenance man to the fault device, should be saved the history database.

At first, a Database of the Microsoft Access should be created, the user data source created by the ODBC data source pointed to the available Access database. Then one OPC client should be defined in the I/O device driver menu of the configuration software of the PCAuto3.6 and make contact with Phoenix Contact OPC server, at the same time, the response variables are defined in Real-time Database\Database configuration and make contact with the server variables, and the real-time data of the I/O module of the interbus system will be gained. In the end, the programming is written in the configuration software of the PCAuto3.6.

3.2.2 Real-Time Monitoring of the Materials and Devices

Every figure of the real-time monitoring graph is corresponded to the relevant material and device, and the dynamic effect is formed by defining the dynamic connection to change the color, position and size of the figure. Therefore, the real-time monitoring graph can show the real-time requirement status of every station and the real-time running status of every device.

3.2.3 Fault Diagnosis of the Interbus Controller System

There are two 16-bit diagnostic registers within the interbus Controller: Diagnostic status register and Diagnostic parameter register. Each bit in the diagnostic status register represents a bus system running status. The data of the Diagnostic status register and Diagnostic parameter register can be gained by the interbus OPC server in the configuration software of the PCAuto3.6, so the running status of the bus can be estimated in the upper computer. If the bus works abnormality, the maintenance man can deal with this fault according to the detail information of the fault and the solution method that can be seen from the fault code of the diagnosis register. Therefore, the reliability of the interbus control system will be enhanced.

4 Transmission Design of Real-Time Monitoring Graph and Data of Database in Ethernet

According to the network technology, the real-time monitoring graph can be published through network and the data of the history database can be transmitted by network. Every management department can access to the remote monitor and management through shared database resources.

4.1 Network Publication of Real-Time Monitoring Graph

The system utilizes the WEB function provided by the configuration software of the PCAuto3.6 to realize the network publication of the graph. Firstly, selects the submenu command Draw of File\Web server configuration to set up Web root directory, the initialization graph and the IP address of the server. Then opening the graph window needs to be published in the command Draw, and selects the menu command File\Publish to Web to make it.

4.2 Network Transmission of History Data

For making the resource sharing and enhancing the managerial level, the material providing information and the device running information that saved in the Access Database should be transmitted to every office of the management through network. Because the configuration software of the PCAuto3.6 is short of the function of transmitting the user designed data directly through network. For realizing above functions, this system designs a program based on VISUAL C++.NET.

The whole program is mainly composed of five classes: CsourceLib class, CworkLib class, CNETBBDYDlg class, CAVIDlg class and CDataDlg class. The CsourceLib class is corresponded to the Access Database of the upper computer, the data of the Database can be operated through its member function. The

CworkLib class is corresponded to the Access Database of the office computer in the Network. The CNETBBDYDlg class has many functions as follows: transmitting data to the destination Database from the information source and querying, displaying, saving and printing of the report data and so on. The CAVIDlg class plays a role to display the real-time data transmitting status. The CDataDlg class is corresponded to the material or device information sheet of the Access Database, it have some functions just as modify, add and save. To begin with, the program achieve some necessary information (for example: the Network IP address of the source) by reading the configuration information in the file of the ini automatically. Then it transmits the data of the Network source Database to the local Database. Finally it carries out some function as enquiring, displaying, saving and printing. The user can dispose the data according to their demand because this data can be saved as Excel file.

5 Conclusion

The management system of material call and equipment call in the process of automobile production are realized by using the technology of the connection of interbus and utilizing the configuration software of the PCAuto3.6 to design the monitoring software in the host computer. Each interbus control system and local area network of factory are interlinked by the Ethernet technology, which realize the function of the network issuing and network transmission of the real-time data. it provides the powerful technical support for enhancing the production efficiency and the management level of the company.

References

1. Jin, J., Li, K.: Interbus-Based Control for Tobacco Leaf Redrying. In: Techniques of Automation and Applications, pp. 57–60 (2006)
2. Zhang, X., Ma, S., Hanxue, F.: The PLC Control System of Sluice Gates on Yellow River based on Interbus. Control & Automation, 45–47 (2005)
3. Yang, Y.-q., Zhang, H.: The Realization and Research of Remote Monitoring and Control in the Loop Net Power Distribution System Based on the Interbus Technology. Journal of Shanghai Institute & Electric Power, 27–31 (2005)
4. Miu, X., Liu, Z.: INTERBUS Field Bus Technology and its Development. Programmable Controller & Factory Automation, 6–10 (2006)

Broadband MMIC Power Amplifier for X Band Applications

Jiang Xia[1,2], Zhao Zhengping[2], Zhang Zhiguo[2], Luo Xinjiang[3], Yang Ruixia[1], and Feng Zhihong[2]

[1] College of Information Engineering, Hebei University of Technology.
300130 Tianjin, China
[2] The National Key Lab of ASIC, The 13 Research Institute, CETC.
050051 Shijiazhuang, China
[3] College of Electronic Information, Hangzhou Dianzi University.
310018 Hangzhou, China
tojiangxia@163.com

Abstract. The large signal model of a PHEMT with total gate width 850μm is achieved by microwave on-wafer test and IC-CAP software, then the design and optimization of circuit are implemented by ADS software. A three-stage broadband power amplifier is designed with above model. At operation frequency from 6 to 18 GHz, the output power is over 33dBm, the power gain is higher than 19dB and the PAE is more than 25%. Otherwise, the power amplifier has better power gain flatness.

Keywords: PHEMT, MMIC, power amplifier, X band.

1 Introduction

The current trend in microwave technology is toward circuit miniaturization, high-level integration, improved reliability, low power consumption, cost reduction, and high volume applications. Component size and performance are prime factors in the design of electronic systems for satellite communications, phased-array radar systems, electronic warfare, and other military applications, while small size and low cost drive the consumer electronics market[1],[2]. The wideband MMIC power amplifiers based on PHEMT technology just agree with above requirements. They play an increasing role in consumer electronics and military applications[3].

In this paper , we described the design and simulation characteristics of a wideband three-stage MMIC power amplifier for X band applications.

2 Design

The amplifier used 85μm×10 devices as a basic cell. Figure 1 shows the schematic diagram of the GaAs PHEMT structure.

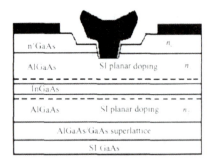

Fig. 1. Schematic diagram of the GaAs PHEMT structure

It is key to establish the non-linear large signal model for PHEMT. By microwave on-wafer test and IC-CAP software, we get the large signal model of a PHEMT with total gate width 850μm[4]. After optimizing these extracted parameters according to Ref.[5], the simulated S-parameters from the equivalent model are fit into the measured data, the results are show in Fig.2.

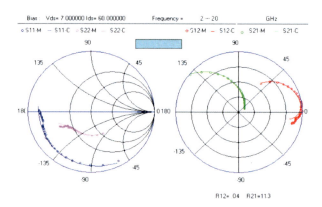

Fig. 2. S parameter comparing curve

Papers not complying with the LNCS style will be reformatted. This can lead to an increase in the overall number of pages. We would therefore urge you not to squash your paper.

Figure 3 shows that a three-stage topology design was adopted to meet the 20dB gain target . The first stage used a distributed amplifier topology to achieve a good input match over the design band and to provide positive gain slope compensation. The second stage used two 1700μm gate width transistors to drive an output stage consisting of four 1700μm transistors.

The first matching network for a power amplifier is the output matching network which is designed to transfer maximum output power from the FET to a 50Ω system. Lossy matching techniques in the interstage network were used to provide additional gain slope compensation and to provide the optimum impedance level for power matching.

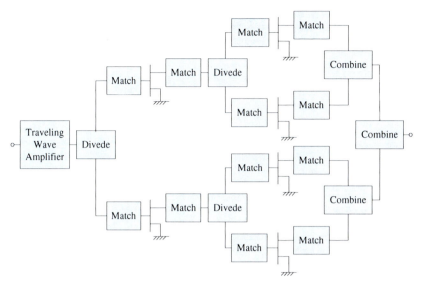

Fig. 3. Topology of there-stage MMIC power amplifier

The layout photograph of the MMIC power amplifier is show in Figure 4. The chip size of the MMIC power amplifier is 5.2×3.6mm.

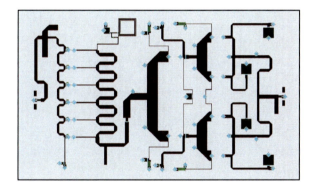

Fig. 4. Photograph of MMIC power amplifier

3 Performance

Based the above model, the three-stage broadband power amplifier was designed and simulated. All the simulation were performed at the fixed DC biases of V_{ds} = 7.0V and V_{gs} = - 0.4V. Figure 5 shows the simulated output power, power gain and PAE performances as a function of input power for various frequencies. At operation frequency from 6 to 18 GHz, the output power is over 33dBm, the power gain is higher than 19dB and the PAE is more than 25%. Otherwise, the power amplifier has better power gain flatness.

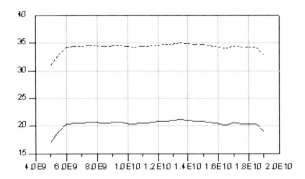

Fig. 5. Simulated P$_0$(*ine*) and G$_p$ (*real line*)

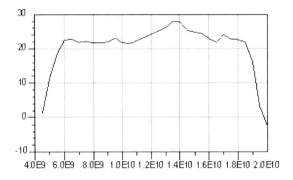

Fig. 6. Simulated PAE

4 Conclusions

We have described the design and simulation performance of a wideband power amplifier operating from 6 to 18GHz for X band applications. Using 0.85μm GaAs PHEMT technology, the three- stage MMIC power amplifier is designed. Over 6 ~ 18GHz, the output power is above 33dBm, the power gain is higher than 19dB and the PAE is more than 25%. These performance are satisfied.

References

1. Barnes, A.F., Moore, M.T., Allenson, M.B.: A 6-18GHz broadband high power MMIC for EW applications. J. IEEE MTT-S Digest, 1429–1430 (1997)
2. Inder, B., Prakash, B.: Microwave solid state circuit design, New Jersey, Hoboken (2003)
3. Lim, J.S., Kang, S.C., Nam, S.: MMIC 1watt wideband power amplifier chip set using pHEMT technology for 20/ 30GHz communication systems. In: J. Asia Pacific Microwave Conference (1999)
4. Kim, Y.G., Maeng, S.J., Lee, J.H.: A PHEMT MMIC broad-band power amplifier for LMDS. In: RAWCON Proceedings (1998)
5. Zhang, S.J., Yang, R.X., Gao, X.B.: The large signal modeling of GaAs HFET/ PHEMT. Chinese Journal of Semiconductors 28, 439–442 (2006)

The Selection of Dry Port Location with the Method of Fuzzy-ANP

Jinyu Wei, Aifen Sun, and Jing Zhuang

Department of Managemen, Tianjin University of Technology
Tianjin, China
wch6021552@163.com

Abstract. With the increasing of maritime container, functional seaport inland access is important for the efficiency of the transportation chain as a whole. The selection of Dry port location bases on the index system in view of the factors. The factors that affect the facilities of the dry port were systematically analyzed and an evaluation model was built. As a result of complexity in assessing Dry port selection performance, it is difficult for decision-makers to provide exact numerical values for the factors. Therefore, a fuzzy ANP method can solve problems in an uncertain condition effectively.

Keywords: fuzzy, analytic network process (ANP), Dry port.

1 Introduction

The establishment of Dry port is a useful method to solving the shortage of the space in the sea port areas. In general Dry port provides services like storage, maintenance, repair for containers, consolidation if individual container flows and custom clearance. At the same time, it can reduce the transport costs and expand rail transport.

Dry port can improve the situation resulting from increased container flows. In the dry port goods can be turned in as if at the seaport. The concept of the dry port is based on a seaport directly connected by rail with inland intermodal terminals. In such dry ports large goods' flows can shift from road to more energy efficient traffic modes. Furthermore, a dry port can provide services such as storage, consolidation, depot, custom regulation and service, maintenance of containers, and customs clearance, so it can relieve seaport cities from some of the congestion, make goods handling more efficient and facilitate improved logistics solutions for shippers in the port's hinterland[1,2].

Dry port means that most containers are moved by rail, road and airport from cities to seaport. Furthermore the development of dry ports is an essential possibility to promote sustainability and effectiveness of goods transport in sea related transport chains. In Oct, 2002, as the first Dry port Chaoyang dry port was built up in Beijing. It means China started the construction of dry port.

To be a dry port, a place must fulfill some conditions[3] such as, it should locate near a developed city, have direct connection to a seaport either by rail or by road, have a high capacity traffic mode, have the abundant human resources.

The concept of the dry port has been described in detail by Andrius Jaržemskis and Aidas Vasilis Vasiliauskas [1].FDT not only gives the concept and background of dry port, but also gives some dry port examples in 5 areas[4]. In Chain, Lv shunjian writes about the development and defects of China's dry port [5]. Xu wei and Lu meng researched about the important role of the dry port [6].

To build a dry port, the location must be carefully evaluated so as to obtain strategic advantages over competitors.

2 Application of Methodology

2.1 The Analytic Network Process (ANP)

The Analytic Network Process (ANP) is a process that allows one to include all the factors and criteria, tangible and intangible that has bearing on making a best decision. The Analytic Network Process allows both interaction and feedback within clusters of elements and between clusters. Such feedback best captures the complex effects of interplay in human society, especially when risk and uncertainty are involved.

a. Supermatrix [7]

The first step of ANP is to compare the criteria in whole system to build up the supermatrix. It is done through pairwise comparisons by asking "How much importance does a criterion have compared to another criterion with respect to our interests or preferences?" The relative importance value can be determined using a scale of 1-9 to stand for equal importance to extreme importance. We postulate network structure is composed of hierarchy $C_k (h = 1,2,\cdots\cdots,m)$. For each hierarchy, C_k assume there exist elements $e_{k1}, e_{k2},\cdots\cdots, e_{km}$, so the influence of $C_k (h = 1,2,\cdots\cdots,m)$ can be denoted as below:

$$W = \begin{array}{c} \\ C_1 \\ C_2 \\ \vdots \\ C_m \end{array} \begin{array}{c} \begin{array}{cccc} C_1 & C_2 & \cdots & C_m \end{array} \\ \left[\begin{array}{cccc} W_{11} & W_{12} & \cdots & W_{1m} \\ W_{21} & W_{22} & \cdots & W_{2m} \\ \vdots & \vdots & \vdots & \vdots \\ W_{m1} & W_{m2} & \cdots & W_{mm} \end{array} \right] \end{array}$$

Which is the general form of the supermatrix. W_{ij} shows the influence of each element of the i hierarchy on j the hierarchy, which is called a block of a supermatrix, whose form is as follows

$$W_{ij} = \begin{bmatrix} W_{i_1 j_1} & W_{i_1 j_2} & \cdots & W_{i_1 j_{n_j}} \\ W_{i_2 j_1} & W_{i_2 j_2} & \cdots & W_{i_2 j_{n_j}} \\ \vdots & \vdots & \cdots & \vdots \\ W_{i_{n_i} j_1} & W_{i_{n_2} j_2} & \cdots & W_{i_{n_i} j_{n_j}} \end{bmatrix}$$

b. Weighted Supermatrix

The priorities of elements in one hierarchy according to a certain criterion can be denoted with a supermatrix, which means every column of every hierarchy in the supermatrix is column stochastic. However, the influence that other hierarchy according to this criterion is not concerned. As a result, each column of the supermatrix is not column stochastic. It is essential to consider the influence between every two hierarchy. The method is: regarding each hierarchy as an element, and pairwise comparing according a certain hierarchy, then computing corresponding priorities. Assume a_{ij} is the influence weight of the i hierarchy on the j hierarchy, let

$$\overline{W}_{ij} = a_{ij} W_{ij} \quad (1)$$

$\overline{W}$ is a weighted supermatrix. In a weighted supermatrix, addition of elements in each column is 1. Matrix has this trait is called column stochastic [8] This step is much similar to the concept of Markov chain for ensuring the sum of these probabilities of all states equals to 1.

c. Limited Supermatrix

What we want to obtain is the priorities along each possible path in a supermatrix, on the other word the final influence an element on the highest goal. This kind of result can be acquired by solving $\overline{W}^\infty$,

$$\overline{W}^\infty = \lim_{k \to \infty} \overline{W}^k \quad (2)$$

The weighted supermatrix is raised to limiting powers like in (2) to get the global priority vector or called weights.

2.2 Fuzzy Set and Fuzzy Number

Zadeh introduced the fuzzy set theory to deal with the uncertainty due to imprecision and vagueness. A major contribution of fuzzy set theory is its capability of representing vague data. The theory also allows mathematical operators and programming to apply to the fuzzy domain [9]. Generally, a fuzzy set is defined by a membership function, which represents the grade of any element x of X that have the partial membership to M. The degree to which an element belongs to a set is defined by the value between zero and one. If an element x really belongs to M, $\mu_M(x) = 1$ and clearly not, $\mu_M(x) = 0$.

A triangular fuzzy number is defined as (l, m, u), where $l \leq m \leq u$. The parameters l, m and u respectively, denote the smallest possible value, the most promising value, and the largest possible value that describe a fuzzy event. (l, m, u) has the following triangular type membership function.

$$\mu_M(x) = \begin{cases} (x-l)/(m-l) & l \leq x \leq m \\ (u-x)/(u-m) & m \leq x \leq u \\ 0 & otherwise \end{cases} \quad (3)$$

A triangular fuzzy number can be shown in Figure. 1

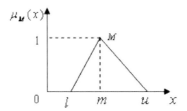

Fig. 1. A triangular fuzzy number

2.3 Chang's Extent Analysis Method [10]

Chang's extent analysis method is one of fuzzy AHP methods, the steps of which are as follows:

Let $X = \{x_1, x_2, \cdots, x_n\}$ be an object set, and $G = \{g_1, g_2, \cdots, g_m\}$ be a goal set. According to the method of Chang's extent analysis, each object is taken and extent analysis for each goal, g_i, is performed respectively. Therefore, m extent analysis values for each object can be obtained, with the following signs:

$M_{g_i}^1, M_{g_i}^2, \cdots, M_{g_i}^m, i = 1, 2, \cdots, n$, where all the $M_{g_i}^j$ $(j = 1, 2, \cdots, m)$ are triangular fuzzy numbers.

The steps of Chang's extent analysis can be given as follows:

a: The value of fuzzy synthetic extent with respect to the i th object is defined as:

$$S_i = \sum_{j=1}^{m} M_{g_i}^j \otimes \left[\sum_{i=1}^{n} \sum_{j=1}^{m} M_{g_i}^j \right]^{-1} \quad (4)$$

where

$$\sum_{j=1}^{m} M_{g_i}^j = \left(\sum_{j=1}^{m} l_j, \sum_{j=1}^{m} m_j, \sum_{j=1}^{m} u_j \right) \quad (5)$$

$$\sum_{i=1}^{n}\sum_{j=1}^{m}M_{g_i}^{j} = \left(\sum_{i=1}^{n}l_i, \sum_{i=1}^{n}m_i, \sum_{i=1}^{n}u_i\right) \quad (6)$$

$$\left[\sum_{i=1}^{n}\sum_{j=1}^{m}M_{g_i}^{j}\right]^{-1} = \left(\frac{1}{\sum_{i=1}^{n}u_i}, \frac{1}{\sum_{i=1}^{n}m_i}, \frac{1}{\sum_{i=1}^{n}l_i}\right) \quad (7)$$

b: The degree of possibility of
$M_2 = (l_2, m_2, u_2) \geq M_1 = (l_1, m_1, u_1)$ is defined as:

$$V(M_2 \geq M_1) = \begin{cases} 1 & m_2 \geq m_1 \\ 0 & l_1 \geq u_2 \\ \dfrac{l_1 - u_2}{(m_2 - u_2) - (m_1 - l_1)} & otherwise \end{cases} \quad (8)$$

c: The degree possibility for a convex fuzzy number to be greater than k convex fuzzy numbers $M_i(i=1,2,\cdots,k)$ can be defined by:

$$V(M \geq M_1, M_2, \cdots, M_k) = \min V(M \geq M_i), i=1,2,\cdots,k \quad (9)$$

Assume that

$$d'(A_i) = \min V(S_i \geq S_k) \quad (10)$$

for $k = 1, 2, \cdots, n; k \neq i$.
Then the weight vector is given by

$$W' = \left(d'(A_1), d'(A_2), \cdots, d'(A_n)\right)^T \quad (11)$$

where $A_i(i=1,2,\cdots,n)$ are n elements.

d: Via normalization, the normalized weight vectors are

$$W = \left(d(A_1), d(A_2), \cdots, d(A_n)\right)^T \quad (12)$$

where W is a nonfuzzy number.

3 A Fuzzy ANP Application in Dry Port

3.1 Modeling in ANP

This paper summarizes the influencing factors of dry port location. According to the function of dry port and logistics center location, this paper intends to decide indicator systems. The indexes of indicator systems are chosen based on the referred literature [11], [12], [13]. The following evaluation indicator system (Table 1) was constructed:

Table 1. Dry port location evaluation indices system

Criteria	Sub-criteria
infrastructure status	traffic
	information infrastructure
	state of public facilities
costs	transport costs
	the environment protection
	local labor wage level
operating environment	labor conditions
	the distribution and quantity of goods
	customer conditions

3.2 Formation of Fuzzy Matrices

When pairing comparisons under each control criterion Delphi method is employed. To make sure the result is more reasonable and exact, more experts are expected to participate in pairwise comparison. The elements in a cluster are calculated by employing fuzzy scale. The fuzzy scale regarding relative importance to measure the relative weights is given in Table2 [9].

Pairwise comparison matrices are computed by the Chang's extend analysis method and local weights are determined. Pairwise comparison matrices of main factors and the local weights for the factors are calculated as shown in Table3.

Table 2. Linguistic scales for difficulty and importance

Linguistic Scales for Difficulty	Linguistic Scales for Importance	Triangular Fuzzy Scale
Just equal	Just equal	(1,1,1)
Equally difficult (ED)	Equally important (EI)	(1/2,1,3/2)
Weakly more difficult (WMD)	Weakly more important (WMI)	(1,3/2,2)
Strongly more difficult (SMD)	Strongly more important (SMI)	(3/2,2,5/2)
Very strongly more difficult (VSMD)	Very strongly more important (VSMI)	(2,5/2,3)
Absolutely more difficult (AMD)	Absolutely more important (AMI)	(5/2,3,7/2)

Table 3. Local weights and pairwise comparison matrix of main factors

Factors	Infrastructure status(IS)	Costs(C)	Operating environment(OE)	Local weights
IS	(1,1,1)	(1,3/2,2)	(1/2,1,3/2)	0.369
C	(1/2,2/3,1)	(1,1,1)	(2/3,1,2)	0.300
OE	(2/3,1,2)	(1/2,1,3/2)	(1,1,1)	0.331

The Selection of Dry Port Location with the Method of Fuzzy-ANP

From Table 3, according to (4), (5), (6), (7),

$$S_{OE} = (3\ 4\ 5) \otimes (1/12.667, 1/9.667, 1/7.576) = (0.237, 0.414, 0.660)$$
$$S_{IS} = (2.667, 3.5, 5) \otimes (1/12.667, 1/9.667, 1/7.576) = (0.211, 0.362, 0.660)$$
$$S_C = (1.9\ 2.167\ 2.667) \otimes (1/12.667\ 1/9.667, 1/7.567) = (0.150, 0.224, 0.352)$$

According to (8),

$$V(S_{OE} \geq S_{IS}) = 1 \qquad V(S_{OE} \geq S_C) = 1 \qquad V(S_{IS} \geq S_{OE}) = 0.891$$
$$V(S_{IS} \geq S_C) = 1 \qquad V(S_C \geq S_{OE}) = 0.377 \qquad V(S_C \geq S_{IS}) = 0.505$$

According to (9), (10), (11), (12), the weight vector from Table 3 is calculated as $W = (0.441, 0.393, 0.166)^T$

Local weights of sub-factors and inner dependence matrix of factors can be computed by the same approach, the result of which are shown in Table 4 and Table 5.

Table 4. Local weights of sub-factors

Sub-factors in	Local weights
infrastructure status	$(0.451, 0.392, 0.157)^T$
costs	$(0.369, 0.300, 0.331)^T$
operating environment	$(0.238, 0.381, 0.381)^T$

Table 5. Inner dependence matrix of factors with respect to each main factor

Main factors	Relative importance weights
infrastructure status	$(0.316, 0.684)^T$
costs	$(0.684, 0.316)^T$
operating environment	$(0.500, 0.500)^T$

In this step, the interdependent weights of the factors are calculated by multiplying the dependence matrix of factors provided in Table5 with the local weights of factors provided in Table3 as follows

$$\begin{bmatrix} 1.000 & 0.684 & 0.500 \\ 0.316 & 1.000 & 0.500 \\ 0.684 & 0.316 & 1.000 \end{bmatrix} \times \begin{bmatrix} 0.369 \\ 0.300 \\ 0.331 \end{bmatrix} = \begin{bmatrix} 0.740 \\ 0.582 \\ 0.678 \end{bmatrix}$$

After normalization, the interdependent weights of the factors are $[0.370\ \ 0.291\ \ 0.339]^T$.

In the final step, the global weights for the sub-factors are computed. Global sub-factor weights are calculated by multiplying interdependent weight of the factors with local weight of the sub-factor which belongs to main factors and the values are obtained in Table6.

Table 6. Global weights of sub-factors

Factors	Sub-factors	Local weights	Global weights
infrastructure status (0.370)	traffic	0.451	0.167
	information infrastructure	0.392	0.145
	state of public facilities	0.157	0.058
costs (0.291)	transport costs	0.369	0.107
	the environment protection	0.300	0.087
	local labor wage level	0.331	0.096
operating environment (0.339)	labor conditions	0.238	0.081
	the distribution and quantity of goods	0.381	0.129
	customer conditions	0.381	0.129

4 Conclusions

In this paper, we have discussed several criteria for Dry port problem. When decision makers are uncertain about their own level of preference, a systematic decision procedure is given in this study. Through the table 6, decision makers can know which factor is the most important and which sun-factor influence the main factor which it belongs to.

References

1. Jaržemskis, A., Vasiliauskas, A.V.: Research on dry port concept as intermodal node. Transport XXII, 207–213 (2007)
2. Rui, Y.: Analyze on inland dry port construction in China, Port Economy (2006)
3. Shunjian, L., Dongyandan: Development of dry port in China. Water carriage management 29 (2007)
4. FDT: Feasibility study on the network operation of hinterland hubs (dry port concept) to improve and modernise ports' connectiond to the hinterland and to improve networking. Integrating logistics center networks in the baltic sea region (2007)
5. Shunjian, L.: The development and defects of China's dry port, China Ports, pp. 13–14 (2007)
6. Wei, X., Meng, L.: Dry port in the role of port development. Water Transport Management 28, 8–9 (2006)
7. Saaty, T.L.: Decision Making With Dependence Feedback: The Analytic Network Process. RWS Publications, Pittsburgh (2001)

8. Tang, X., Feng, J.: ANP Theory and Application expectation. Statistics and Decision-making 12, 138–140 (2006)
9. Kahraman, E., Ertay, T., Büyüközkan, G.: A fuzzy optimization model for QFD planning process using analytic network approach. European Journal of Operational Research 171(2), 390–411 (2006)
10. Chang, D.Y.: Application of the extent analysis method on fuzzy AHP. European Journal of Operational Research, vol 95, 649–655 (1996)
11. Wen-jin, Y., Kai, L.: The study on logistics center location based on ANP. Journal of Transportation Systems Engineering and Information Technology 6 (2006)
12. Min, Z., Yangjun, Y.c.: Location of logistics distribution center based on AHP/DEA. Chinese Journal of Management 2 (2005)
13. Jie, X., Kai, Z., Yuan, T., Yi-hong, R.: Research of logistics center location and case analysis. Journal of Northern Jiaotong University 25 (2001)

Growing and Declining of Aged Population Asymmetric Real Estate Price Reactions: The Proof of China

Jinqiu Xu

School of Management, Huazhong University of Science and Technology
Wuhan, China, 430070
xjq315@126.com

Abstract. The population of China will be to undergo considerable permanent reduction because of aging significantly and a low fertility rate by 2050. After discussing different factors which induced the real estate price fluctuation of China , this paper proposes a mixed model to analyze the relations between the change of aged population with housing price. The empirical result indicates that asymmetric price reactions: growth in aged population numbers has not significantly lowered prices, whereas declining in aged population has a significant effect on price .The paper gives the guide about how to choose the balance between aged population with real estate price.

Keywords: Aged population, real estate, asymmetric price reaction, regional analysis.

1 Introduction

Almost all the studies on the demographic development of Chinese societies come to the general conclusion that by the year 2050 the populations will considerably decline , and The decline is due to both a high aging rate and a low fertility rate, and that is not compensated particularly dramatic. Whilst the population projection assumes that a medium scenario would see a 10% reduction in population to some 85 million inhabitants by the year 2050, the projected reduction is 20% in the minimal scenario. The pessimistic prediction is as low as 5.7 million inhabitants every year, a decline of 15% from the present. Furthermore, these average figures conceal that population reduction will affect the Chinese regions to differing degrees. In general, rural areas will be less affected, and metropolitan areas will bear the brunt of the decline. These demographic developments will have a significant effect on the pension, health and nursing care insurance systems. Although the emphasis is often the ageing phenomenon rather than population decline, there is little awareness of the effects of projected population decline and ageing on other areas such as regional real estate markets.

In the first empirical studies of demographics and real estate, Mankiw (1989) find a significant positive relationship between housing prices and demand in the United States. His study motivated a series of other works. Engelhardt and Poterba (1991) find no relationship for Canada. In a different approach, DiPasquale and Wheaton (1996)perform a cross-section analysis for the United States. Meese and Wallace (2003) elaborate on the Paris real estate market, Terrones and Otrok (2004) estimate the growth in house prices using a multivariate model, and find a significant influence of population growth at a highly aggregated national level. As far as we are concerned, there are no models of both population growth and decline and their effects on real estate prices. Leaving aside regional restrictions on land availability, the chronic under-utilization of building production capacity means that increased demand could be satisfied without increasing prices.

By contrast, if demand declined there is the possibility of an inelastic supply reaction due to typical construction methods. in spite of considerable levels of unoccupied buildings – up to 5% in some areas ,The rental and the property sectors are characterized by low price elasticity of demand, indicating that significant real estate price decreases can result. As real estate assets dominate private household portfolios in most western economies, price reductions could have significant complications for consumption and growth. This work supplements existing studies by examining real estate prices on a disaggregated level of Chinese metropolitan areas and studying the effects of both aged population growth and decline. At the same time, checks are made for other potential factors that influence prices, such as household income and building costs.

2 Methodology and Data

House prices in the metropolitan areas is examined by the log–log approach used by e.g. Engelhardt and Poterba (1991), DiPasquale and Wheaton (1994).The real estate prices (PRICE) base on the real estate price index for residential real estate, using total property prices for ready-to-inhabit detached houses of medium housing quality. The level of residential property is around 34%, with most homes of medium housing quality. The arithmetic mean of house prices is ￥205789. The cross-section analysis covers 100 of the 118 metropolitan areas that population number is above 1 million, No price data is available for the 18 metropolitan areas not included. Taking the equation of DiPasquale (1996) as a starting point, it is appropriate to test the aged population of the Chinese metropolitan areas as determinants. Given that no data is available at district level for the number of households, the aged population numbers are taken from the INKAR PRO database (BBR, 2006). Generally, population in the cities fell by just under 5%. In order to accommodate for this peculiarity and to estimate the potential consequences of a future reduction in aged population, two dummy variables are introduced that are multiplicatively linked with the Pop Growth (percentage growth of aged population) variable. The variable INCREASE=1, if a city increased in population in 1995–2005, and otherwise=0. Analogously, the variable SHRINK=1, if a city had reduced aged population in 1995–2005, and otherwise=0.The variable COST (ln of construction costs) is composed of the regional cost of new residential buildings per square meter (BBR, 2004b). The arithmetic mean of construction costs per square meter is 1200. Finally, for China it is appropriate to test the regional annual per capita income (INCOME) as an influencing factor. Regional per capita income data is supplied by regional Offices of Statistics, the arithmetic mean value is ¥ 8875.40 The estimating equation is:

$$Ln(PRICE) = \beta_0 Ln(POP) + \beta_1 INCREASE * POPGROWTH + \beta_2 SHRINK * POPGROWTH + \beta_3 Ln(COST)$$
$$\beta_4 Ln(INCOME) + \mu \_(1)$$

where μ is the random error term.

3 Results

Because the White test rejected homoscedasticity, we use the White correction in the following regressions. Table 1 shows the results using the model of Di-Pasquale and Wheaton (1996) which, due to the cross-section approach, is most similar to the present study (model 1). Population size, building costs, aged population, and demographic changes have significant effects with the expected signs. The estimates from Eq. (1) are in Table 1 (model 2). The highly significant aged population growth of model 1 is only statistically significant in model 2 when it is negative, thus a growing aged population has no significant influence on prices.This asymmetry or ratchet effect can be explained by a sufficient level of construction capacity producing adjustment to increasing demand in the medium term, without price effects. Adjusted R2 (R2adj)=0.66, indicating increased goodness of fit compared to model 1.11 For the purpose of sensitivity analysis the estimation was subjected to several modifications. Firstly, to account for possible influences of residential development structures on real estate purchase prices (BBR,2004b, 103), the urban districts were categorised; the 100 cities are distributed among district types as follows:

25 core cities in agglomeration areas (Type 1); 20 districts with high aged population density in agglomeration areas (Type 2); 0 dense districts in agglomeration areas (Type 3); 16 rural districts in agglomeration areas (Type 4); 14 core cities in urban areas (Type 5); 11 dense districts in urban areas (Type 6); 0 rural districts in urban areas (Type 7); 9 dense rural districts (Type 8); 5 low-density rural districts (Type 9).

From the above tables we find that only core cities in agglomeration areas (Type 1) have a significant influence on housing prices. The results of Model 2 are confirmed: a decline in aged population significantly decreases prices, whereas population increase has no influence. Construction costs and income are highly significant. In order to test for a possible price-increasing influence connected with the seat of a regional government, a dummy variable CAPITAL is introduced (Model 4). This variable is not significant. To test whether land regulation and other building restrictions are important determinants of price, the fraction of land zoned for residential purposes is used (Model 5) Although the variable SHARE

Table 1. Determinants of (ln of) house prices in Chinese metropolitan areas

Model no.	1	2	3	4	5	6	7	8
CONST	4.350***	−0.641	−0.780	−0.456	0.46	0.029	0.606	1.543
	-0.146	-2.036	-2.045	-2.037	-2.45	-2.146	-2.181	-2.274
	-1.544	-2.231	-2.218	-2.245	-2.59	-2.215	-2.272	-2.303
lnPOP	0.175***	0.151***	0.123**	0.127***	0.118***	0.123***	0.109**	0.104***
	-0.034	-0.032	-0.041	-0.039	-0.047	-0.044	-0.041	-0.035
	-0.027	-0.029	-0.028	-0.036	-0.041	-0.037	-0.035	-0.041
POPCHANG	2.32***							
	-0.37							
	-0.376							
SHRINKPOPCHANGE		2.455***	2.331***	2.271***	2.536***	2.326***	1.836***	
		-0.425	0.432	0.465	0.668	0.447	0.623	
		0.446	0.443	0.449	0.605	0.455	0.598	
INCREASEPOPCHAN		-0.176	-0.004	-0.017	0.675	0.677	0.119	
		-1.228	-1.254	-1.193	-1.098	-1.134	-1.125	
		-1.167	-1.152	-1.154	-1.178	-1.194	-1.136	
ln COST	0.778***	0.669***	0.712***	0.709***	0.617***	0.669***	0.726***	0.679***
	-0.218	-0.2	-0.187	-0.189	-0.199	-0.187	-0.189	-0.179
	-0.184	-0.178	-0.182	-0.186	-0.198	-0.178	-0.179	-0.187
ln INCOME		0.644***	0.610**	0.626***	0.609***	0.614***	0.526***	0.408***
		-0.189	-0.18	-0.184	-0.219	-0.186	-0.187	-0.219
		-0.214	-0.21	-0.213	-0.242	-0.212	-0.22	-0.2
CORE			0.109**	0.101**	0.114**	0.118**	0.109**	0.128**
			-0.05	-0.049	-0.057	-0.053	-0.05	-0.054
			-0.054	-0.055	-0.062	-0.057	-0.055	-0.0501
CAPITAL				-0.055				
				-0.067				
				-0.063				
SHARE_HOUSING_AREA					−0.475			
					-0.628			
					-0.545			
SHARE_HOUSE_AND_BUSIN_AREA						−0.118		
						-0.324		
						-0.304		
R^2	0.6	0.64	0.65	0.66	0.61	0.66	0.65	0.65

** = significant at the 5% error level.   *** = significant at the 1% error level.
Numbers in parentheses are t-values: first values with, second without White correction.

HOUSING AREA has the expected sign, it is not significant. However, the relevant data for 16 of the 100 cities is not available, which might explain the reduced goodness of fit. As an alternative the fraction of land zoned for residential and commercial purposes is used (Model 6), which has only two missing values. This determinant has the expected sign, but is also non-significant. Finally a regression with the interaction between the dummy and the asymmetric effect of aged population growth is run (Model 8). The asymmetric effects of aged population growth and population shrinkage are not lost.

4 Conclusions

This study confirms empirical results that aged population size and being one of the district-type leads to higher purchase prices. Significant positive effects are also shown for disposable income and construction costs. There is an asymmetrical supply reaction to aged population increase or decrease. Aged population growth and the resulting increases in demand have no significant effect on price; however, a declining aged population leads to significantly reduced prices. Of the 100 cities in the present study, the greatest decline is 15% by 2050. Construction costs will probably fall (BBR, 2004a, 10). Whilst these foreseeable price reductions vary widely between regions, they are highly likely to occur.

Further research on these developments could be fruitful. It is also worthwhile examining whether the asymmetries noted also occur in other contexts, e.g. housing rents. To determine potential consequences for economic policy in particular, an analysis could be made of tax-related factors. Attention should also be paid to the financing conditions. This was not possible in the cross-section analysis in the present study since – in as much as it is possible for regionally differing values to exist at all – no data was available for lending limits etc. at district level.

References

[1] Mankiw, N.G., Weil, D.N.: The baby boom, the baby bust, and the housing market. Regional Science and Urban Economics 19, 235–258 (1989)
[2] Malkiel, B.: Returns from investing in equity mutual funds 1971–1991. Journal of Finance 50, 549–572 (1995)
[3] Mueller, G., Pauley, K.: The effect of interest rate movements on real estate investment trusts. The Journal of Real Estate Research 10(3), 319–326 (1995)
[4] Peterson, J., Hsieh, C.: Do common risk factors in the returns on stocks and bonds explain returns on REITs? Real Estate Economics 25(2), 321–345 (1997)
[5] Poterba, J., Summers, L.: The persistence of volatility and stock market fluctuations. American Economic Review 76, 1141–1151 (1986)
[6] Hendricks, D.: Evaluation of value-at-risk models using historical data. Federal Reserve Bank of New York Economic Policy Review 2(1), 39–69 (1996)
[7] Hull, J., White, A.: Value at risk when daily changes in market variables are not normally distributed. Journal of Derivatives 5(3), 9–19 (1998)
[8] Jackson, P.J., Maude, D., Perraudin, W.: Bank capital and value at risk. Journal of

Mobile Learning Application Research Based on iPhone

Zong Hu and Dongming Huang

College of Education, Ningbo University,
No. 818 Fenghua Road, Ningbo, Zhejiang 315211, China
nbshiny@gmail.com

Abstract. In recent years, as a new mode of learning, more and more domestic and foreign scholars pay attention to mobile learning. This paper use iPhone act as mobile learning carrier, learners can get and interact information through iPhone. Application research based on iPhone will maximize the potential advantages of iPhone, so that learners can participate in learning at anytime, in anywhere.

Keywords: Mobile Learning, iPhone, Cocoa.

1 Introduction

With the development of science and technology, 3G network communication technology has opened a new era, and has greatly contributed to the rapid development of wireless communications. Mobile learning (m-learning) will become the new mode of learning after the e-learning learning mode, people can access the appropriate learning resource information and be participated in learning process through the wireless communication devices. Therefore, mobile learning can provide an anytime, anywhere learning environment to any people to fit for their special learning habits, and people can select different learning content that they need, have real-time and non real-time discussion with other learners and teachers, this will greatly promote the development of the education.

The iPhone is an Internet and multimedia enabled smartphone designed and marketed by Apple Inc. The iPhone functions as a camera phone (also including text messaging and visual voicemail), a portable media player (equivalent to a video iPod), and an Internet client (with email, web browsing, and Wi-Fi connectivity) — using the phone's multi-touch screen to render a virtual keyboard in lieu of a physical keyboard. Here we use iPhone to get the needed learning resource, and can be participated in the learning process through the iPhone.

The rest of the paper is organized as follows. Section 2 describes related theories of mobile learning. In section 3, we give the detail of the iPhone, and design the corresponding application software to improve the learning process. Finally, the conclusion is given in section 4.

2 Background

At present mobile learning research mainly focus on Europe and North America, based on the research purpose, mobile learning research divide into two categories, one is sponsored by educational institutions, which based on their school education, try to improve teaching and management through the new technologies. The other is initiated by the e-learning providers, who learn from the experience of e-learning, put mobile learning into market, but more for business training. International distance education specialist Desmond • Keegan[1], presided over the "From the e-learning to mobile learning" project, proposed that the e-learning will change to the mobile learning in the future, designed a virtual learning environment for mobile technology and provided the appropriate learning environment model. At this time, many other foreign scholars conducted in-depth study and practice in this area, including:

- Stanford University learning lab (SLL) Study[2]: SLL has developed a initial model fit for the mobile learning, chosen foreign language learning as a mobile education curriculum, allowing users to practice new words, do quizzes, access to words and translate phrases, users can use the mobile phones, PDAs, wireless Web to help to send and receive e-mail, and maintain real-time contact to ensure that mobile devices can work in a safe, trusted, personalized, and immediate needs of the environment, to provide the learning opportunities of review and practice.
- Mobile learning in rural of Africa: taking into account the students who participate in the learning in rural of Africa have no PDA, no E-mail and other digital-based learning facilities, but 99% of them who have cell phones, so the research is based on mobile phone to learn, through mobile phones to receive and send text messages to learn to ensure that students can receive learning anytime, anywhere information and thus go to learn.
- Ericsson and other commercial companies to develop the "mobile learning" project[3]: to study the integration of the mobile learning and traditional classroom teaching methods, through the 3G communications applications technology to achieve information sharing and communication exchanges.

3 Mobile Learning Application Research Based on iPhone

3.1 iPhone

The first-generation phone (known as the Original) was quad-band GSM with EDGE; the second generation phone (known as 3G) added UMTS with 3.6 Mbps HSDPA;[4] the third generation adds support for 7.2 Mbps HSDPA downloading but remains limited to 384 Kbps uploading as Apple had not implemented the HSPA protocol.[5] iPhone have the following features:

- The iPhone run an operating system known as iPhone OS. It is based on a variant of the same Darwin operating system core that is found in Mac OS X. Also included is the "Core Animation" software component from Mac OS X v10.5 Leopard. Together with the PowerVR hardware (and on the iPhone 3GS,

OpenGL ES 2.0),[6] it is responsible for the interface's motion graphics. It is capable of supporting bundled and future applications from Apple, as well as from third-party developers. Software applications cannot be copied directly from Mac OS X but must be written and compiled specifically for iPhone OS.
- The interface is based around the home screen, a graphical list of available applications. iPhone applications normally run one at a time, although most functionality is still available when making a call or listening to music. The home screen can be accessed at any time by a hardware button below the screen, closing the open application in the process. By default, the Home screen contains the following icons: Messages (SMS and MMS messaging), Calendar, Photos, Camera, YouTube, Stocks, Maps (Google Maps), Weather, Voice Memos, Notes, Clock, Calculator, Settings, iTunes (store), App Store, and (on the iPhone 3GS only) Compass, as shown in Figure 1. Almost all input is given through the touch screen, which understands complex gestures using multi-touch. The iPhone's interaction techniques enable the user to move the content up or down by a touch-drag motion of the finger.

Fig. 1. The Home screen of iPhone

- The iPhone allows audio conferencing, call holding, call merging, caller ID, and integration with other cellular network features and iPhone functions.
- The layout of the music library is similar to that of an iPod or current Symbian S60 phones. The iPhone can sort its media library by songs, artists, albums, videos, playlists, genres, composers, podcasts, audiobooks, and compilations. Options are always presented alphabetically, except in playlists, which retain their order from iTunes. The iPhone allows users to purchase and download songs from the iTunes Store directly to their iPhone. The feature originally required a Wi-Fi network, but now can use the cellular data network if one is not available.
- Internet access is available when the iPhone is connected to a local area Wi-Fi or a wide area GSM or EDGE network, both second-generation (2G) wireless data standards. The iPhone 3G introduced support for third-generation UMTS and HSDPA 3.6, but not HSUPA networks, and only the iPhone 3GS supports

HSDPA 7.2. The iPhone will ask to join newly discovered Wi-Fi networks and prompt for the password when required. Alternatively, it can join closed Wi-Fi networks manually.[7] The iPhone will automatically choose the strongest network, connecting to Wi-Fi instead of EDGE when it is available. The iPhone 3G has a maximum download rate of 1.4 Mbps in the United States. Safari is the iPhone's native web browser, and it displays pages similar to its Mac and Windows counterpart. Web pages may be viewed in portrait or landscape mode and supports automatic zooming by pinching together or spreading apart fingertips on the screen, or by double-tapping text or images. The iPhone supports SVG, CSS, HTML Canvas, and Bonjour. The maps application can access Google Maps in map, satellite, or hybrid form. It can also generate directions between two locations, while providing optional real-time traffic information.

- The iPhone also features an e-mail program that supports HTML e-mail, which enables the user to embed photos in an e-mail message. PDF, Word, Excel, and Powerpoint attachments to mail messages can be viewed on the phone.
- The iPhone and iPhone 3G feature a built in fixed-focus 2.0 megapixel camera located on the back for still digital photos. It has no optical zoom, flash or auto-focus, and does not support video recording, however jailbreaking allows users to do so. Version 2.0 of iPhone OS introduced the capability to embed location data in the pictures, producing geocoded photographs. The iPhone 3GS has a 3.2 megapixel camera, with auto focus, auto white balance, and auto macro (up to 10 cm). It can also record VGA video at 30 frames per second, although compared to higher-end CCD based video cameras it does exhibit the rolling shutter effect. The video can then be cropped on the device itself and directly uploaded to YouTube, MobileMe, or other services.

From the above iPhone features, we can know iPhone has powerful capabilities, and use the features of telephone, text messaging, pictures, and mail to carry out the appropriate information in the discussion or interaction between the learners[8], and thus better support in-depth learning and interaction between learners, as shown in Figure 2.

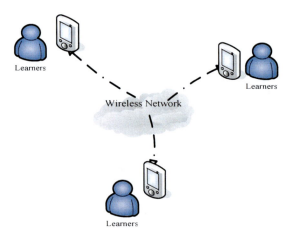

Fig. 2. iPhone learning mode

3.2 iPhone Application Environment–Cocoa

In the Mac OS X operating system, Cocoa, Carbon, and Java act as an application environment. It consists of a set of object-oriented software libraries and a runtime environment component, its integrates development environment is as same as other applications environment. Cocoa using the MVC design pattern, which divided into three objects: the model object, the view object, and the controller object.

Model object is responsible for packing the data and the basic behavior. Model object reflect the special knowledge and professional skills, they are responsible to maintain the application's data and logic definition of the operational data. A well-defined MVC application, will encapsulate the all important data in the model object. Any data represent the application retained status (regardless of the status is stored in a file, or stored in a database), once loaded into the application, it should reside in the model object. Because they represent a particular problem domain with the knowledge and professional skills, so there may be reused. In the ideal case, the model object does not responsible for establish an explicit connection with the user interface. For example, if there is a representative of a person's model object (assuming you are in the preparation of an address book), you may want to store the person's birthday, then you will store birthday on Person model object, this is a better way. However, the date format string, or other relevant date on how that information may be stored in other places better.

View object is responsible for that information to the user: View object knows how to display the application's model data, and may allow users to edit it. View object should not be responsible for storing the data it displays (which, of course not to say that the view object that it is never stored display data. For performance reasons, the view object may be cached the data, or use a similar technique). A view object may be responsible for showing part or all of model object, or even many different model objects. View object may have many changes. View object should be reusable and can be configured as far as possible, they can be in different applications to provide a consistent display.

Controller object link model object with view object: Controller object is the co-ordinator between the application's view object and model object. Under normal circumstances, they are responsible for ensuring view access to their display model, and act as a communication channel, so that make view to understand model changes. Controller object can also be configured for the application program execution and coordination tasks, managing the life cycle of other objects. In a typical Cocoa MVC design, when the user through a view object to enter a value or to make a choice, the value or choice of the object passed to the controller. Controller object may be interpretation the user input by the application-specific way, and then tell the model object, or how to deal with this type - such as "add a new value" or "delete the current record," or make the model object react the changed value in one of its attributes.

Compose: We can compose multiple MVC roles, so that an object at the same time as multiple roles, such as while serving as a controller and the view of the role of the object - in this case, the object is known as the View - Controller. Similarly, you can model - controller objects. From Figure 3, we can see in this compose model, the controller combines the arbiter and strategy model of the object model, two-way co-ordination the data flow between model object and view object. The change of the model state can pass to the view object through the controller object. In addition, the view object adopt command mode in the target - action mechanism.

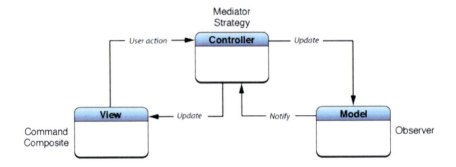

Fig. 3. Composite Design Pattern

3.3 iPhone Application Case

Learners can get the appropriate information through the iPhone, and interact information with other persons, not face to face, which will greatly stimulate their enthusiasm for learning, so that they can participate in the learning process anytime ,anywhere, such as animal knowledge software based multi-lingual, as shown in Figure 4, learners can first see one animal picture in the iPhone, and then learners would be to consider the animal's name, the animal's living habits, and then

Fig. 4. Animal knowledge software based multi-lingual

will go to recall the animal sounds, the last they will compare these information with the answer stored in iPhone, which ultimately deepening learner knowledge understanding. Other case is the language learning system based Multi-language, as shown in Figure 5, more suitable for foreigners learning Chinese, for example a foreigner want to learn word '学', the system will first show the corresponding meaning of the word in foreign language, so that learners can understand, and then tell the students how to write this word. Finally this software make knowledge simplification, learners can master the knowledge without the help of the teachers.

Fig. 5. Language learning system based Multi-language

4 Conclusion

With the development of the 3G wireless network communication technology, mobile learning will become a new learning mode, play a great role in the education filed and gradually become a hot spot in the future, however as mobile learning carrier—iPhone, it has its powerful features to ensure that learners will be able to achieve information and interact with other learners, while the application research based on iPhone will be to maximize the potential advantages of iPhone, so that learners can participate in learning at anytime, in anywhere.

Acknowledgements. This study was supported by Scientific Research Fund of Zhejiang Provincial Education Department(Y200804422).

References

1. Keegan, D.: The future of learning: From E-learning to M-learning[DB/OL], http://learning.ericsson.net/mlearning2/project_one/thebook/chapter1.html
2. Stanford Learning Lab, http://sll.stanford.edu/projects/mobilelearning/

3. Ericsson Education Online,
   http://learning.ericsson.net/mlearning2/resources.shtml
4. Apple Inc. Apple Introduces the New iPhone 3G. Press release,
   http://www.apple.com/pr/library/2008/06/09iphone.html
5. Macworld.co.uk. iPhone 3GS upload limited to 384 Kbps upstream,
   http://www.macworld.co.uk/mac/news/index.cfm?newsid=26559
6. Apple Inc. Apple Announces the New iPhone 3GS—The Fastest, Most Powerful iPhone Yet, http://www.apple.com/pr/library/2009/06/08iphone.html
7. Apple Inc. iPhone: About Connections Settings,
   http://docs.info.apple.com/article.html?artnum=306249
8. Hu, Z., Gao, Q.: Ubiquitous Learning Application Environment based on JXTA. In: The First IEEE International Conference on Ubi-media Computing, pp. 355–358 (2008)

Research of Improved Frame-Slot ALOHA Anti-collision Algorithm[*]

Shuo Feng, Fei Gao, Yanming Xue, and Heng Li

School of Information and Electronics, Beijing Institute of Technology,
Beijing, China
fengshuolove2008@gmail.com, gaofei@bit.edu.cn
xueym@bit.edu.cn, liheng1109@163.com

Abstract. In order to further improve identification efficiency of tag anti-collision algorithm in RFID system, we analyze the performance the frame-slot ALOHA anti-collision algorithm, apply the idea of grouping in binary search algorithm to frame-slot ALOHA anti-collision algorithm, and propose an improved frame-slot ALOHA anti-collision algorithm combining frame-slot ALOHA anti-collision algorithm and binary search algorithm, identifying the collided tags in time slots with the binary search algorithm. The simulation results show that, the improved algorithm does improve the anti-collision performance and the tag identification efficiency in RFID system.

Keywords: RFID, frame-slot ALOHA, binary search.

1 Introduction

In RFID system, to avoid the tags collide with each other in the data transmission process is a key technology. Domestic and foreign scholars have proposed many tag anti-collision algorithms, such as binary tree search algorithms with adaptivity [1] algorithms aiming to environment with high-density labels, anti-collision algorithms based on inquiry binary tree search, anti-collision algorithms based on energy-aware.

Frame-slot ALOHA algorithm and binary tree search algorithm are widely used in RFID system because of their simple design and practicality. However, there are shortcomings in each of them, so we combine them to get better performance.

2 Analysis of Frame-Slot ALOHA Anti-collision Algorithm

The main idea of frame-slot ALOHA algorithm is that, divide time-domain into multiple discrete time slots of the same size, then packages N time slots into a

---

[*] Fund project: 863 project: Research of Communication Test Technology in RFID (2006AA04A106).

frame, and the tag randomly send a message for only one time in every N time slots.[2] The algorithm needs synchronization operation between the reader and tags, and it is a random time-division multiplexing address (TD-MA) anti-collision algorithm which is controlled by the reader (Reader-Driven). Identification Process of frame-slot ALOHA algorithm is shown in Table1.

Table 1. Identification process of Frame-Slot ALOHA Algorithm

	TS1	TS2	TS3
Tag1	10100011		
Tag2		11011010	
Tag3		11001010	
Tag4	11100111		
Tag5			11101101

In table 1, tag1 and tag4 collide with each other in TS1, tag2 and tag3 collide in TS2, only tag5 succeeds in sending data in TS3. Since collisions occur, tag1, tag2, tag3 and tag4 will continue to send data controlled by the reader in the following frames, until all the tags are correctly identified by the reader.

When there are only a few tags, with the increase in time-slot, it's easy to solve the anti-collision problem. However, too many time slots will not be used, wasting the valuable channel resources in vain, this situation will not be considered [3].

When there are quite a lot of tags, this algorithm is much better than ALOHA algorithm, its maximum throughput is twice as much as ALOHA. However, because of the large amount of tags, the tags 1 to 4 collided in TS1 and TS2 will still need to send data to the reader. Therefore, there is still large collision probability, and perhaps there would be data collisions in almost every time slot. The amount of the frame slots is fixed, so the collisions will occur successively in the following frames. In the end, in a very long period of time, the readers can't complete the recognition of the tags within the coverage area. So, frame slot ALOHA algorithm is not useful when there are large amount of tags.

3 Improved Frame-Slot ALOHA Algorithm

Frame-slot ALOHA algorithm has the shortcomings above, so it needs improve. The binary search algorithm uses a series of instruction sequences, being able to control the tags which respond, so we consider to apply the idea of binary search algorithm to the frame slot ALOHA algorithm.

3.1 Binary Search Algorithm Theory

Binary search algorithm uses the reader control anti-collision method, and its basic idea is by defining a group of required instruction sequences between the reader and a number of responders, select only one responder and finish the exchange of data between them.[4] Now we describe the instruction sequences of binary search algorithm in detail.

1. REQUEST (SNR): This command sends a sequence number as a parameter to the responders, who will compare their own SNR with the SNR they received. If less than or equal to, then the responder will send its own SNR back to the reader, thus reducing the scope of the pre-selection of the responder.
2. SELECT (SNR): Send a certain (pre-determined) SNR as a parameter to the responder, then the responder with the same SNR will be selected. While the responders with other SNRs will only respond to REQUEST command.
3. READ-DATA: The responder selected will send the data stored to the reader (in the actual system, there are also identification or input, cashier registration, cancellation, etc. commands)
4. UNSELECT: Cancel pre-selected responders, the responder goes into the 'silence' state. In this situation, the responder is totally non-active, doesn't respond the received REQUEST command. In order to re-activate the responder, the respond must temporarily leave the range of the reader (equivalent to no supply voltage), in order to implement reset.

Below we will give an example to describe the binary search algorithm.

In the range of the reader, there are two responds with the unique serial number of their each own are: 10110010 and 10101010. First, readers send REQUEST (1111 1111) command, and the two responders in the range of the reader all respond, then the reader will receive a serial number : 101* *010. There are collisions in the 3rd and 4th bits, and the highest level the collision occurred is the 4th bit, so the reader send REQUEST (1010 1111) (the variation of parameters of SNR is that, set the highest collision location 0, and following bits all 1), at this time, only 10101010 will respond, and send its own serial number back to the reader. Then the reader send SELECT (1010 1010) command to select the respond, and send READ-DATA to read responder data, and at last the reader sends UNSELECT command to make it in a 'silent' state. Then take go-back strategies, the reader continue to send REQUEST (1111 1111) command, at this time, there is only one responder 1011 0010 is in the range of the reader and will respond. When the reader send REQUEST (1111 1111) command, while there is no response, then the reader can go into hibernation. Now collisions have been effectively solved, and data in responders are fully read out.

3.2 Improved Frame-Slot ALOHA Algorithm

Apply the above binary search algorithm to the frame-slot ALOHA algorithm, because tag1 and tag4 collide in TS1, and tag2 and tag3 collide in TS2, in the following frames, tag1,tag2,tag3 and tag4 will continue to send data controlled by the reader [5].

In TS1, tag1 and tag4 collide with each other. First, readers send REQUEST (1111 1111) command to tag1 and tag4, making the two tags respond, while UNSLECT command to the tags collide in other time-slots, making them 'silent'. Then the reader will receive the serial number: 1*10 0*11, detecting there are collisions in the 2nd and 6th bits. The highest bit collide is 6th, then the reader will send REQUEST (1011 1111), at this time, only 1010 0011 will respond and send its own serial number back to the reader. Then the reader selects the responder, and sends READ-DATA to read the responder data, and at last the reader sends UN-SELECT command to make it in a 'silent' state. Then take go-back strategies, the reader continue to send REQUEST (1111 1111) command, at this time, there is only one responder 1110 0111 is in the range of the reader and will respond. When the reader send REQUEST (1111 1111) command, while there is no response, then the reader can go into hibernation. Now collisions have been effectively solved, and data in responders are fully read out.

Similarly, use the same algorithm in TS2, its anti-collision process is as follows: First, readers send REQUEST (1111 1111) command to tag2 and tag3, making the two tags respond, while UNSLECT command to the tags collide in other time-slots, making them 'silent'. Then the reader will receive the serial number: 110* 1010, detecting there is collision in the 4th bits. The highest bit collide is 6th, then the reader will send REQUEST (1100 1111), at this time, only Tag3 (1100 1010) will respond and send its own serial number back to the reader. Then collision in TS2 is solved.

3.3 Algorithm Performance Analysis

When the amount of the tag is very large, because all the tags are distributed into N time slots, the probability to occur collision in each slot is significantly reduced. Thus, dividing all the tags into groups according to different time slots collision occurred, and then using the binary search algorithm for each group to detect collision, the performance of the system will be improved. Using MATLAB to carry out simulations, assuming that the length of the EPC tag is 64 bits, each tag ID is randomly distributed, then the Figure 1 shows that, as the increase in the number of tags, the variety of time slots used in frame-ALOHA algorithm and an improved frame-ALOHA anti-collision algorithm. We can see that, when the amount of the tags is small, the advantage of improved algorithm is not obvious. However, with the increase in the number, the improved algorithm of grouping tags will show its apparent advantages, that is, the number of time slots used is obviously much fewer.

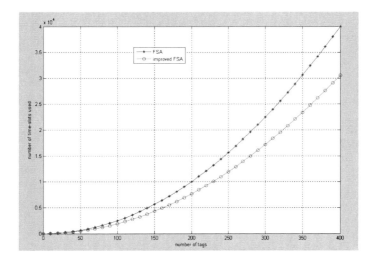

Fig. 1. Contrast between FSA and improved FSA

4 Prospect

The algorithm analyzes the lack of frame slot ALOHA algorithm, basing on it and combining the grouping ideas of binary search algorithm, when there are a large amount of tags, grouping the tags according to the different time slots they collided, and then use the binary search algorithm in each group, so the probability to collide in each group is much smaller than before, and so we can use fewer time slots to achieve the purpose of collision avoidance. This algorithm is a significant improvement than frame slot ALOHA algorithm.

References

1. Finkenzeller, K.: RFID-HANDBOOK, fundamentals and applications in contactless smart cards identification, 2nd edn. Wiley& Sons LTD., Chichester (2003)
2. Burdet, L.: RFID multiple access methods, Technical Report ETH Zurich (2004)
3. Vogt, H.: Mutiple Objet Identification with Passive RFID Tags Systems, Man and Cybernetics. In: 2002 IEE International Conference, pp. 6–9 (2002)
4. Kalinowski, R., Latteux, M., Simpwt, D.: An adaptive anti-collision protocolfor smart labels (2001)
5. Marrocco, G., Fonte, A., Bardati, F.: Evolutionary design of miniaturizedmeander- line antennas for RFID applications. In: IEEE Antennas andPropagation Society International Symposium, vol. 1(2), pp. 362–365 (2002)

Research on Transmission of Power Telecontrol Information Based on IEC 61850/OPC

Changming Zhang, Yan Wang, and Zheng Li

Department of Computer Science and technology,
North China Electric Power University, Yonghuabei Street,
Baoding, Hebei, P.R. China
{zcm19740405,wanyan1206}@126.com, yeziperfect@163.com

Abstract. The traditional means of power telecontrol communication must be changed because of the use of IEC 61850 standards in Substation communication. Considering the integration of IEC 61850 standards and IEC 61970, this paper presents an IEC 61850/OPC-based telecontrol communication model. The telecontrol information can be exchanged through ACSI/MMS services between IEDs in substation and OPC server in control center. The OPC server performs the mapping of IEC 61850 information models and OPC information model, and publishes information to SCADA applications which run as OPC clients. A prototype system is built to verify the feasibility of the model. It is proved that this system can meet the real-time requirements. It is different from gateway and need no complex protocol conversion.

Keywords: power telecontol system, IEC 61850, OPC, manufacturing message specification, Component Interface Specification.

1 Introduction

With the advent of IEC 61850 standard series, more and more substation automation equipment manufacturers begin to develop IEDs based on IEC 61850. Many substation have been recast according to IEC 61850 standard. It's an inevitable trend for substation automation system to comply to IEC 61850 standard. According to the report of IEC TC 57 SPAG, IEC 61850 has been chosen as foundation of the future seamless telecontrol communication architecture and be applied for modeling control center view of substation and for communication between substations and control centers. In addition, IEC 62445-1 (Use of IEC 61850 for the communication between substations), IEC 62445-2 (Use of IEC 61850 for the communication between control centers and substations), and IEC 62445-3 (Mapping of IEC 61850 based Common Data Classes, information addressing, services onto IEC 60870-5-104/101) are preparing in IEC TC57. IEC61850 will be the exclusive communication protocol used in substations and control center.

For the standards related to telecontrol have not been published, some transitional method are used to exchange data between substations that based on IEC 61850 and control center. Converting IEC 61850 standards to traditional telecontrol protocol by gateway is the most popular solution. When there are a variety of telecontrol protocols, multiple protocol conversion should be implemented. In addition, when IEC 61970 applies to EMS, this approach has to be changed again. So a power telecontrol communication model that based on IEC 61850/OPC is presented in this paper. It avoids the complex protocol conversion, and it is accordance with the IEC 61970 that adopts OPC as part of its component interface specification (CIS).

2 Communication Model of Power Telecontrol System Based on IEC 61850/OPC

With the development of computer network and communication technology, the power data communication network based on Asynchronous Transfer Mode (ATM) and Synchronous Digital Hierarchy (SDH) has taken shape in China. Intelligent Electronic Devices (IEDs) in substations and SCADA and computers in control center are connected through ATM, SDH and Ethernet in a rapid and reliable manner. Therefore, the communication speed bottleneck problems caused by the traditional front-end communication computer can be solved, and the communication reliability is much higher than that of traditional telecontrol system. All of this made the transmission of telecontrol data through network possible. Although the data exchange between substation and control center is not defined in IEC 61850, IEC 61850, as the only substation communication standard based on network platform in the world, will become the only communication protocol between substation and control center.

A set of object-oriented abstract communication service interface (ACSI) is defined in part 7 of IEC 61850, which is independent of concrete communication protocol. The core of the ACSI is mapped to the Manufacturing Message Specification (MMS). The WAN scoped transmission of power telecontrol information can be realized using IEC 61850 and MMS. Meanwhile, IEC 61970 standard series is proposed by IEC TC 57, which is used for the power system management and information exchange. The Common Information Model (CIM) and Component Interface Specification (CIS) defined by IEC 61850 make the Energy Management System (EMS) applications component-based and open that make it easy to implement plug-and-play and interconnection, and it is helpful to reduce the cost of system integration. At present the coordination of IEC 61850 and IEC 61970 standards is in progress. So the possibility to integrate IEC 61850 and IEC 61970 is discussed in this paper, and a power telecontrol communication model based on IEC 61850/OPC is presented, as shown in Fig. 1.

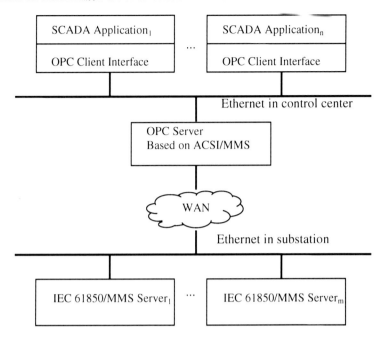

Fig. 1. IEC 61850/OPC-based telecontrol communication scheme

2.1 Network Structure of IEC 61850/OPC-Based Power Telecontrol System

At present, Ethernet has become the mainstream network to implement IEC 61850 protocol. MMS+TCP/IP+Ethernet have become the preferred mode to realize the communication between substation and control center. This architecture can combine OSI and TCP/IP organically. The hierarchy OSI has great clarity of style and it makes the resource subnet can be changed easily for the reason of development of technology and increase of function. On the other hand, TCP/IP is easy to implement and efficient. So it is adopted in many fields and is regard as de facto communication standard.

In the IEC 61850/OPC-based communication model, Ethernet or Optical Network, which is based on TCP/IP, is adopted to connect substation level and bay level. The core services of ACSI are mapped to MMS to exchange real-time data. The network between bay level and process level is one-way Ethernet and used to achieve analog sampling. SDH or Optical Network is used to connect substation and control center. The OPC server located in control center can exchange data with the IEDs in substation directly with ACSI/MMS API. The SCADA applications based on OPC can access the OPC server in a uniform method.

2.2 Real-Time Communication Standards

The core services of ACSI are mapped to MMS to realize data exchange that based on Client/Server mode. While MMS is an application layer communication protocol which is used in industrial control systems. It provides real-time data and control information exchange capability for interconnected IEDs or computers. MMS is selected as the underlying real-time communication protocol. This specification is independent of the concrete realization of application functions so that the devices from different vendors are interoperable and such character may meet the requirement of real-time of the telecontrol communication perfectly. Also it makes system integration easier and convenient.

OPC is a popular interface standard in industrial control field, which provides real-time data access, event alarm and historical data access, etc. OPC DA is adopted by WG 13 of TC 57 for real-time data access.

2.3 ACSI/MMS-Based OPC Server

There are many kinds of automation device in electric power system and the related protocol is various. Therefore, no matter what kind of model, master control units are indispensable in power telecontrol system. They gather the information from field devices and publish the information to the SCADA applications. OPC, as a communication standard defined for industrial control, can provide uniform communication service interface for applications in control center, and shield the differences of information model of different communication protocols.

The ACSI-based OPC server is the core in the above model. OPC server, as a software module of master control unit, receives the real-time information from the IEDs based on IEC 61850, and published it to the SCADA applications in the control center through uniform interfaces. An OPC server can be consist by three kinds of objects-server object, group object and item object. The OPC server should have the ability to map the ACSI/MMS objects to OPC item objects. Then the current OPC-based SCADA applications can support IEC 61850 standard without change. The design of OPC server will be discussed in the rest of this paper.

3 Design of ACSI/MMS-Based OPC Server

In the real-time communication process, the IEC 61850 data model is mapped to MMS information and transported to the OPC server located in control center using MMS service. Then the IEC 61850 data is mapped to OPC item. The SCADA applications can access the data from substation as an OPC client. The software structure of ACSI/MMS-based OPC server is shown as Fig. 2.

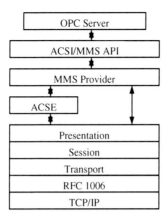

Fig. 2. Software architecture of ACSI/MMS-based OPC server

3.1 MMS Service Provider

The exchange of primitives between MMS client and MMS server is implemented by MMS service provider. The provider receives MMS service request and sends the request to proper object. The selected object executes the corresponding service. The service result is sent to MMS client by MMS service provider. When a service request arrives, the service provider creates a transaction object and checks the service execution conditions. If the conditions are not met, the transaction object is put into the queue of task, and then the other arrived service request will be handled.

The most important part of MMS service provider is MMS protocol machine (MMPM). The layered approach was used in the development of MMPM. The whole protocol stack can be separated into several Dynamic Link Library (DLL) according to their different functions, such as transmission layer, session layer, presentation layer, Association Control Service Element (ACSE) layer, MMS layer, and the protocol stack can also provide APIs of MMS and ACSI for the OPC server. ACSE layer, presentation layer and session layer can be develop under ISO development environment(ISODE) and the transport layer can be implement by the use of RFC 1006 specification over the TCP/IP protocol suite.

3.2 ACSI/MMS Service API

The service mapping from ACSI to MMS is implemented in ACSI/MMS service API and the API is provided to OPC server. There is no one-one relation between ACSI service and MMS service and the mapping is very complex. The simplest case is mapping an ACSI services to a unique MMS services. For example, Associate services of ACSI can be mapped to Initiate of MMS directly. In many cases an ACSI services need to map to multiple MMS service. For example, GetServerDirectory services of ACSI should be mapped to different MMS service when one of its parameter called ObjectClass has different value. When the value is

LOGICAL-DEVICE, it is mapped to the GetNameList services of MMS. When the value is FILE, it is mapped to the FileDirectory services of MMS. In addition, an ACSI services may also be mapped to one MMS service, but when the ACSI service is called once, the MMS service need to be executed many times. One ACSI service may also be mapped to a combination of several MMS services.

3.3 OPC Server

The interface IDs and input/output parameters of member functions are defined strictly in OPC specification. Many development tools can be used to implement an OPC server, such as OPCTEST (OPCDA Server Tool), which encapsulates the kernel functions of OPC server. The data source mapping module should be realized to develop a domain-specific OPC server. When OPC technology is used in power telecontrol system, the data source is IEC 61850-based IEDs located in substation. So the OPC server should have ability to identify the IEC 61850 data information. The mapping of information model from IEC 61850 to OPC can be achieved by parsing IEC 61850 configuration document.

With regard to hierarchy of information model, IEC 61850 mainly defines five-layer structure which consists of server, logical device (LD), data, data attributes (DA). As for the above-mentioned five kinds of objects, the former always contain the latte. One or more services are defined for each kind of objects. In hierarchy of OPC, three kinds of objects are defined. So more than one kind IEC 61850 objects may be mapped to the same OPC object class, as shown in Table 1.

Table 1. Mapping IEC 61850 objects to OPC objects

IEC 61850 Objects	OPC Objects
Server	Server
LD	Group
LN	Item
Data	Item
DA	Item
Data Set	Item List
Control Block	Item
Log	Log
File	File

Substation Configuration Language (SCL) is used to describe the information and service model of substation IEDs in electric power dispatch system. The configuration document consists of XML and XML Schema document. Therefore, in order to make OPC server understand the information from underlying devices, a XML parser is necessary in the OPC server. The configuration file written in SCL is parsed and a DOM tree is created according to the analysis results. Then the information of IED can be read and used to configure the variable storage space of OPC. Then the information structure of OPC can be built by traversing the DOM tree and then corresponding items can be added to the OPC server.

4 A Prototype System and Tests

A prototype system has been built based on proposed model and scheme to in laboratory environment. Firstly, an IEC 61850 server is built that based on ACSI/MMS to simulate the behavior of an IED. Then a simple ACSI/MMS-based OPC DA server is used as the master control unit of control center. Since any well defined OPC client can access OPC servers in the same way, OPCTest, which is developed by Rockwell Software for the purpose of testing OLE for Process Control (OPC), is used as an SCADA application.

This model has proved feasible and it can provide bi-direction communication between control center and substation. The uniform interfaces of OPC server have the ability to provide real-time data exchange service for the OPC-based applications conveniently. The experimental results observed from OPC client is shown as Fig. 3.

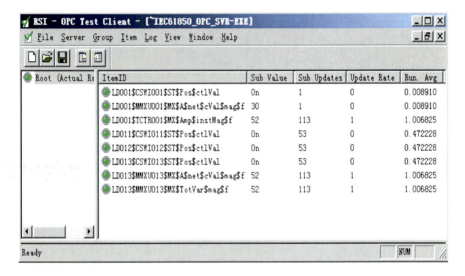

Fig. 3. Experimental results observed from OPC client

5 Conclusion

The IEC 61850/OPC-based telecontrol communication model presented in this paper combines the advantages of IEC 61850 and OPC and makes the system extension easier. A prototype system is given to verify the feasibility of the model. IEDs in substations and OPC server in control center are connected through ATM, SDH and Ethernet in a rapid and reliable manner and the communication reliability is much higher than that of traditional telecontrol system. Therefore, this architecture can meet the system requirements of real-time data exchange.

Acknowledgment. This research is supported by the youth research fund of North China Electric Power University (200811018).

References

1. Wenshu, T.: Seamless telecontrol communication architecture. Power System Technology 25, 7–10 (2001)
2. Zaijun, W., Minqiang, H.: Research on a substation automation system based on IEC 61850. Power System Technology 27, 61–65 (2003)
3. IEC61850, Communication networks and systems in substations. IEC Standard (2004)
4. Kostic, T., Preiss, O., Frei, C.: Towards the Formal Integration of Two Upcoming Standards: IEC 61970 and IEC 61 850. In: Power Engineering, 2003 Large Engineering Systems Conference, pp. 24–29 (2003)
5. Bingquan, Z., Yanming, R., Jianning, J., et al.: Strategy for implementation of IEC 61850 in substation automation system during transitional. Automation of Electric Power Systems 29, 54–57 (2005)
6. Energy Management System Application Program Interface (EMSAPI), Part 401: Component Interface Specification (CIS) Framework, http://webstore.iec.ch/webstore/webstore.nsf/artnum/034755?opendocument
7. Liu, J., Lim, K.W., Ho, W.K., Tan, K.C., Tay, A., Srinivasan, R.: Using the OPC standard for real-time process monitoring and control. IEEE Software 22(6), 54–59 (2005)
8. Wang, D., Zhu, Y., Liu, P., Huang, J., Zhao, W.: Research on distributed transmission of power telecontrol information based on ACSI-MMS. In: 3rd IEEE Conference on Industrial Electronics and Applications, ICIEA 2008, pp. 670–674 (2008)
9. Mercurio, A., Di Giorgio, A., Cioci, P.: Open-Source Implementation of Monitoring and Controlling Services for EMS/SCADA Systems by Means of Web Services— IEC 61850 and IEC 61970 Standards. IEEE Transactions on Power Delivery 24(3), 1148–1153 (2009)

E-Commerce Leading Development Trend of New Sports Marketing

Zhongbin Yin[1], Binli Wang[1], and Lina Wang[2]

[1] Hebei University of Engineering
Handan, China
Liyunsheng337@sina.com
[2] No. 2 middle school of Handan city
Handan, China
wanglinaenglish@126.com

Abstract. The traditional sports marketing is based on sport used as a carrier to promote their products and brands which is a marketing activity as well as a means of marketing. E-commerce sports marketing is based on the traditional sports marketing, adding e-commerce support to re-arrange the advantages of e-commerce and traditional sports marketing, which, accordingly makes traditional sports marketing present more flexibility, communication and high-profit.

Keywords: E-commerce, sports marketing, trend.

1 Introduction

Since 2000, with the extensive application of electronic technology and the popularization of Internet, e-commerce and the development of industrial has achieved the deep integration. People's understanding of e-commerce gradually extends to E concept from e-commerce level [1]. As a business, e-commerce does not exist in isolation, which also exerts a strong impact on kinds of aspects, such as people's way of life, the government's economic policy, enterprise's operation and management, and social economic efficiency. In June 2007, the National Development and Reform Commission, the State Council Information Office jointly issued China's first e-commerce development planning ---"Eleventh Five-Year" plan of E-commerce development. " It clearly put forward the overall goal of e-commerce development in the period of China's "Eleventh Five-Year" is: By 2010, the environment of e-commerce development, support systems, technical services and promotion of the use of coordinated development pattern will have basically taken shape, as a result, e-commerce services will become an important new industry. Thereby national economic and social development level of e-business applications in various fields will have achieved a substantial increase with the apparent results [2].SAIC is currently under an active investigation in preparation for the introduction of a "network market supervision and management approach." With the continuous improvement of the external environment of e-commerce, through the combination of government, society and the various vendors, the problems of the integrity, security, payment and logistics that have been plaguing the e-commerce are gradually being resolved and have achieved initial success. It also prompts a growing number of businesses to begin to understand the role of e-commerce, and then to get it started.

According to China Internet Network Information Center (CNNIC), in January 2009 the report of the development of China's Internet shows: by the end of 2008, the number of China's online shopping users has reached 74 million, accounting 24.8% of the proportion of Internet users. The Internet is regarded as a emerging market with tremendous untapped potential consumption. Nowadays with the increasing number of on-line users and the development of network market, as well as mergers, acquisitions and integration the Internet industry, e-commerce has entered a healthy development track. Traditional industry has also been widely applied, and e-commerce is to become more prevalent in our country. The Internet is not only associated with a particular industry, which can also affect every industry. Because the world has entered a new stage —"the stage of network society ". United Nations Trade and Development Conference said in "report": e-commerce and Internet will be widely used in the next few years. In Asia-Pacific region, China will become the dominant force in e-commerce development. On the other hand, China is building an e-commerce platform with Chinese characteristics.

2 Sports Marketing and Sports E-Commerce

2.1 The Concept of Sports Marketing

There are two aspects of the concept of sports marketing. From the sports itself, it refers to how to promote the industrialization and marketization of sports more deeply and effectively, so that it will gradually become operating mechanism with self-development and self-improvement ,which, will lead it be an independent operating body. While as far as the enterprises are concerned, as a promotional tool or image background, sports activities and sport itself make use of the title, sponsorship and other forms to promote their brand or establish a corporate image through the sponsorship of sports. Therefore their marketing purposes can be achieved [3].

2.2 Sports Concept of E-Commerce

Brenda G. Pitts said: Synchronization effects generated by the Internet on the sports industry have become a major commercial power. It creates a timely manner to exchange and develop sports trading. As one of the sport new products and new branches of the sports industry, it is added to the sports industry, which is known as the e-sports business (i.e., e-sport business) [4].

3 E-Commerce in the Development Trend of Sports Marketing

3.1 The Development Trend of E-Commerce Sports Marketing Application in Sporting Eevents

E-commerce sports marketing is based on the traditional sports marketing, adding e-commerce support to re-arrange the advantages of e-commerce and traditional sports marketing, which, accordingly makes traditional sports marketing present more flexibility, communication and high-profit. Because the traditional sports marketing is needed in multiple channels to promote sporting events themselves and bedding in order to achieve the purpose of interesting sports fans. When the sports fans gain the information, and sporting events are focused on, the corporate sponsors will be attracted to come to sponsor the events to complete their own marketing purposes. Meanwhile, in a special sports marketing e-commerce web site, it can also be used to sponsor sporting events, as well as sales of related products to combine the traditional sense of the sports marketing so as to maximize the interests of superposition. Because of the nature of widely spreading of e-commerce itself (based on the developed Internet), the emergence of e-sports marketing can also allow events promotion and sponsorship to achieve synchronization to form a one-stop sales results. As soon as the information is released, a huge number of first Internet users will access to this information, including the target sponsors. This will enable two information destinations access to information at the same time, which will make marketing efforts more efficiently, improve sports marketing efficiency and reduce the cost of sports marketing.

The advertising of e-commerce site includes the value of traditional advertising as well as the nature of widely spreading. Enterprises get the enterprise promotion and achieve the exchange of intangible assets and enterprises in the process of sponsoring a variety of events, of which the most important is to achieve the sports promotion by means of e-commerce marketing. That is to say, some enterprises are willing to pay for corporate sponsorship events, so that a variety of events can own the sources of market funding. For example, Samsung, Coca-Cola enterprises make use of the sports marketing to fully integrated brand positioning and sporting events. They integrate customers, consumers, and other multi-media resources in the minds of consumers to shape a clear brand image. In particular, they are influenced by large-scale international events and take the "brand internationalization" of the road to enhance corporate brand image. Therefore enterprise marketing cost of inputs is very impressive, and naturally, the corresponding gains are unknown. A variety of events take advantage of e-commerce in the application of sports marketing to reduce and mitigate the marketing inputs of costs and capital investment. Accordingly financial resource utilization in the unit is enhanced and the overall sales revenue is also raised.

3.2 The Development Trend of Application of E-Commerce Sports Marketing in Sports Business

E-commerce can help sports enterprises to open international markets. E-commerce breaks the boundaries of space. The product created by sports enterprises via the Internet is closely linked with consumers, manufacturers, suppliers and vendors to complete the entire information flow, logistics, and capital flow operation. Because the Internet is global in nature, companies that are engaged in sports products can timely access to the trade price information, supply and demand information, as well as the production material information, technical information and human resources information, in international markets. So these companies can become truly multinational. Namely, employees, partners, markets, and even competitors are internationally oriented, who come from around the world. Its direct impact is that they can rapidly turn around the business enterprise decision-making, and adjust the development direction and strategies, integrate various resources efficiently, which will effectively improve the ability of sporting goods companies' responding to and participating in international market competition. Sports enterprises will also have access to high profits and more market opportunities for development. Sports enterprises will also have access to high profits and more market opportunities for development. Meanwhile the strengths and weaknesses of business are also expected to be doubly enlarged and the large-scale production and management will focus on the advantageous enterprises.

E-commerce can promote sports enterprises and consumers to communicate directly. E-commerce also enables enterprises to make full use of the mass storage technology and resources of Internet to build customer relationship management system (CRM). Thus a kind of low-cost channel of communication is provided, which is closely linked to the direct communication with consumers, and then it will save the cost of communications without going through a middleman. Sports enterprises are able to track, query, maintain client relationships and business relationships at any time, which will realize personalized customer service effectively and increase customer satisfaction. It becomes possible for enterprises to develop tailor-made and one-to-one marketing. The interactive features of e-commerce can achieve cost-effective customer complaint handling system (CRM subsystem). Besides the problems can be detected timely and the quality of service for sports enterprise can also be improved. The use of e-commerce platform can create a virtual community, and increase the customer loyalty at the same time according to feedback information to adjust business programs. The nature of global of e-commerce makes the virtual community, or the customer relationship management (CRM) systems have a direct response to the global consumers, which will be directly involved in international competition. Sports enterprises thus can effectively enhance the international image and increase their international competitiveness [5].

E-commerce can effectively induce changes in the internal management of sports enterprises. The progress of sports enterprises in areas such as the information collection, dissemination, treatment makes inter-departmental co-operation more coordinated. The organizational structure will become flatter at the same time. SOHO business owners will become a common phenomenon in the future, and a large number of task-oriented virtual sports enterprises will emerge. For example, the company of Li Ning is a management innovation. Two special sectors in two peer companies are set up in Li Ning's organizational structure in the system, namely digital marketing and e-commerce department. The two departments are responsible for the operation of virtual communities, and Li Ning Direct Shop and online agents operating management. Therefore Li Ning's importance degree of the network market is evident. In April 2008, Li Ning established e-commerce and launched the directly managed brand's flagship stores and direct brand discounters in Taobao. There are more than 1000 sale of Li Ning products shops in Taobao and other e-commerce platforms at this time. Li Ning's e-commerce section makes use of dealer system to incorporate the existing storefronts to solve the mixed network and difficult situation. In 2009, all on-line shops will use CI and VI provided by Li Ning, which will make a unified marketing theme.

4 Problems of E-Commerce in the Development of Sports Marketing

4.1 Sports Industry Model Lacking Innovation

In e-commerce of recent years, China's sports industry is also appearing a lot of e-commerce sites. However, most e-commerce sites are taking such a path "great hype to attract the public, the fight for advertising, market means for misappropriating". Because the online sports trading volume is too low, its income is insufficient to maintain the daily operation. Most sites have to rely on foreign capital to bring in invest. Therefore, since the second half of 2000, under the influence of the rapid decline in the U.S. Nasdaq index, many Chinese sports enterprises have been facing the surviving crisis, layoffs, and closure. Faced with the profound lesson, Internet companies need to reconsider their position and return to the track of "profit center".

4.2 Low Level of Information of Sport Enterprise

Enterprise information is the basis of development of e-commerce. Enterprise Information develops backward seriously so that it has hampered the development of e-commerce in China. As the main body of e-commerce, enterprises' business processes and information of management process is a necessary prerequisite for enterprises to develop e-commerce. At present, about 70% of the Internet sporting goods businesses have a certain degree of information tools or proceed to realize enterprise information.

4.3 Shortage of High-Level Talent in Sports Industry

Although the Tsinghua University, Zhejiang University and other colleges and universities begin to develop specialized e-commerce professionals, however, at present an acute shortage of talent with innovative thinking of the theory of e-commerce, planning and management, which of course is another problem of the development of electronic commerce. Particularly the design of professional subjects of e-commerce is an interdisciplinary field, which is related to computers, economics, management and legal aspects. Therefore, the shortage of human resources in this field leads to stagnation in e-commerce of China's sports industry.

Acknowledgments. We are grateful to the support of Hebei University of Engineering and the Science and Technology Bureau of Handan City.

References

1. Jian, J., Ting Zhuo, C.: E-Commerce Introduction, p. 13. People's Posts & Telecom Press, China (2009)
2. Chen, Y., Xiaoming Meng, C.: E-commerce and Internet marketing, p. 17. Electronics Industry Press, China (2008)
3. Huiqiao Duan, T.: Digital-era sports consumption and marketing, vol. 11, pp. 60–61. Sports Culture Guide, China (2006)
4. Baojun Gheng, T.: On the Modern Olympic Movement and sports E-commerce. Journal of Coal Management Institute, China 2, 208–209 (2009)
5. Chen, H., Chen Zhang, T.: Based on e-sports industry information countermeasures, vol. 4, pp. 48–50. Technology Square, China (2009)

A Modified Robust Image Hashing Using Fractional Fourier Transform for Image Retrieval

Delong Cui, Jinglong Zuo, and Ming Xiao

College of Computer and Electronic Information, Maoming University, Maoming China
delongcui@163.com, oklong@gmail.com, xiaoming1968@163.com

Abstract. In order to improve the security and robustness of image retrieval, a novel image hashing algorithm based on fractional Fourier transform (FRFT) for content-based image retrieval (CBIR) application is proposed in this paper. By employing FRFT and normalized Hamming distance, a robust hashing sequence is obtained by preprocessing, feature extracting and post processing. The security of proposed algorithm is totally depended on the orders of FRFT which are saved as secret keys. For illustration, several benchmark images are utilized to show the feasibility of the image hashing algorithm. Experimental results show that the proposed scheme is robust against perceptually acceptable modifications to the image such as JPEG compression, mid-filtering, and rotation. Therefore, the scheme proposed in this paper is suitable for CBIR application.

Keywords: image hashing, FRFT, CBIR, LDO, normalized Hamming distance, encryption keys.

1 Introduction

With the rapid development of information-communication and personal computers, copyright protection of digital media as image, video and audio becomes a more and more important issue. Image hash as one of content-based image authentication techniques has become an important research topic recently. Image hash can be used in authentication, content-based image retrieval and digital watermarking.

Security and robustness are two important requirements for image hash functions. By security, it means that one image should have different function values according to the different applications. By robustness, it means that the hash function should keep invariable by common image processing operations such as additive noise, filtering, compression, etc. Traditional cryptographic hashes such as SHA-1 and MD5 produce uniformly distributed hash values. They are not applicable in the aforementioned multimedia applications because they are extremely sensitive to the message being hashed; i.e., even a one bit change in the input media will change the output hash dramatically.

In this paper, an image hashing algorithm based on fractional Fourier transform (FRFT) is proposed for content-based image retrieval (CBIR) application. The scheme proposed in this paper extracts a robust feature vector to generate a content-based hash sequence, which includes three-step preprocessing, feature generation and post processing. To improve the security of the proposed scheme, the orders of FRFT are used as the encryption keys. And the normalized Hamming distance is employed to measure the performance of the proposed image hash arithmetic. Experimental results show that the proposed scheme is robust against common image processing such as JPEG compression, mid-filtering, and rotation. Therefore, the scheme proposed in this paper is suitable for CBIR application. The remaining of this paper is organized as follows. The concept of FRFT is briefly introduced in section 2. In section 3, the details of the proposed scheme and some experimental results are presented. A conclusion is drawn in section 4.

2 Fractional Fourier Transform

The fractional Fourier transform (FRFT) is the generalization of the classical Fourier transform (FT). It depends on a parameter α and can be interpreted as a rotation by an angle α in the time-frequency plane or decomposition of the signal in terms of chirps. FRFT is not only richer in theory and more flexible in application, but is also not expensive in implementation. Recently the application of FRFT has attracted so many researches. The application areas include optics, wave propagation, signal analysis and processing, etc.

If FT can be considered a linear differential operator (LDO) acting on a signal, than FRFT generalizes LDO by letting it depend on a continuous parameter α. The definition of FRFT can be given as follows [1,2]:

$$x_p(u) = F_p(x)(u) = \int_{-\infty}^{+\infty} x(t) K_p(t,u) dt \quad (1)$$

where p is the rank of transform; $\alpha = p\pi/2$ is LDO; $Kp(t,u)$ is transform kernel.

$$K_p(t,u) = \begin{cases} \sqrt{(1-j\cot\alpha)} \exp j\pi \cdot \\ (t^2 \cot\alpha - 2ut\csc\alpha + u^2 \cot\alpha), \alpha \neq n\pi \\ \delta(t-u), \alpha = 2n\pi \\ \delta(t+u), \alpha = (2n+1)\pi \end{cases} \quad (2)$$

The inverse FRFT can be given as:

$$x(t) = F_{-p} x_p(t) = \int_{-\infty}^{+\infty} X_p(u) K_{-p}(t,u) du \quad (3)$$

Let $f(x)$ be a sampled periodic signal with a period Δ_0, the pth order discrete fractional Fourier transform (DFRFT)[3] of $f(x)$ can be obtained by using (1), given

$$f_p = \sum_{k=-\frac{N}{2}}^{\frac{N}{2}-1} f(k\frac{\Delta_0}{N}) \sum_{n=-\infty}^{\infty} K_p(x,(n+\frac{k}{N})\Delta_0) \tag{4}$$

The forward and inverse two-dimensional discrete fractional Fourier transform (2D-DFRFT) [3] of the image signal are computed as

$$F_{\alpha,\beta}(m,n) = \sum_{p=0}^{M-1}\sum_{q=0}^{N-1} f(p,q) K_{\alpha,\beta}(p,q,m,n) \tag{5}$$

$$f_{\alpha,\beta}(p,q) = \sum_{p=0}^{M-1}\sum_{q=0}^{N-1} F_{\alpha,\beta}(m,n) K_{-\alpha,-\beta}(p,q,m,n) \tag{6}$$

where (α,β) is the order of 2D-DFRFT, $K_{\alpha,\beta}(p,q,m,n) = K_\alpha \otimes K_\alpha$ is the transform kernel, and K_α, K_β are the one dimensional discrete fractional Fourier transform kernel.

3 Image Hashing Algorithms

In general, an image hash can be constructed by preprocessing, extracting and post processing appropriate image features. In order to improve the property of feature extracting, the preprocessing of image is always used. The common image preprocessing includes applying a low-pass filter, rescaling, or adjusting the components of image, and so on. To achieve robustness, security, and compactness, the feature extraction is the most important stage of constructing an image hash. A robust image feature extraction scheme should withstand various image processing that does not alter the semantic content. Various image hashing schemes mainly differ in the way randomized features and extracted. For post-processing, the aim is compression the length of hash sequence and without less the magnitude feature. In this paper, in order to improve the validity and veracity of image retrieval, a novel robust image hashing algorithm based on FRFT for CBIR application is proposed. The framework of proposed hashing algorithm is shown in Fig.1, which includes the following steps:

Preprocessing: rescale the input image to $N \times N$ and normalize the pixel value, than apply a low-pass filter on the image, finally block the image and the number of blocks is dependent on the length of image hash.

Feature generation: FRFT of every block using angles of (α, β), and the orders of FRFT are saved as a secret key. An image feature vector is obtained by the following steps: the bigger FRFT domain coefficients which contain the most energy of every block image are selected, and then sum up the coefficients which is only dependent on the block image content.

Post processing: normalize the vector, and quantize the resulting statistics vector according to the block energy and the size of block hash length, obtain the binary hash sequence finally.

Fig. 1. The framework of generating a hash

3.1 Performance Study and Comparison

Performance metrics and experiment setup: to measure the performance of image hash, the normalized Hamming distance between the binary hashes is employed. The defined of normalized Hamming distance is:

$$d(h_1, h_2) = \frac{1}{L}\sum_{k=1}^{L} |h_1(k) - h_2(k)| \qquad (7)$$

where $h_1(k)$, $h2(k)$ are different image hash sequence values; L is the length of image hash. The normalized Hamming distance d has the property that for dissimilar images, the expected of d is closed to 0.5, else the expected is closed to 0.

3.2 Robustness Analysis

Several benchmark images (such as Lena, Baboon, Peppers, F16, Cameraman, etc) are used to test the performance of the proposed scheme. The results under various attacks are drawn respectively in table 1. It can be observed that the performance of the proposed scheme is robust to common image processing.

Table 1. Normalized Hammming Distances of proposed algorithm between several benchmark image

Image	Boboon	Cameraman	F16	Fishingboat	Lena	Peppers
Boboon	0	0.4805	0.4453	0.4375	0.5	0.4570
Cameraman	0.4805	0	0.3633	0.3945	0.4258	0.5
F16	0.4453	0.3633	0	0.3359	0.4609	0.5273
Fishingboat	0.4375	0.3945	0.3359	0	0.4531	0.5430
Lena	0.5	0.4258	0.4609	0.4531	0	0.5117
Peppers	0.4570	0.5	0.5273	0.5430	0.5117	0

For comparison, three existing robust image hash schemes by Venkatesan et al. [4], by Fridrich [5], and by Mihçak [6] are chosen. The hash lengths in various schemes are shown in table 2. The comparison curves under various attacks are

A Modified Robust Image Hashing Using FRFT for Image Retrieval 313

drawn respectively in Fig.2-Fig.4. It can be observed that the performance of the proposed schemes under these attacks is comparable to the existing schemes, and the lengths of image hash are shortened.

Table 2. Hash lengths for various Hashing schemes

Hashing Method used	Hash Length
Venkatesan's scheme [4]	805
Fridrich's scheme [5]	420
Mihçak's Algorithm B[6]	1000
Proposed scheme	256

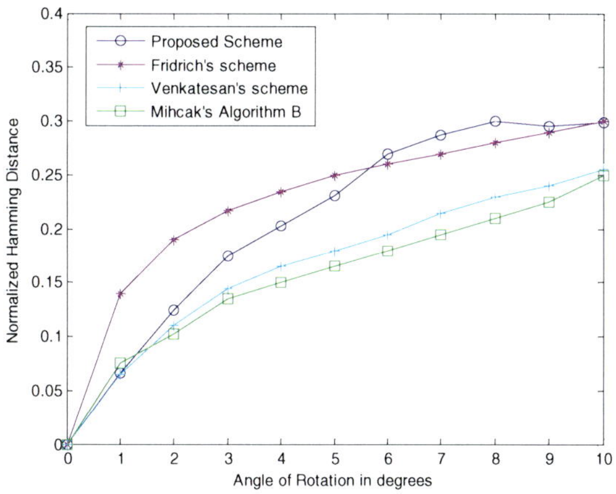

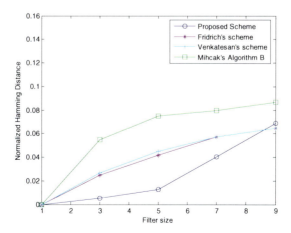

Fig. 2. Performance of various hashing schemes under rotation

Fig. 3. Performance of various hashing schemes under filtering

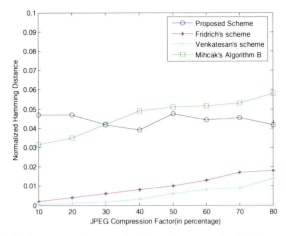

Fig. 4. Performance of various hashing schemes under JPEG compression

3.3 Security Analysis

The security is also measured in terms of the normalized Hamming distance of the image hash bits. About 400 transform orders and secret keys are generated from interval [0,1], and the 200th value is set equal to the original transform order and secret key. The security of proposed algorithm is total depended on the transform orders. The results are shown in Fig.5. It can be observed that the proposed algorithm provides a high level of security to image hashing for a given application.

Although proposed algorithm is robustness and security, but the method is not sensitive enough to detect the small-area tampering. This is because the feature vector extracted from the image is a global-based feature, which is unable to capture small changes in the image.

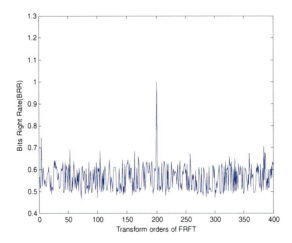

Fig. 5. Right bit rate under various FRFT orders

4 Conclusion

In this work, a novel robust image hash scheme for CBIR application is proposed. The FRFT is used for constructing robust image hashes, and the orders of FRFT are used as the encryption key. The image is first re-scaled to a fixed size and normalized to interval [0,1], then after the FRFT of every block image, feature vector is extracted from the transform field coefficients, finally the resulting statistics vector is quantized and the binary hashes sequence is obtained. Experimental results show that the proposed scheme is robust against common image processing such as JPEG compression, mid-filtering, and rotation. Therefore, the scheme proposed in this paper is suitable for CBIR application.

Image hashing techniques can also be extended to audio, video and other multimedia data. Further research on the FRFT-based image hash is in order. This includes reducing the length of hash sequence meanwhile the improvement of the hash performance, the relation between various parameters and the hash performance, improvement of detection robustness against less obvious tampering, and so forth.

Acknowledgments. The work presented in this paper was supported by Maoming Municipal Science & Technology Program (No.203624) and the PHD startup fund of Maoming University (No. 208054).

References

1. Pei, S.-C., Ding, J.-J.: Closd-form Discrete Fraction and Affine Fourier Transforms. J. IEEE Trans on Signal Processing 48(5), 1338–1353 (2000)
2. Ozaktas, H.M., Arikan, O.: Digital Computation of the Fractional Fourier Transform. J. IEEE Transactions on signal processing 9, 2141–2149 (1996)
3. Pei, S.C., Yeh, M.H.: Two Dimensional Discrete Fractional Fourier Transform. J. Signal Processing 67, 99–108 (1998)
4. Venkatesan, R., Koon, S.M., Jakubowski, M.H., Moulin, P.: Robust Image Hashing. In: IEEE Conf. on Image Processing, pp. 664–666 (2000)
5. Fridrich, J., Goljan, M.: Robust Hash Functions for Digital Watermarking. In: IEEE Int. Conf. Information Technology: Coding and Computing, pp. 178–183 (2000)
6. Mihcak, K., Venkatesan, R.: New Iterative Geometric Techniques for Robust Image Hashing. In: ACM Workshop on Security and Privacy in Digital Rights Management Workshop, pp. 13–21 (2001)

Short-Circuit Current Calculation of Distribution Network Based on the VDNAP

Yungao Gu[1], Yuexiao Han[2], Jian Li[3], and Chenghua Shi[4]

[1] Department of Physics and Electrical Engineering, Handan college, Handan, China
guyungao@126.com
[2] College of Information Engineering, Handan college, Handan, China
hanyuexiao@yahoo.cn
[3] Handan Polytechnic College, Handan, China
hdxykyc@sohu.com
[4] College of Economics and Management, Hebei university of engineering, Handan, China
shichenghua@163.com

Abstract. This article elaborates the establishment mechanism and realization method of distribution network analysis platform VDNAP (Visual Distributed Network Analysis Platform) based on geographic information systems relatively deeply, based on describing the method of short-circuit current calculation on superposition principle and then use the method to carry out short-circuit current calculation in VDNAP with taking a single line of certain urban areas for example, which achieve the effective combination of VDNAP and short-circuit current algorithm, and avoid the tedious work of user manual modeling, using the method of short-circuit current calculation based on superposition principle overcome the difficulties encountered when the traditional symmetrical component method is applied to distribution system short-circuit current calculation.

Keywords: electric distribution network, VDNAP, short-circuit current calculation, superposition principle.

1 Introduction

Short-circuit is a common and very serious fault in power system. Short-circuit fault makes the system voltage lower and current increase significantly in the loop of short circuit current, which seriously affect the stable operation of power system and cause electrical equipment damage[1]. Short-circuit current calculation is often performed in power system planning design and operation. If using the hand calculations, it is time-consuming and labor-intensive and prone to error. The initial conditions of short-circuit current calculation software practically applied required the user to manually input in the way of graphic or data, consuming a lot of effort. For the above issues we propose that the short-circuit current calculation based on the superposition principle [2] algorithm will be applied to distribution network model analysis platform VDNAP, and it does not require to model or input data as its initial calculations are completely extracted from the database, bringing great convenience to the operator and having a high accuracy.

2 Establishment of Distribution Network Model Platform VDNAP Based on Gis

VDNAP (Visual Distributed Network Analysis Platform) as the electrical distribution network graphics output window shows users the Graphical Distribution Network Electrical Wiring Diagram generated by Power GIS Conversion, which have set up the electrical graphical browsing, power flow calculation, short-circuit current calculation, Distribution network reconfiguration and other functions, and on this platform the user can achieve the conversion from Geographic graphic to electrical Graphics ,and the analysis and calculation of a variety of applications of Distribution network, and its operating process is easy and intuitive.

2.1 The Establishment Methods and Significance of VDNAP

GIS is in store for geographically graphics, but for the short-circuit current calculation what the user care about is the electrical graphics, so to achieve the conversion geographical map graphics to electrical requires to create a platform for electrical diagram. Platforms need to set up graphics primitives corresponding with the equipment of GIS (such as substations, lines, circuit breakers, transformers, surge arresters, etc.), and the equipment that electrical calculation does not need can be used as nodes to deal with. VDNAP using VC + + language implementation, concrete realization is as follows [3];

(1) use VC + +6.0 to create a MFC multi-document projects.

(2) Set pixel electrical components in it to achieve image element in the document display.

(3) Each electrical components is Object,and all have inherited the CComponent derived from the CObject class. Line is another type, and has no inheritance CComponent class.

(4) Respectively set for a single endpoint 𝟙 One or more pairs of endpoint endpoint pixel, according to the characteristics of electrical components,and establish connections among each other through its endpoints.

(5) Each document corresponds to one view,and each document has three lists: CObList (parts list) used to store electrical components graphics; CNodeList (node list) used to store electrical nodes in the graph; CLineList (line list) used to store electrical circuit graph.

(6) Add the components one by one to the appropriate list in the painting for later use of querying device .

(7) Each component has a CRectTracker, which is used for mobile use. Achieve it by calling the drag of CRectTracker. For the drag line , the line is divided into a number of box lines, and then each box has a CRectTraker tracking, used to drag, and then move other parts accordingly.

(8) Querying and modifying its properties can be realized by achieving double-clicking electrical components in the graph .

(9) We can carry out a second model of the electrical graph and save it.

VDNAP establishment has created a stand-alone graphics platform, and the platform provides abundant electrical parts distribution network primitives, thus the user can alone achieve Electrical Distribution Network Graphics Modeling (single-line diagrams, station house plans, etc.)on this platform. We can realize querying and modifying device Properties in the VDNAP with the best graphics processing capabilities, and VDNAP can also achieve the judgment of connectivity between the various components through the three (COblist, CNodelist and CLinelist) , laying the foundation for the future distribution network application analysis, while the establishment of VDNAP provides the premise for achieving graphics conversion from geographical graphic to electrical.

2.2 VDNAP Interconnects with GIS

GIS system contains a lot of data information, from which transforming the data we need is the basis of the establishment of distribution network model in VDNAP. As the original GIS data is SQL Server and DB2 attributes graphics library database,we can extract data from the GIS and put it into two parts graphics transformation and properties transformation, and the transformed data is put into a unified intermediate database we have designed, as shown in Figure 1.

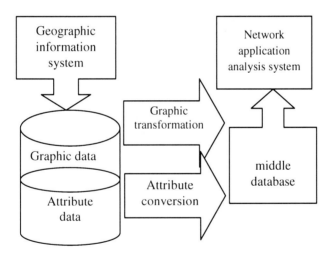

Fig. 1. Graphical transformation and properties transformation based on GIS

Grap Transformation

The central theme of Graph transformation is to read logo names and coordinate information of graphics components from the GIS graphics library so as to model in drawing platform.

Graphics library information is stored in graphical form, so first of all we should turn them into graphics files (shape files)that are easy to read, and then get elements of the document name and coordinates and put them into the database. Turn all single-line diagrams and station plans in GIS System out into shape files, and show the shape file through the MapObjects (control) and read the coordinate value of each element of the shape file.

The coordinates of the different elements indicate different methods: point coordinate values (x, y), the line coordinates value $((x_1, y_1), (x_2, y_2))$, surface coordinate value $((x_1, y_1), (x_2, y_2), (x_3, y_3), (x_4, y_4))$, combining the ID number of each element generates a corresponding record ▯ and the table structure got is: ID x_1 y_1 x_2 y_2.

In general, the load, capacitor, current / voltage transformers belong to the form of point , and lines, switches, transformers, etc are in the form of lines, and what are the form of transformer substations, SSP and so on is surface. However, in the original shape file in the GIS, the switches, transformers, etc are all in the form of points, because in the GIS, we concerned only about the geographical location of these components rather than the connection between them. However, the graphics we had converted as the electrical wiring diagram, must be able to vividly expressed the connection and status of various electrical components in the circuit, such as opening and closing state of the switch. Therefore, you need to convert these points into the form of line ,that is to say you need to change the coordinate representation of the electrical components ,expand the coordinates of the point into coordinates of line, at the same time, generate a new line direction according to the direction of both ends of the line.

From the coordinate values of electrical components we can draw the electrical topological relation, which is based on the principle of topological connection that elements with electrical topological relationship must have coordinates of the same value.

Property Conversion

The ID of each element corresponds with their basic properties, so we extract the corresponding attribute data in the properties databases of GIS, according to the ID number of electrical components converted out of the graph.

What stores geographical connection associated with all devices and equipment attribute information is the GIS attribute database, which mainly draws the data included in single-line diagram and stations map of the GIS databases and required by establishing electrical connection between line and a variety of computing, including device properties online, line name equipment belonging to, device name, impedance models and the subordinate relationship of various equipment and stations, equivalent impedance information of substation and so on.

3 Superposition Principle Short-Circuit Current Calculation Method Introduction

Short-circuit current calculation is a basic calculation of the grid. The structural characteristics of distribution network is its radiation type and that the slip parameter will appear the situation ▯ r>x. Short-circuit current calculations of the above 220kV transmission network generally uses the node admittance matrix, and the use of sparse techniques can significantly improve the efficiency of short-circuit current calculation [4]. However, the method applied to short-circuit current calculation of the distribution network is not the most effective way, because: (1) From the node number optimization to the formation of complex factors table, it is time-consuming and has more trouble when calculating; (2) It needs to re-establish or modify the sequence of the mathematical model of the original network when the network topology changes.

The traditional short-circuit current calculation generally uses the symmetrical component method, and its approach is: break it down into three sequence networks and list their equations respectively, calculate voltage and sequence components current of the fault point combining with its boundary conditions, and if you want to analyze the voltage, current of any point in computing network, you must first of all obtain each sequence components of the voltage, current in each sequence network, and then synthesize them into three-phase voltage and current. The advantage of symmetrical component method is that it can separately deal with three sequential network matrix and equations, so symmetrical component method has been widely use in the past few decades. However, the symmetrical component method applied to short-circuit current calculation of the distribution network encounters difficulties, which is determined by characteristics of the distribution feeder circuit structure: the distribution feeders are generally three-phase parameters asymmetry three-phase load asymmetry and the network is extremely large.

This article uses short-circuit current calculation method based on the superposition principle and the method uses abc three-phase model, which sees short-circuit fault as a sudden injection of additional current suddenly superimposed under normal operation conditions and each branch current and node voltage can be calculated though a pushed back to the previous generation calculating. Additional injection current is the short-circuit current of the faults, and its calculation can be obtained by pre-fault three-phase power flow calculation results and fault boundary conditions.

Please note that, if your email address is given in your paper, it will also be included in the meta data of the online version.

4 Short-Circuit Current Calculation Method Is Implemented in the VDNAP

4.1 Short-Circuit Point Positioning

Short-circuit current calculation is to be achieved in VDNAP, be converted by the GIS-generated electrical power to set graph the failure point (VDNAP set the failure point in the pixel, the user can click in the graphics settings), while setting the failure point attributes (two-phase short-circuit or three-phase short-circuit). Short-circuit current calculation of the implementation steps are as follows.

Set Failure Point
Click the task bar on the failure point primitive, set in a graphical failure point at the same time set the failure point properties, need to be the failure point of the pixel location of the endpoint and the wish to set the short-circuit node connected to the color change when the two nodes immediately upon the phase connection

Network Topology
This part of the contents with the author in his paper "The distribution network based on VDNAP flow calculation" discussed the same, the difference lies on short-circuit current calculation in the network topology be an array of information on the short-circuit set points. Searching the current document the equipment list (CObList) to obtain a pointer to the failure point, while extracting the failure point in the corresponding pointer information, to achieve fault location. If not found the failure point, you can not short-circuit calculations, while the failure point of the node array of message settings. As the author in his paper "On the basis VDNAP flow calculation of the distribution network," discussed, the node array defined as follows:
struct strnode

{
int nodeID; The original serial number of the node
int flowcalID; The serial number of the node after topology
int IsFault; Fault node identifiers
int FaultType; Fault type identifiers
    double U[3][2] ; Three-phase voltage of the node
double P[3] ; Injection three-phase active power to the node
double Q[3] ; Injection three-phase reactive power to the node
double G[3] ; Three-phase conductance of components parallel connection with the node
double B[3] ; Three-phase susceptance of components parallel connection with the node
double S[3][2] ; Three-phase injection apparent power of the node
double Y[3][2] ; Three phase bus impedance matrix of components parallel connection with the node
double I[3][2] ; Three-phase injection current of the node
};
Topology failure point in the node array IsFault = 1; FaultType = 0 (three-phase short-circuit) / 1 (two-phase short-circuit), and the remaining nodes in the array IsFault and FaultType are -1.

4.2 Short-Circuit Impedance Search

The Search Method Proposed
Impedance value is calculated as graphic as the platform, electrical short-circuit current calculation of the focus and difficulty, this paper adopts the reverse search method to solve this problem the better, according to distribution network of

radiation during normal operation of network characteristics, available brought to triangular form of the network diagram, as shown in Figure 2:

Assume that node8 short circuit, the purpose of calculating short-circuit current value is between the need to obtain node1 to node8 impedance. To avoid the initial node node1 Search the node will be qualifying disaster issues, proposed to node8 as the initial node by node and the connection between devices and circuits reverse search, namely:

$$node8 \longrightarrow node5 \longrightarrow node3 \longrightarrow node1$$

You can get:

$$Z_{sc} = \rho_{L5} * l_{L5} + \rho_{L3} * l_{L3} + \rho_{L2} * l_{L2}$$

Z_{sc} is the short-circuit impedance, ρ is the resistivity, l is the length of the line. l is extracted from the property line. By line attribute the middle of the line models to GIS database query and get the corresponding ρ values.

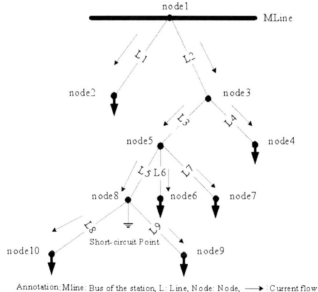

Fig. 2. Schematic diagram impedance search

Determine of the Search Direction

To ensure that by node8 Start Search only along the L5 to L2 to L3 bus and through the suspension and will not search for L8, L6, etc. slip road, take control of the search direction that is along the current flow to reverse search, specific methods are as follows:

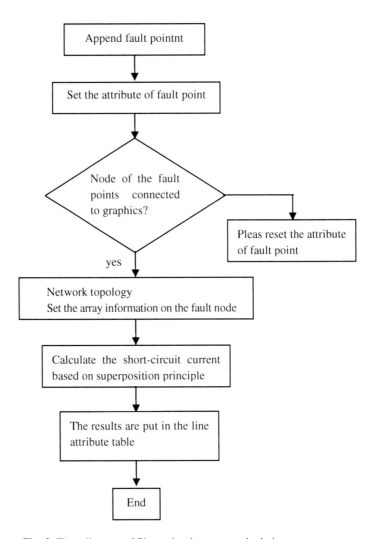

Fig. 3. Flow diagram of Short-circuit current calculation

Between points from the bus to short-circuit equipment (components, circuits) are double-end components, each device class is defined by two nodes node1 and node2, respectively, at both ends of the device on behalf of the graph nodes, node1 and node2 derived from the base class CComponentnode, CComponentnode defines a pointer variable Linknode, Linknode recorded in the device-side nodes and the rest of the graphic device-side connections between nodes.

Short-circuit impedance in the calculation carried out before the first circuit topology:

The definition of two-dimensional array StoreNode [100] [100].

Graphic in the busbar Linknode as the initial node, StoreNode into the first layer, through the devices and circuits of Linknode list Search and bus the same equipment, and its other endpoint Linknode into StoreNode the second tier.

If it is the same bus of the Linknode endpoint for the equipment, node2, the need to swap node1 and node2.

Longer StoreNode the second tier of data as the initial node, continuing their search, will be equally obtained by means of an array into the first layer of Linknode.

This cycle until the search to the Linknode number 0.

Circuit topology to achieve the electrical diagram of the device along the current flow endpoints sorted, topology after the end of the device node node1 beginning of the end for the current flow, node2 for the current flow at the end. After circuit topology of Linknode after node8 initial node, search CobList, CLineList equipment in node2 in Linknode, if it is the same as the initial node, then record the device information, and this Linknode as the initial node, continue to search until Linknode and Linknode the same end of the bus. At this point the search to achieve the reverse direction to maintain the current flow and, based on device information records obtained Z_{sc}. Flow chart is shown in Figure 3.

5 Conclusion

This paper presents the combination of VDNAP and distribution network short circuit calculation, electrical graphics generated by conversion instantly reflect the operation state, while avoiding the tedious modeling and the input of a large number of raw data. Short-circuit current calculation method is simple, practical and high efficiency. The constructed distribution network electrical model platform provides the basis for application development for dispatch automation, simulation, and network reconfiguration of distribution network, and it has broad application prospects.

References

1. Yang, H.: Improvement for power system short-circuit current calculation procedures. Qinghai Electric Power 23(2), 37–39 (2004)
2. Lu, J., Li, P.: The development and key functions of power GIS. Mapping and spatial geographic information 27(2), 31–34 (2004)
3. Zhong, F., Xu, T., Zhang, H.: The Modeling of distribution network based on geographical information system. Jiangsu Electrical Engineering 24(37), 37–39 (2005)
4. Che, R.F., Li, R.J., Li, Y.Z.: Short circuit calculation of distribution network based on superposition theorem. Automation of Electric Power Systems (12), 22–25 (2001)
5. Li, G.Q.: Transient-state analysis of power system. China electric power press, Bingjing (1995)
6. Hong, Z.: Study on practical algorithm for short circuit current in radial distribution networks. North china electric power (3), 20–23 (2001)

People's Event Features Extraction

Wen Zhou, Ping Yi, Bofeng Zhang, Jianfeng Fu, and Ying Zhu

School of Computer Engineering and Science, Shanghai University,
No. 149, Road Yanchang, 200072, Shanghai, China
{Zhouwen,pyi,bfzhang,jffu,yzhu}@shu.edu.cn

Abstract. In this paper we describe the system architecture and method of people's event feature extraction. Although the individual techniques are not new, but assembling and applying those to extract person's feature are new and interesting.

Keywords: Information Extraction, People's Event Feature.

1 Introduction

Previous work in information extraction has shown two types. One is the rule based method[1]. Another is based on machine learning techniques. There are two major machine learning techniques: hidden Markov models [2, 3] which learns a generative model over input sequence and labeled sequence pairs; the other is the technique which based on SVM classifiers [4].

However, for the machine learning method, a labeled corpus is necessary. This corpus must big enough and contain not only the positive example but also negative example. The rule based method need to use the knowledge from human who has the knowledge of language expression.

There are some important information extraction conference which are important in promote these technique's development. They are past MUC (Message Understanding Conference) [5] and now the ACE (Automatic Content Extraction) [6].

In this article, we present system architecture of people's event features extraction from web. The extracted person's event features are not only can be used for supplying the information but also used for the construction of people ontology.

The rest of the paper is organized as follows. First the system architecture for the extraction of the people's event features in Sections 2. In Section 3 gives the experiment results and result discussion, then conclude the paper in Section 4.

2 The Extraction of the People's Event Feature

The event features of people are extracted from webpage which contain the information of the person. First, the biographies in Chinese are collected from the Internet. Then, after analyzing the biographies, there are eight basic information categories of a person: date of birth, gender, education, nationality, birth place, address, occupation, and political status. Then, the biographies are labeled by those eight categories to form the corpus. For each category, the trigger words are recognized. Then according to the triggers, the extraction rules of the character information are set.

The new webpage in Chinese about the person is collected. After word segmentation, POS tagging and named entity, the triggers identification is conducted. Then according to the extraction rules, people's features are extracted. The process of the rule-based extraction method is shown in Figure 1.

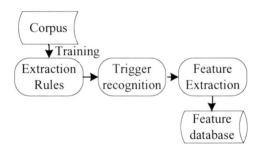

Fig. 1. The framework of people's event feature extraction

2.1 Extraction Rule

Using the annotated corpus, the people's feature is represented by a five-tuple <Left, tag1, Middle, tag2, Right>. Tag1 represents the name of the person. Tag2 is one of the person's features. Left contains the content of a two-vocabulary window in the left side of tag1. Right contains the content of a two-vocabulary window in the right side of tag2. Middle contains the content between tag1 and tag2.

There are a lot of words in Left, Middle, and Right, which don't typically reflect the people's features. So, stop word list are used for removing the "Auxiliary", "name", "punctuation" from the Left, Middle, and Right. After stop word removal, Left, Middle, Right values may be empty. When the Middle is empty, that indicates tag1 and tag2 adjacent to each other. Their relationship may be more closely. We used a special symbol "&" to mark Middle is null.

2.2 Trigger Recognition

Trigger recognition is the key step before feature extraction. There are two steps for trigger recognition: lexical analysis and trigger word recognition.

The ICTCLAS is used for Chinese lexical analysis. The lexical analysis in our method includes word segmentation, part of speech, and the recognition of named entity.

People's Event Features Extraction

The key step is establishing trigger list. In order to finding the properties of people from the corpus without annotation, the key is to establish a trigger list. The method of trigger list establishment is shown in Table 1.

Thus, a new sentence is analyzed by lexical analysis. If the trigger is found in the sentence, then began to determine whether the property is a specific person's event features.

Table 1. Trigger list establishment

People properties	Trigger word
Date of birth	Part of speech is "t". phrase compose the digitals, "years", "month"
Gender	"Male" or "female". A number of general and confusing words are removed.
Education	high school, undergraduate, Bachelor, Master's, Ph.D
Nationality	The name of 56 nation in China
Origin	the words with part of speech as a "ns"
Address	the words with part of speech as "ns" or start with "live in"
Position	"hold a position" ,"position" or the words expressing a specific position
Political landscape	Chinese Communist Party, democratic parties, the Communist Youth League, the masses

2.3 People's Event Feature Extraction

Our method uses the results of lexical analysis to identify the features corresponding to the largest feature phrase. Table 2 shows the rule for identifying the phrase including people's feature.

In addition, we also try to identify the position phrases and the names phrase. If those two phrases use slight pause mark or connect with "and", then the relationship between them is side by side. Parallel phrases share the same feature vectors.

Table 2. The Rules of People's feature identification in the phrase

People Feature	Identify rule of people character
Date of birth	from the trigger word for the next part of speech in order to find "t" and the Center for "Month", "Day" Words
Gender	Gender is only words can be triggered
Education	alone can trigger the word, pick the highest academic qualifications
nationality	Words alone can trigger the national
Origin	from the trigger word to find the words with part of speech as "ns", "f", "a", "n", "b", or "s" , until meet with "people"
Address	from the trigger word to find the words with part of speech as "ns", "f", "a", "ng", "m", "q", "n", "b" or "s"
Position	from the trigger word to find the words with part of speech as "n", "nt", "ns", "nz", "nl", "ng", "b", "f", "vn", "m", "q", "s", "nsf", "ag", or "a"
Political landscape	The trigger alone

3 Implementation

The system development environment is Visual Studio 2005. The operating environment is of CPU P4 and Memory 1G. Operating system is Windows XP. The interface of this system is shown in Figure 2.

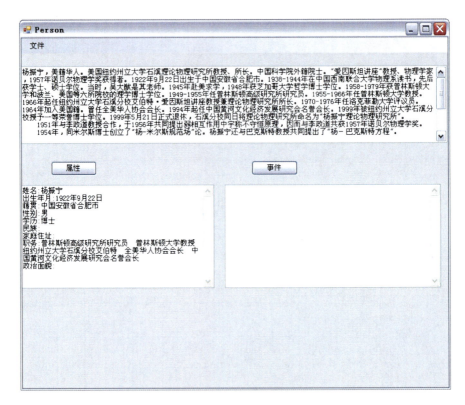

Fig. 2. The system interface of people event feature and event extraction

4 Conclusion

This paper establishes a method for people's event feature extraction. Name recognition have been directly used in Chinese lexical analysis system ICTCLAS, the part of speech tag for the "nr", "nrf" the words as a person's name, as a person information mining trigger words. Using a number of simple and effective rules to carry out the names refer to the handling of the characters corresponding to the determination of discourse.

There are some technique problems left for the future research, such as the rule need to refine, the evaluation method can be improved.

Acknowledgement. This work is supported by six projects of National Science Foundation of China (NSFC No.60975033), Shanghai Scientific Special Funds for Cultivation and Selection of Excellent Young Teachers (shu-07027), Shanghai University innovation Funding Project, Shanghai Leading Academic Discipline Project(J50103), and Shanghai Undergraduate student innovation project.

References

1. Soderland, S.: Learning information extraction rules for semi-structured and free text. Machine learning 34(1), 233–272 (1999)
2. Seymore, K., McCallum, A., Rosenfeld, R.: Learning hidden Markov model structure for information extraction, vol. C42, p. 37
3. Takasu, A.: Bibliographic attribute extraction from erroneous references based on a statistical model, p. 60
4. Han, H., Giles, C., Manavoglu, E., et al.: Automatic document metadata extraction using support vector machines, pp. 37–48
5. Marsh, E., Perzanowski, D.: MUC-7 evaluation of IE technology: Overview of results. In: Proceedings of the Seventh Message Understanding Conference, MUC-7 (1998)
6. National Institute of Standards and Technology, ACE (Automatic Content Extraction) Chinese Annotation Guidelines for Events (2005)

A Server-Based Secure Bootstrap Architecture

Qiuyan Zhang, Chao Chen, Shuren Liao, and Yiqi Dai

Department of Computer Science and Technology, Tsinghua University
100084 Beijing, China
{shuimuren,chaochen.thu,liaoshuren}@gmail.com,
dyq@theory.cs.tsinghua.edu.cn

Abstract. The computer terminal plays an import role in the security of whole Local Area Network. However, the uncontrolled way of bootstrap brings about difficulties of providing sufficient trustworthiness to the LAN. To enforce the terminal security of the LAN and especially its ability of resisting ill-meaning tampering, this paper puts forward a server-based bootstrap architecture, based on the trusted computing technology. By verifying the integrity of the terminal before booting the OS, this architecture can effectively prevent the terminal from booting into a tampered OS, and the recovery module meanwhile enforces the robustness of the system. We present an implementation of the architecture, which extends the Trusted GRUB by adopting an attestation process between the GRUB level and the attestation server. The performance analysis shows that at a low time delay, the security of the system has been improved, and the proposed architecture can also provide server with stronger control and management ability towards the whole LAN.

Keywords: Attestation server, Trusted Computing, bootstrap, Trusted GRUB, Recovery Module.

1 Introduction

As basic units of Local Area Network (LAN) computer terminals' security has become more and more important, especially when the LANs with high security requirements are concerned, e.g.: in militaries or government[1]. However, most clients are mainly under the arbitrary control of their users and if the user boots the terminal into untrusted or unsecure status unconsciously or intentionally, it may bring potential security problems into the terminal as well as the LAN. Thus it is very important to ensure that terminals are booted into a trusted Operating System (OS). Since it is the base of the security of terminal, the security of the bootstrap has raised more concern from people.

The bootstrap process consists of several discrete levels: BIOS, boot loader, OS kernel [2]. If an adversary manages to gain control over any particular level no subsequent level can be trusted. For example, consider a client system with a compromised Boot Loader. The Boot Loader can modify the OS kernel before it is executed, where backdoor can be inserted before the OS gains control.

1.1 Related Work

The problem with bootstrap described above is well known, and several solutions have been put forward. AGES [2] proposed by Arbaugh et al. improved the security of the bootstrap process by ensuring the integrity of bootstrap code. But their work was built on some strong assumptions, e.g.: the ROM won't be compromised. However, the ROM maybe replaced if an internal attacker is able to access the PC physically, which will break the assumptions easily. On the other hand, the use of digital signatures introduces a key management problem that is amplified by the requirement to store the initial public key in ROM.

The trusted computing technology [3-5] provides another solution which introduces a non removable secure device Trusted Platform module (TPM) [4] into bootstrap process. The TCG-based Integrity architecture has been implemented in [6]. However, it only measures the code of the boot entity and saves its checksum into PCR, no attestation action is taken before booting the OS kernel. Only after the OS is launched, terminal offers a trusted attestation for remote systems. This is not secure enough when the client system is taken advantage by adviser, especially when the terminal is in a high security requirements LAN. Additionally, TCG assumes the OS is in security architecture which at this moment can not be satisfied, for current Operating System is monolithic, too complex to provide a sufficiently small Trusted Computing Base (TCB) [7-8] and hence OS is prone to security vulnerabilities [9].

Considering all the points mentioned above, we introduce a server in our bootstrap architecture, which verifies the integrity of all boot entities before launching the OS.

1.2 Outline of Paper

The rest of this paper is organized as follows. In section 2, we focus on functionality provided by Trusted Computing platforms. Section 3 gives an overview of the server-based secure bootstrap architecture. Section 4 presents details of its implementation. The complete bootstrap performance evaluation is given in Section 5, discussions and future works are described in Section 6, and the paper is concluded with Section 7.

2 Technical Background

In this section, we present the background of trusted computing in these aspects: Trusted Platform Module, integrity measurement and integrity reporting and verify.

2.1 Trusted Platform Module

Trusted Platform Module (TPM) proposed by Trusted Computing Group is the main component of Trusted Computing. In addition to a physical true random number generator, a TPM offers cryptographic functions such as SHA-1, RSA encryption/decryption and signing/hashing. It stores hashes of the boot sequence in a set of Platform Configuration Registers (PCRs) which cannot be directly written but only be modified using the extending operation (1) with PCR[k] the previous register value, PCR'[k] the new value, m a new measurement and ‖ denoting the concatenation of values.

$$PCR'[k] = Hash(PCR[k] \| m) \qquad (1)$$

2.2 Integrity Measurement

The initial state of the computer system is measured by computing hashes of every components loaded during bootstrap process before the components' execution. As shown in figure 1, the dotted line denotes that the lower level measures the upper level and extends it into the PCR, while the solid line dedicates the transferring control between levels. The main entities during bootstrap are Core Root of Trust Measurement (CRTM)[3-5], BIOS, Boot Loader and OS. CRTM is the root of trust which will be trusted unconditionally after measuring code and parameters of BIOS, it extends the hashes into some referenced PCR and then passes control to BIOS. BIOS still need to measure Boot Loader before transferring control to it. The measurement and transfer process will continue until OS get control. Thus a chain of trust is established from the root of trust to the OS.

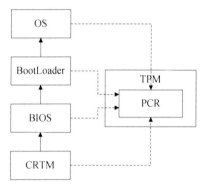

Fig. 1. the dotted line denotes that the lower level measures the upper level and extends it into the PCR, the solid line dedicates the transferring control between levels

2.3 Integrity Reporting and Verification

Integrity reporting and verify is a challenge-response protocol. By offering PCR values signed with an Attestation Identity Key (AIK) [3-4] to a remote party, a client can present its integrity to the remote party. A trusted third party called

Privacy Certification Authority (Privacy CA) is used to certify the AIK. If one of the client's components is compromised, the related PCR value will be different from the hash of the original one computed in the remote party and then will be verified as not secure.

3 Architecture

The system architecture is shown in figure 2. It consists of Trusted terminal, Attestation Server and some related system modules. The trusted terminal equipped with the security chip TPM includes four bootstrap entities, which are CRTM, BIOS, Boot Loader, and OS. Attestation server with verifying policy is protected by LAN with special strategy. Our system architecture consists of four major function modules:

Measurement Module. The measurement module is responsible for the measurement of every bootstrap entity and save the measurement value securely.

Integrity Reporting Module. The Integrity Reporting Module ensures that the integrity information about the client's bootstrap process is reported to the remote attestation server before the OS is launched.

Verification Module. The Verification Module called by attestation server makes decisions about whether the client can be given the permission of launching OS.

Recovery Module. The Recovery Module forces the client to recovery into a suitable verified replacement module if any integrity failure is detected, which enforces the robustness of the system.

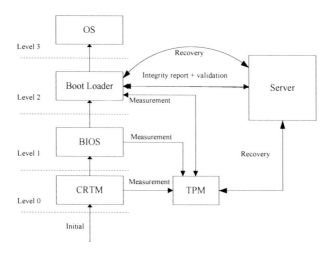

Fig. 2. The server-based bootstrap architecture

Before each control transfer, every boot entity of terminal, BIOS for example, has to call the measurement module to measure the next entity (Boot Loader as for BIOS). Especially, after Boot Loader measures the OS kernel and related key files, it does not pass control to the next boot entity (OS) directly, but reports the integrity information of the bootstrap system to the attestation Server. According to the validation policies, server attests the integrity of the client, and then returns 'permission' to the terminal if the verification of every component is successful or 'forbidden' to the client if any integrity inconsistency is detected. In case of entity inconsistency, the server may enforce the recovery module which triggers a recovery process causing the client to boot into a verified kernel contained on the attestation server.

As described above, the validation process happens before The OS is launched, which can prevent terminal boot into a tampered status.

The proposed architecture improves the security of trusted terminal in two aspects. The first is that no tampered OS is booted unless its verification is successful; the second is that when an integrity failure is detected the recovery module will be triggered which forces tampered components be replaced by consistency ones.

The Measurement module is similar to the current work presented in [6]. Chapter IV will focus on the implementation of integrity reporting module, validation module and recovery module.

4 Implementation

By introducing grub as boot loader into our bootstrap process, we implement the system architecture, whose work flow is shown in figure 3. The Arabic numerals in figure 3 denote the work sequence of the bootstrap.

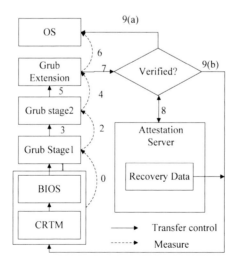

Fig. 3. The Work flow of System

4.1 Integrity Reporting and Verification Module

GRUB is a multi-Operating System boot loader, which is able to load a wide variety of free operating systems, as well as proprietary operating systems with chain-loading [10]. In fact, Grub is a micro-operating system which mainly consists of Grub stage1 and Grub stage2 shown in the figure3. GRUB stage1 including the initial sequence of system which is 512 Bytes is loaded into the Master Boot Record (MBR) when installing. Its major role is loading the code of Grub stage2. As the core of GRUB, Grub Stage2's major role is loading and booting the OS kernel.

Besides, several open source projects have patched GRUB to make it a trusted boot loader [11-13]. Trusted GRUB implemented in [12] is one of them. Grub Extension module as an extension of Trusted Grub stage2 is developed by us. We enable it to communicate with the remote server before the OS being launched. The extended module first measures the OS kernel and the key files of the OS that will be booted, reports the measure value to the remote server, and then boots the OS if "permission" is returned from the remote attestation server.

In the communication process of integrity reporting and validation module between the GRUB extension module and server, we adopt an improved protocol presented in [6] which is shown in figure 4.

T denotes the trusted terminal and S the attestation server. In step 1, T first sends a verification request to the server. In step 2, S creates a random through TPM's random generator, and sends it to back to T. In the next step, T contacts with TPM using quotes command which is defined in specification [14]. In this stage, T gets PCR values signed with AIK from the TPM. In step 4, PCR values and AIK certification are sent to server, and both of them are encrypted by the public key of server. In step 5 and 6, S verifies the AIK certification through trusted third party, and attests the integrity of values of measurement.

```
1) T—>S: Verify_Req
2a) S: Creat(Ns)
2b) S—>T: Ch_Req(Ns)
3a) T—>TPM: TPM_Quote_Req(Ns)
3b) TPM—>T: Quote_Res(AIK_{pvt}sig(PCR,Ns))
3c) T: Fetch(SML),E_{Spub}(SML)
4) T—>S:Ch_Res(Quote,E_{PUBs}(SML),AIKcert)
5a) S:validate AIKcert
5b) S: verify Ns
5c) S: validate the Stored Measurement Log
6) S—>T: Verify_Res
```

Fig. 4. Integrity reporting and verification protocal

This protocol provides security to the integrity reporting and verification modules in three parts: first, the random number Ns guarantees the freshness of the TPM quotes, which reflects the actual system status. The second is the bidirectional authentication which can prevent masquerading attacks. The third is sending the stored measurement log to the remote server which is able to prevent tampering attacks [15-16].

If the server finds any unexpected measurement value, then according to its rules, it may require the client to block the booting process, and initiate the system recovery process.

4.2 Recovery Module

The goal of the recovery module is to recover and correct the inconsistency module from the trusted sources when necessary. Here the trusted sources refer the attestation server.

If the component is detected to have been malicious tampered then it must be recovered by using the TPM chip. The recovery process is similar to a memory copy from the address space of the Non-volatile Memory (NV) [5] of TPM, where original BIOS code is saved.

If the component that fails its integrity check is a portion of OS, the recovery process begins between GRUB extension module and the server. By taking advantage of the Preboot Execution Environment (PXE) module [17] of the GRUB boot loader, recovery module enables terminals to boot into a verified kernel loaded from the server.

5 Performance Evaluation

Since we introduce server validation into the bootstrap system, it will bring some extra time cost into the boot process. Besides, the recovery process will also slow the boot process while the verify operation fails. Now, taking the measure and verify operation into consideration, we test the performance of the process of bootstrap in a 100Mb/s LAN. The client system is equipped with Pentium Dual 1.60GHz, 1GB memory and a TPM that complies with TCG TPM v1.2 specification. The Operating System we boot is Fedora 9.0.

Let ΔT denote the whole extra time while the system boot into OS normally. Then we have equation (2), where M_t denotes the time cost of all the measure operations, V_t the attestation operation's time cost.

$$\Delta T = M_t + V_t \qquad (2)$$

By testing, we get M_t 1.8 seconds and V_t 1.42 seconds, and thus we have $\Delta T = 3.22$ seconds, which is insignificant compared to the length of time currently cost to bootstrap a normal PC.

6 Discussion and Future Work

The proposed server based architecture ensures that the OS is in a trustworthy state after the secure bootstrap is processed. If any damage has been done by the malicious applications to the system in their runtime, the damage will be detected in the next boot process during the integrity reporting and verify steps, hence system recovery can be applied to guarantee that the trustworthiness of system.

By introducing the attestation server into the bootstrap process, this architecture also enables central administration of terminals, which makes deployment and change of security rules relatively easy. Additionally, in the LAN environment, the server monitoring mechanism is very necessary to improve the overall security of the LAN system.

However this architecture is not perfect since the application level security monitoring is not addressed in this paper, which may results in running time security problems [18]. Our future work aims to provide execution and runtime security of applications on terminals.

7 Conclusion

In this paper, we present a server-based bootstrap architecture to enhance client system's security. The attestation server and recovery module are introduced to enforce the client boot the OS into a trustworthy status. Based on the integrity measurement architecture, we construct a trusted chain from BIOS up to OS.

The Trusted GRUB Extension module is then developed which reports the terminal integrity measurement value to the attestation server and helps ensure compromised OS will not be launched. The recovery process enforced execution while verify failed ensures the usability and security of system. Performance evaluation of the system shows that the time cost is acceptable, while the security of the system has been improved.

References

1. Sandhu, R., et al.: Client- side access control enforcement using trusted computing and PEI models. Journal of High Speed Networks 2006 15, 229–245 (2006)
2. Arbaugh, W.A., Farber, D.J., Smith, J.M.: A secure and reliable bootstrap architecture. In: Proceedings of the 1997 IEEE Symposium on Security and Privacy, pp. 65–71. IEEE Computer Society, Los Alamitos (1997)
3. Trusted Computing Group. TCG Specification Architecture Overview. Revision 1.4 (August 2007)
4. Trusted Computing Group. TPM Specification, v1.2 (March 2006)
5. Challener, D., Kent, Y.: A practical guide to trusted computing, pp. 16–17. IBM press (2008)
6. Sailer, R., et al.: Design and Implementatin of a TCG-based Integrity Measurement Architecture. In: Thirteenth Usenix Security Symposium, August 2004, pp. 223–238 (2004)
7. Dinh, T.T.A., Ryan, M.D.: Trusted Computing: TCG proposals (2006), http://www.cs.bham.ac.uk/~mdr/teaching/modules/security/lectures/TrustedComputingTCG.html

8. Dvir, O., Herlihy, M., Shavit, N., Dvir, O., Herlihy, M., Shavit, N.: Virtual leashing: Internet-based software piracy protection. In: 25th International Conference on Distributed Computing Systems, ICDCS 2005, Columbus, OH, USA, June 6-10. IEEE Computer Society, Los Alamitos (2005)
9. Schellekens, D., Wyseur, B., Preneel, B.: Remote attestation on legacy operating systems with trusted. In: Science of Computer Programming (April 2008)
10. The multiboot specification, http://www.Gnu.org/software/grub/manual/multiboot/multiboot.html
11. http://tboot.sourceforge.net
12. Applied Data Security Group. What is trusted GRUB (2006), http://www.elinux.org/upload/28/Trusted_Boot_Loader.Pdf
13. TrustedGRUB, http://www.rub.de/trusted_grub.html
14. TCG PC Client Specific Implementation Specification. Revision 1.2 (July 13, 2005)
15. Bell, D., La Padula, L.: Secure computer systems: Mathematical foundations, Tech. Rep. MTR-2547, vol. I, Mitre Corporation, Bedford, Massachusetts (1973)
16. Liu, J., Jia, Z.: A Remote Anonymous Attestation Protocol in Trusted Computing. In: Proceedings of the first ACM workshop on Scalable trusted computing, pp. 7–16 (2006)
17. Pxeboot Execution Environment, PXE (2009), http://en.wikipedia.org/wiki/Preboot_Execution_Environment
18. Trusted Computing Group, TNC Architecture for Interoperability Specification v1.3, Revision 6 (April 2008)

Design and Implementation of the Integration Platform for Telecom Services Based on SOA

Xiaoxiao Wei, Xiangchi Yang, and Pengfei Li

School of Management Engineering
Xi'an University of Posts and Telecommunications, 710061, Xi'an, China
`wxxsally@163.com`

Abstract. In the trend of telecommunication network and Internet towards convergence, it is important to learn the success experiences of existed service provision and service development technologies from telecommunication domain and introduce the new technologies of SOA from software domains, for convergent network oriented next generation service provision. According to the demand characteristic of business process integration for telecom enterprises, an integration platform for telecom services is designed and implemented based on SOA, which will provide the flexibility for enterprises to change their business process quickly and swiftly. It makes the new and old system couple with better and cut down the cost in development and maintenance.

Keywords: Services Oriented Architecture, Business Process Execution Language, Business Process Integrated Service, Component.

1 Introduction

With the deepening study of software development method and wide applications of distributed technology, a software system which based on a variety of development platforms and development tools has been large-scale application, making the enterprise software systems become increasingly large and complex.

Most IT systems of telecom enterprises are faced with many problems, firstly, enterprises have been in existence for information islands and information can not be shared because of many scattered isolated application systems; Secondly, as IT resources can not be re-used, it causes the waste of resources; Thirdly, IT system is unable to respond business demand quickly; The Fourth, there is lack of the overall control of enterprises IT systems and business processes, the new demand of the business is increasing progressively constantly, as a result , rising IT construction costs, the corresponding business process management has lagged far behind, and there is no foster core business processes for the management of its core concept, and no perfect form the process of strategic management system. Reasonable ways to put Web services into the telecom enterprise IT systems and processes into commercial activities, and bring direct economic benefits to enterprises so that enterprises of the IT infrastructure quickly and efficiently adapt to business changes, to meet market and customer's personality the demand-oriented, so that timely response, that has become the key to the success of enterprise information. According to the demand characteristic of business integration for telecom enterprise and the advantage of SOA technology, an integration platform for Telecom services is presented and implemented based on SOA.

2   Service-Oriented Architecture SOA

SOA is an architecture that unifies business processes by structuring large systems as a loosely coupled collection of smaller modules called services. These services may be used by all the systems and applications (both internal and external) to achieve better unification and flexibility [1]. SOA concentrates the migration problem on the middleware layer. It allows replacing custom code by processes that can be designed and re-factored as required without writing code. It also integrates message mediation into the infrastructure that is provided by the middleware layer. That is to help different systems with different message formats to be combined in one business process. SOA is a growing trend to help business units to be less reliant on IT departments to actually make changes to IT infrastructure when the business changes. SOA model is shown in Figure 1.

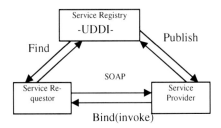

Fig. 1. The SOA model

Among them, a service provider is a Network node that provides a service interface for a software asset that manages a specific set of tasks. A service provider node can represent the services of a business entity or it can simply represent the service interface for a reusable subsystem. A service requestor is a network node that discovers and invokes other software services to provide a business solution. Service requestor nodes will often represent a business application component that performs remote procedure calls to a distributed object, the service provider. In some cases, the provider node may reside locally within an intranet or in other cases it could reside remotely over the Internet. The conceptual nature of SOA leaves the networking, transport protocol, and security details to the specific implementation.

The service broker is a Network node that acts as a repository, yellow pages, or clearing house for software interfaces that are published by service providers. A business entity or an independent operator can represent a service broker.

These three SOA participants interact through three basic operations: publish, find, and bind. Service providers publish services to a service broker. Service requesters find required services using a service broker and bind to them. The interactive process among these 3 agents calls/centers on the service components (rather than objects which characterizes object paradigm).

At present, because business processes are essential to the SOA approach, the model must also integrate an additional standard for this aspect: the Business Process Execution Language (BPEL) [2]. BPEL provides a language for the specification of Executable and Abstract business processes. By doing so, it extends the Web Services interaction model and enables it to support business transactions. BPEL defines an interoperable integration model that should facilitate the expansion of automated process integration both within and between businesses, and through Web Service interfaces; the structure of the relationship at the interface level is encapsulated in what called a partner link [3].

3 Design of the Integration Platform for Telecom Services Based on SOA

In the telecommunications industry, operation support systems how to achieve application integration and business process integration has been the telecom operators focus of attention. In order to meet the increasingly fierce market competition, to meet customer needs rapidly changing environment, the telecom operators needs the flexibility to quickly adjust their business models and business processes, and needs to integrate the data sources and business processes across the multiple distributed business systems.

This system is the integration platform model for the telecom Services based on SOA framework, and selected IBM WebSphere as the underlying support platform of the system. The whole system using open standards and integration technology take control of the business processes. The development of the new system, following a service-oriented architecture and component-based development approach in ensuring the application of the new structure which is clear and easy to expand, at the same time, the new system can be guaranteed the ability to integrate. System framework is shown in Figure 2.

The system mainly includes two parts: one is to achieve data integration through a variety of adapters and CRM, SPS, Billing, RPS and other systems; the other realizes telecom business processes logic control through the process control, at the same time, it carries out effective monitoring and analysis for processes such as service fulfillment.

4 Integration Platform for Telecom Services Based on SOA

While the business treatment of telecommunications is in various types, but the ways in dealing with have some similar characteristics. Therefore, it can abstract and assemble some of the typical business processes into different atomic processes, and the business processes logic is packaged together and independently administered. It can provide the flexibility and maneuverability for the demand of telecom services [4].

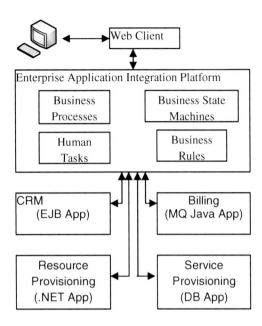

Fig. 2. System Framework

4.1 Business Process Choreographer

A business process is a specific process for a class of the corresponding business. It is mainly based on certain principles derived from the atomic assembly processes. In order to achieve the control of business processes and business rules definitions, you can use the BPEL description of the components for development which includes business processes, business state machine; business rules, etc [5]. Using BPEL describe the atomic processes, which mainly include the information of atomic processes nodes and task activities. The description of business processes is a business process that contains the atoms process information. According to the formation of specific business rules database, we will design the corresponding scheduling engine and business rules engine of the atomic processes.

This system is the integration platform model for the telecom Services based on SOA framework .The system will integrate existing IT systems to the platform, and the business processes of the various systems are integrated into this platform too. It choreographs reusable processes into a complete end to end business processes. Take an order processing for example, this thesis has explained the process of business processes choreographer and implementation.

The business processes describe as follows: the operator submits customer orders through the web client, and orders contain one or more requirements for service fulfillments.Dependencies may not exist between several service fulfillments, for example the service fulfillment of ADSL requests that the installation of telephone is complete first, and necessary resources of the order have already preempted. Then CRM stores the order and confirms all worksheets of the order have been completed, if so, the entire order is completed. After all worksheets have been completed, the resources which have been pre-accounted preempt actually in the resource provisioning system. Then add new billing items in the billing system, and update the status of orders in the CRM system, and inform the operator that customer's orders have been completed and the operator notifies the customer that service has been opened.

Thus, the business process contains a number of interactions between business systems. After these business systems are integrated into the enterprise application integration platform, and the integration platform will interact with these systems, and call the services which they provide [6]. Figure 3 shows how to organize the order processing in the integration platform.

4.2 Business Process Development

Order processing as a service, it has interface OrderHandling based on WSDL standard, interface operation including submit, modify, cancel and enquire orders. To achieve the service components of the order processing interface will realize all of the operations of the interface. After export the interface, the other SCA modules and non-SCA programs can call these services, such as call submitOrder to operate, and the orders can be submitted, and then the integration platform drives business processes running to handle the submitted orders.

In this case, we can use business processes to achieve the multiple sub-business processes, and call various external services. Sub-processes include storage, submit, update and finish orders, etc. Each sub-process is a short process, to facilitate reuse of sub-processes. To achieve an internal interface SalesOrder by business processes, its operation that includes is the function that each sub procedure will be realized; it contains the operation that each sub procedure will be realized.By using the business state machine (BSM) to achieve control over the order status, and call the business process components, drive sub-processes to achieve specific business functions and achieve an external interface OrderHandling. Using the Business Rules components to achieve business dependencies, the business logic is spun off from the business process to facilitate the customization and modification.WorkOrderDependency as a business rule interfaces determine the existence of dependencies between services according to the input type of service.

As shown in Figure 4, in accordance with service design,the various modules is assembled together and connected the call relation between services. OrderHandlingBSM is a component to achieve business state machine, and SalesOrderBPEL is a component to achieve sub-business processes. WorkOrderDependencyRuleGroup is a component to achieve business rules,and OrderHandlingExport is the export of the Order Handling service. WorkOrderResponseExport is the export of the WorkOrderResponse services for the callback of worksheets update. The rest of the import module BillingImport, CRMImport, ResourceImport, WorkOrderImport and CustomerNotificationImport is the import of external services.

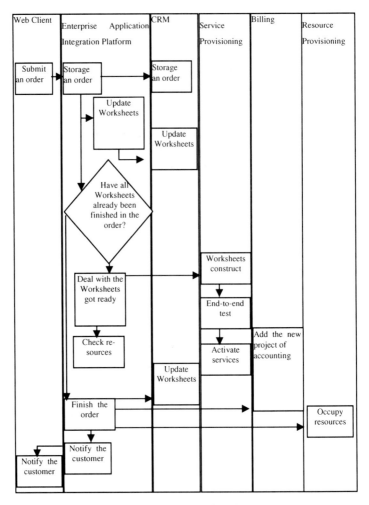

Fig. 3. Order processing flowchart

OrderHandlingBSM is one of the main modules in the whole business process, which maintains the state of order processing. OrderHandlingBSM has realized the service that OrderHandling interface and WorkOrderResponse interface describe. Business Process SalesOrderBPEL call services provided by other modules to complete several business sub-processes.

Using IBM WebSphere as the underlying support platform of the system for development of atomic processes, using Business Process Execution Language to establish different activities that can finish carrying out the jobs, such as Web service call, operation data, the termination of a process, etc, and then they are connected. Through the preparation of Java classes and a systematic tool to generate the corresponding WSDL file to describe it in the business process, it will need to be used in process which called to them.

Design and Implementation of the Integration Platform

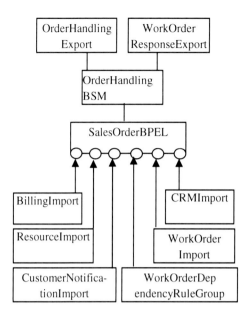

Fig. 4. Assembly Drawing of Order processing Module

Ways to generate work orders, for example, the code of its partial implementation WSDL file is shown below.

```
<defmitions name="SalesOrderBPEL"
targetNamespace="http://services.sps.neusoft.com/Sales
OrderBPEL/"
 xmlns="http://schemas.xmlsoap.org/wsdl/"
 xmlns:tns="http://services.sps.timeson.com/S
WorksheetBri/"
 xmlns:xsd="http://www.w3.org/2001/XMLSchema"
 xmlns:xsd2="http://xmlapache.org/xml-soap">
 <import namespace="http://xml.apache.org/xml-soap"/>
 ……
 <message name="WorkOrderRequest">
 <part name="WorkorderFlow" type="xsd:string"/>
 <part name="WorkorderType" type="xsd:string"/>
 ……
 </message>
 <message name="WorkOrderResponse">
 <part name="Order" type="xsd:string"/>
 </message>
 ……
 <portType name="SalesOrderBPEL">
 <operation   name="CreateWorkOrder"   parameterOrder=
"WorkorderFlow Workorder Type ">
```

```
......
<input message="tns: WorkOrderRequest" name="
WorkOrderRequest"/>
<output message="tns: WorkOrderResponse" name="
WorkOrderResponse"/>
</operation>
</portType>
......
</definitions>
```

5 Cconclusions

This thesis presents an integration platform framework for telecom business based on SOA, and discusses the key techniques in the model. The Design and Implementation of the program significantly increases the efficiency of software development, and makes good performance in functionality, usability, scalability, maintainability, security, and reliability. It achieves to work together in an optimum way between enterprise applications, internal applications and external applications, and enhances the flexibility in operation support systems of telecom enterprises, and increases operational efficiency.

References

1. Krafzigd, B.K., Slama, D.: Enterprise SOA service-oriented architecture best practices. Prentice Hall, New York (2005)
2. Hui, W., Bei-en, S.: Using SOA-BPM Combination Framework to Build Real-time Enterprise. Journal of Computer Application 6, 220–223 (2007)
3. Nam, N., Chrisw, M.: An algorithm forvendor-independent BPEL debugger in a Java integrated development environment. Research Disclosure 4, 362–363 (2006)
4. Xiang-chi, Y., Peng-fei, L.: Design of Post Logistics Information System Based On SOA. Computer Engineering and Design 28, 4825–4827 (2007)
5. Web services business process execution language (WS-BPEL), http://www.oasis-open.org/committees/wg_abbrev=wsbpel
6. Understand Enterprise Service Bus scenarios and solutions in Service-Oriented Architecture, http://www-900.ibm/developerworks

An Online Collaborative Learning Mode in Management Information System Experimental Teaching

Hanyang Luo

College of Management, Shenzhen University, P.R. China, 518060
Shenzhen Graduate School, Harbin Institute of Technology, P.R. China, 518055
hanyang@szu.edu.cn

Abstract. Nowadays, the teaching reformation based on network is one of the new trends of higher educational research. Taking the experimental teaching reformation practice of Management Information System, which is the common fundamental subject of managerial specialties, as an example, this paper reviews the constructivism learning theory, which is the theoretical basis of online collaborative learning, and then expound the implementation process of the action research of online collaborative learning under the network environment. At last, the author presents an operable online collaborative learning mode, and analyzes some limitations of this research.

Keywords: Online Collaborative Learning, Experimental Teaching, Management Information System, Action Research.

1 Introduction

A Management Information System is a planned system of the collecting, processing, storing and disseminating data in the form of information needed to carry out the functions of management. It is an interdisciplinary subject based on IT technology, management, economics, business, mathematics and so on, which need a wide range of theoretical basis. The cultivation of students' application and practice ability through effective experimental teaching is a focus of such a subject. With the development of modern computer technology, communications technology and network technology, people launched educational and teaching activities based on Web environmental. By means of online discussion and collaborative learning, E-learning can help to improve students' information literacy, strengthen their team working consciousness and train their innovative ability in practice. The plan, analysis, design and implementation of management information system is a typical project that especially needs communication, collaboration and team spirit.

Based on the above understanding, we start the action research on an online collaborative learning mode in Management Information System experimental teaching since 2008.

2 Literature Review

2.1 Constructivism Learning Theory

Since the 1990s, constructivism learning theory has gradually become the core theory in the field of educational technology. Constructivism is the further development of modern learning theories after cognitivism. Behaviorism and a part of the cognitivism emphasizes on the teachers' teaching, while constructivism particularly on the students' learning.

According to constructivism theory, knowledge is not mainly acquired by learning from teachers, but by means of meaning construction, in which learners make use of necessary learning resources, interact and collaborate with teachers and learning partners in certain circumstances or social culture background [1]. Scene, collaboration, conversation and meaning construction are the four elements of the learning environment. Scene is the social culture background in which learners study. Through out the whole learning process, collaboration is the interaction between learners and teachers or classmates in the learning process. As an important means to realize meaning construction, conversation is an indispensable tache in collaborative process during which each learner shares his/her thinking achievement with the whole learning group. Meaning construction is the ultimate goal of the whole learning process. The meaning to be constructed includes the nature and law of things, and the intrinsic relation between different things.

Social constructivists believe that meaning making is a process of negotiation among the participants through dialogues or conversations [2]. The learner is central to the learning process. Learning is a social activity and learners make meaning through dialogue, communication, collaboration and interaction. The opportunity to interact with other learners in sharing, discussing, constructing and negotiating meaning leads to knowledge construction. Swan suggests that while learners are constructing knowledge they should have support from more knowledgeable people (e.g., educators, peer mentors or experts from the field). These individuals can provide additional expertise, different perspectives and scaffolding in support of the co-construction of knowledge [3].

2.2 Online Collaboration

Harris defines online collaboration as an educational endeavor that involves people in different locations using Internet tools and resources to work together. Much educational online collaboration is curriculum-based, teacher-designed, and teacher-coordinated [4].

Harris specifies four key benefits to online collaborative learning. First, online access to multiple people beyond the learners' immediate classmates and educator gives exposure to differing opinions, perspectives, beliefs, experiences and thinking process. The online arena also provides multiple interactive opportunities with other learners, educators, experts and content. Second, the use of asynchronous communication facilitates learning anywhere and anytime. Third, it enables learners to move from their private to the public world and dialogue to create a shared understanding of meaning through comparing, contrasting, and/or combing similar information collected in dissimilar locations. Fourth, online collaborative learning experiences can create learning communities at local, national or global levels expanding participants' global awareness [4].

According to Hathorn and Ingram, various researchers have identified four critical attributes of the discussion patterns within an online collaborative group: participation, interdependence, synthesis of information, and independence [5]. Of these four attributes, participation is the most fundamental requirement of a collaborative group because it is impossible to collaborate without individual contributions to problem solving. The second attribute of interdependence requires interaction between group members to bring about active responses to one another. The third attribute of synthesis of information requires the product of collaboration to reflect input from every group member. Finally, a collaborative group should be independent of the instructor, which means that whenever a question occurs, group members should attempt to collaborate with each other rather than turning to the instructor for answers [6].

3 Research Design

In our research, we adopt the methodology of action research. Action research is particularly pertinent to the developmental needs of all societies, communities and nations, especially where people struggle with the dramatic changes induced by modern technology and economic activity, and with the deep-seated poverty induced by these changes [7]. Action research is a process of spiral development, each of which includes four interrelated and interdependent parts: plan, action, observation and reflection, shown as Figure 1:

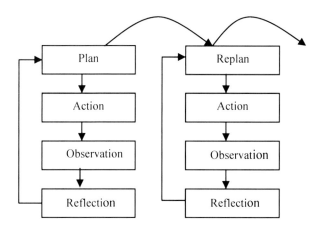

Fig. 1. The spiral cycle model of action research

We took about 160 undergraduates from Grade 2007 in Management School of Shenzhen University as the research object in 2008. We carried out three rounds of action research.

3.1 The First Round Action Research

(1) Plan. The objective in this round research is exploring the feasibility of an online collaborative learning mode in Management Information System experimental teaching on the basis of the utilization of network resources. In action designing, we mobilize test teaching and excite student's interest in network teaching and course study. Finally we put forward an online collaborative learning mode on the basis of the utilization of network resources.

(2) Action. The teacher assigns the task, takes task as the direction and instructs students to utilize online experimental resources to study independently at first. Then students study in groups, discuss online, and then the teacher answers questions. We utilize teaching resource module to carry on expanding knowledge and training ability. Through online investigation, we can understand student's learning adaptability under network environment, and we can also know the students' opinion on this kind of learning mode, activity schedule, learning content and teaching evaluation etc.

(3) Observation. Observe the relation among key elements of the learning mode. Observe the function of various kinds of experimental resources. Observe the students' adaptability of network learning and the ability of utilization of resources.

(4) Reflection. According to the online speech, classroom video and online investigation, we revise and perfect the teaching plan design. In fact, as the main body, most of the students applaud the online collaborative learning mode and participate in the experimental teaching activities actively. At the same time, student's experimental and practice ability has been improved in some degree. Therefore, the online collaborative learning mode in the professional experimental teaching of Management Information System through the Web is feasible. But, we also have found some problems too. For instance, the student's lasting attention are still not enough, the resources suitable for utilizing need to be perfected further; the online collaborative learning mode remains to be improved, and the Web-based teaching evaluation system has not been set up yet.

3.2 The Second Round Action Research

(1) Plan. The objective in this round research is to construct the online collaborative learning mode based on utilization of Web experimental resources, and analyze the key elements and characteristics of this mode. As for action design, we put forward a task-oriented online collaborative learning mode based on utilization of network resources. We guide students start learning with questions and taseks, which will improve the purpose and pertinence of online learning, and then further enhance the students' ability to learn autonomously and collaboratively.

(2) Action. The teacher puts forward some questions, assigns the tasks, encourage and instruct students to utilize online experimental resources to learn independently at first and then collaborate in groups. The students use the network course to learn online, and utilize the little online notebook and the interactive BBS module to exchange idea and discuss with group partners on relevant problems. The teacher will make a scene inspection or give an online guidance. At the end the teacher analyses, appraises and summarizes the situation in which students solve the practical problems and complete the experimental tasks.

(3) Observation. Observe the teaching implementation details and the cooperation among key elements of the online collaborative learning mode which takes questions and tasks as orientation. Observe students' activities and abilities of learning collaboratively to solve the problems and complete the tasks, utilizing online experimental resources.

(4) Reflection. Through the two rounds of teaching reform action research, the teachers have a better understanding of the key elements and the relationship among them in the online collaborative learning mode on the basis of the utilization of network experimental resources. The students take part in online discussion and collaboration more actively, and their initiative and enthusiasms are strengthened further. The network classroom teaching evaluation system promotes the standardization of web-based experimental teaching. But some problems still exists, for example, though we have paid more attention to following and appraising the student's speech and discussion during their online collaboration, the evaluation of works which reflect the students' integrated practical capability is not enough. We still lack the evaluation indexes of students' learning results under the network environment. And the students' collaborative ability and team spirit need to be improved further. Students' practical innovation ability still needs to be trained further.

3.3 The Third Round Action Research

(1) Plan. The objective in this round action research is to explore the teaching tactics and operation procedures of online collaborative learning mode on the basis of the utilization of online experimental resources. According to the result of the second round action research, we make some adjustment to the online collaborative learning mode, which is question-and-task-oriented, and emphasizes the autonomous and collaborative learning, group cooperation, exploration and finding, practical innovation, and pays more attention to teaching evaluation based on the students' works.

(2) Action. First, put forward the questions and define the tasks. Students participate in the online collaborative learning under the double driving force of questions and tasks. Second, students collaboratively learn, communicate and discuss. Students learn autonomously, utilize the network course little web-based notebooks or the interactive BBS to present doubt, comment, express view and discuss with each other. Third, expand knowledge and train ability. Utilizing the relevant online experimental resources in teaching resource module, expand students' knowledge and cultivate their practical and experimental abilities. Fourth, the group numbers cooperate and form the experimental report. Taking a group as a unit, the teacher guide students to carry on deep discussion on the above-mentioned questions, tasks and topics. Then the group leader puts discussion and collaborative experiment results in order and forms the group's experimental report. Fifth, discuss each group's report among groups and in the whole class. Sixth, contrasting with the task aim, appraise and summarize student's achievement of online collaborative learning.

(3) Observation. Observe the implementation course and operating procedure of the online collaborative learning mode during the third round action research. Further observe students' utilization of online experimental resources to solve problems and complete experimental tasks, by means of cooperation in a group and among groups. Through the evaluation indexes based on students' works and the teaching video, teachers can explore some teaching tactics of the online collaborative learning mode based on utilization of online experimental resources.

(4) Reflection. Students are more active and aggressive in this round action research of experimental teaching reform. They can solve the practical problems and finish the experimental tasks better through group cooperation and self-exploration. The teacher should offer proper guidance to students in the utilization of relevant experimental resources, for example, recommending several focal teaching resources for students, leading them to discuss thoroughly, helping them to improve the ability to explore, practice and innovate.

4 Research Results and Limitations

Through the untiring efforts and diligent cultivation, we have found out an online collaborative learning mode based on network environment during the practice of Management Information System experimental teaching reform. It is built upon the basis of constructivism learning theory, and reflects the discipline characteristics of Management Information System and has strong maneuverability, as shown in Fig. 2. During the practice of experimental teaching reform in exploring and using this mode, we have explored the concrete mode which integrates experimental courses with information technology in business or management schools in universities. We also have trained students' learning interest to professional courses of the specialties of Information Management and Information System and E-commerce, improve their ability to utilize the Web to explore, find, collaborate in groups, and innovate in practice. Meanwhile, we cultivate a team of vigorous teachers who are accomplished in the Web-based experimental teaching environment.

However, there are some limitations in our online collaborative learning mode. This mode lays particular emphasis on organization and implementation of the experimental teaching activities, further investigation on the connections and interactions between the five elements, namely teacher, student, network classroom, online course, teaching skill in the learning mode, is not in-depth. In addition, except for the online collaborative learning mode, other modes, for instance, the self-exploration learning mode and researchful learning mode still await follow-up research.

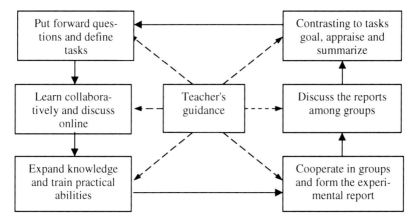

Fig. 2. An online collaborative learning mode in experimental teaching

Moreover, there are some deficiencies in Web-based experimental teaching. For instance, Web-based teaching resources generally adopt the organization of nonlinear hyperlink, which makes it easy for learners to distract their attention from the former page. The virtuality of network often makes it ineffective to track and supervise the learners' study activities. As for the application of technology, for instance, streaming media based video teaching can obtain ideal seeing and hearing effect, but compared with traditional face-to-face teaching, it lacks adequate emotional communication. So, we should combine Web-based experimental teaching with traditional experimental teaching and learn from other's strong points to offset one's weaknesses.

Acknowledgments. This research is supported by Shenzhen University Laboratory and Apparatus Management Research Foundation (No. 2009028).

References

1. He, K.: About the educational thinking and philosophical basis of constructivism—the reflection of constructivism. China University Teaching (7), 16–18 (2004)
2. Jonassen, D.H., Peck, K.L., Wilson, B.G.: Learning with technology: A constructivist perspective. Prentice Hall, Inc., Upper Saddle River (1999)
3. Swan, K.: A constructivist model for thinking about learning online. In: Bourne, J., Moore, J.C. (eds.) Elements of Quality Online Education: Engaging Communities, pp. 13–30. Sloan-C, Needham (2005)
4. Harris, J.: Structuring internet-enriched learning spaces for understanding and action. Learning and Leading with Technology 28(4), 50–55 (2001)
5. Hathorn, L.G., Ingram, A.L.: Cooperation and collaboration using computer-mediated communication. Journal of Educational Computing Research 26(3), 325–347 (2002)
6. Thompson, L., Ku, H.: A case study of online collaborative learning. The Quarterly Review of Distance Education 7(4), 361–375 (2006)
7. Stringer, E., Guhathakurta, M., Masaigana, M., Waddell, S.: Guest editors' commentary: Action research and development. Action Research 6, 123–127 (2008)

Statistical Properties Analysis of Internet Traffic Dispersion Networks

Cai Jun[1,2] and Yu Shun-Zheng[1]

[1] Department of Electronic and Communication Engineering Sun Yat-Sen University,
Guangzhou 510275, P.R. China
gzhcaijun@gmail.com, syu@mail.sysu.edu.cn
[2] School of Electronic and Information GuangDong Polytechnic Normal University,
Guangzhou 510665, P.R. China

Abstract. Recent frequent emergence of applications affecting Internet traffic dynamics have made it imperative to develop effective techniques that can make sense of different applications. In this paper, we build an unweighted and directed complex networks, which a vertex representing a simple IP address, and edge represents the exchange number of flows among these IP addresses on a specific application. Then we analyzed the statistic properties of these networks. According to the experiment result, we find the statistical properties diversity for different applications and can label p2p application.

Keywords: complex networks, statistical parameter, p2p.

1 Introduction

As the Internet continues to grow in size and complexity, the challenge of effectively provisioning, managing and securing it has become inextricably linked to a deep understanding of Internet traffic. There has been research on instrumenting data collection system for high speed networks at the core of the Internet, while the structure of Internet traffic network received little attention.

During the last decade, the study of complex networks has attracted a large amount of attention and works from several domains: computer science, sociology, arts and biology. Consequently, a large set of statistical parameters have been proposed to get a deep insight on structure characteristics. In this paper, we generate network, which a vertex represent a simple IP address, and edge represents the exchange number of flows among those IP address on different applications, which is denoted ITDN and based on the theory of complex networks, we analyzed ITDN statistical parameters for different applications, and label P2P application. We stride the first step to model the ITDN.

The remainder of this paper is organized as follows. In Section 2, we review the related work. In section 3, we describe the methodology in this paper, including empirical traces collected, generating the ITDN, the statistical properties value to analyze in this paper. In section 4 we present the experiment result. Finally, we present conclusions and future work in Section 5.

2 Related Works

Currently, three particular developments have contributed to the complex network theory: Watts and Strogatz's investigation of small-world networks [1], Barabasi and Albert's characterization of scale-free models [2], and Girvan and Newman's identification of the community structures present in many networks [3]. Complex network structures, generally modeled as large graphs, have played an important role in recent real networks. Complex network structures, generally modeled as large graphs, have played an important role in recent real networks. A series of applications to real networks, including social networks [4,5], the Internet and the World Wide Web[6], Metabolic, protein, genetic networks[7], and brain networks[8], have attracted increasing attentions.

Graph based techniques are used by Aiello et al. in the field of communities of Internet (CoI) [9]. CoI research focused on extracting communities and modeling "normal behavior" of hosts. Deviation from normal behavior can then used to trigger an alarm. The profiling of "social" behavior of hosts was studied in BLINC for the purpose of traffic classification. BLINC, however, focused only at host level. In BLINC, the notion of "graphlets" models a single host's flow patterns. A TDG is an aggregation of the graphlets of all hosts in a network for particular key. To simplify the aggregation, TDG use edges only between source and destination IP address unlike "grphlets" which also have edges to nodes that represent port numbers. Our work is similar to TDG, but is not the same to it. The graph metrics we employ in this paper is different from that of TDG.

3 Methodology

3.1 Empirical Traces

To perform the analysis presented in this paper we used the network traces from two geographically different locations. One is a publicly available WAN traffic trace provided by MAWI Working Group [10], which is recently collected at a trans-Pacific line (150Mbps link). The traces we use consist of all traffic measured form sample F during 48 hour period on March 18, 2008 at 00:00:00 to March 19, 2008 at 23:59:59, which is denoted as Dataone in this paper. The other traffic traces is that we collected two weeks from a single site in a large campus environment connected by a private IP backbone and serving a total user population in excess of 16000 hosts in January 2008. The flow records were collected from a boundary router using the Wireshark [11] and storing traffic in one day intervals, which is denoted as Datatwo in this paper. During the two week period we collected flow records corresponding to more than 400TByte of network traffic, not include weekend data. These traces contain up to 16 bytes of payload from each packet thereby allowing the labeling of flows using signature matching techniques similar to the ones described in [12]. Our traces with payload information increase the confidence of our analysis compared to using only port-based traffic.

3.2 Graph Generating

Graph theory is the natural framework for the exact mathematical treatment of complex network and, formally, a complex network can be represented as a graph. In IP networks, a node of ITDN corresponds to IP address, and an edge captures the exchange the exchange of flows number between various senders and destination IP address and the direction. In this paper, the flows is based on the well-known five-tuple (the source IP address, destination IP address, source port, destination port, and protocol fields).

3.3 Statistical Parameters in Networks

Each IP address corresponds a node $i \in V$ in a graph $G = (V, E)$. Edges in our graph $(i, j) \in E$ indicate flows relations between IP address, which are consequence of their communication behavior. Let n_{ij} be the number of flows that IP address i sends to IP address j. We consider two types of graphs. One is the unweighted and directed graph, which can be represented by an adjacency matrix $A = [n_{ij}]$. The adjacency $n_{ij} = 1$ if nodes i and j are connected 0 otherwise. The other is the directed and unweighted network, a directed edge existing from source IP address i to destination IP address j. At the same time, we divided nodes V in three sets V_{Ino} (with incoming and outgoing edges), V_{snk} (with only incoming edges) and V_{src} (with only outgoing edges). Then we defined eight statistic parameters in this paper, such as: the average degree<Avgdeg>, the incoming degree <Avgindeg>, the outgoing degree <Avgoutdeg>, the Clustering coefficient<C4>, the percentage of nodes that have both incoming and outgoing edges<VIno%>, the percentage of nodes that are sinks (only incoming edges)<Vsnk%>, the network density<netdensity> and the network centralization<netcent> can be calculated by Eqs. (1)(8), respectively [13],

$$Avg\,deg = \frac{2E}{V} \quad (1)$$

$$Avgin\,deg = \frac{2E}{Vsnk} \quad (2)$$

$$Avgin\,deg = \frac{2E}{Vsrc} \quad (3)$$

$$C_4 = \frac{the\ number\ of\ squares}{the\ total\ number\ of\ possible\ squares} \quad (4)$$

$$VIno(\%) = \frac{VIno}{V} \times 100\% \quad (5)$$

$$V\,sn\,k(\%) = \frac{V\,sn\,k}{V} \times 100\% \quad (6)$$

$$netdensity = \frac{\sum_i \sum_{j \neq i} n_{ij}}{V(V-1)} \quad (7)$$

$$netcent = \frac{V}{V-2}(\frac{k_{max}}{V-1} - netdensity) \quad (8)$$

4 Experiment

In this section, we present a series of fundamental statistical parameters computed over real-traffic. A set of experimental results of Dataone is summarized in Table1. The analysis results are as follows:

(1) Degree. The degree is the most widely used parameter for distinguishing the nodes of network. As describe in table 1, P2P applications have high average degree. We measure the $VIno(\%)$ and $Vsnk(\%)$. P2P file sharing applications (BitTorrent, eDonkey, fasttrack, Gnutella) have high $VIno(\%)$ value.

(2) Clustering coefficient.
The clustering coefficient of a node gives the probability that its neighbors are connected to each other. Because the data was collected from a single spot, the probability that nodes form groups of 3 is 0. The cluster coefficient is measured the probability that nodes form groups 4, which is given by Eq. (4). It is used to measure the extent of module structure present in the ITDN.

Table 1. Statistical value for different applications

Num	Proto.	Avgdeg	Avdindeg	Avgoutdeg	C4	VIno (%)	Vsink (%)	Netdensity	netcent
1	Web	2.547	1.080	3.798	0.06	1.553	66.276	0.0001	0.291
2	https	1.975	2.223	1.655	0.00	5.276	39.619	0.0017	0.173
2	FTP-data	1.269	1.481	1.111	0.00	1.587	41.269	0.0205	0.162
3	FTP	1.824	3.685	1.184	0.00	2.765	22.119	0.0085	0.220
4	IRC	1.880	1.757	1.756	0.01	10.445	44.776	0.0285	0.189
5	NTP	2.160	2.492	1.765	0.00	4.722	38.603	0.0044	0.189
6	POP3	1.342	1.428	1.190	0.00	0.684	41.095	0.0186	0.179
7	SMTP	3.030	2.232	4.215	0.00	5.131	20.324	0.0003	0.211
8	DNS	3.810	2.949	4.843	0.02	4.104	60.597	0.0001	0.553
9	BitTorrent	2.008	1.556	2.739	0.03	14.163	59.369	0.0087	0.036
10	eDonkey	3.695	1.348	2.207	0.05	3.809	60.544	0.0023	0.0189
11	fast-tracks	3.446	1.051	2.250	0.06	7.4857	66.964	0.0130	0.017
12	Gnutella	2.343	0.939	1.728	0.09	4.975	59.203	0.0067	0.023
13	Slammer	2.041	1.038	60.382	0.00	0.018	98.309	0	0.072

(3) Network centralization. The network centralization (also known as degree centralization) is given by Eq.(8). The centralization is 1 for a network with star topology; by contrast, it is 0 for a network where each node has the same connectivity. The centralization of the ITDN is close to 1, if one IP address connect with all others that in turn strongly communicate with each other, a centralization of 0 indicates that all IP address are equally popular. As described in table 1, the network centralization value of P2P (BitTorrent, eDonkey, fasttrack, Gnutella) is obviously larger than that of other applications.

5 Conclusion

Extracting significant events from vast masses of Internet traffic has assumed critical importance in light of the emergence of new and disruptive applications. In this paper, we have used complex network's theory-based techniques to computer the statistical parameters of the Internet traffic dispersion network. From the experiment result, we have discovered that different applications have different statistical parameter values and we have label P2P application according to them. The next work we shall design efficient thresholds that can be used to identify application from the statistical parameters of ITDN.

Acknowledgements. We would like to thank all the members in our research group, for the valuable discussions about the ideas presented in this paper. Funding from Project supported by the National High Technology Research and from Development Program of China (Grant No.2007AA01Z449) and the Key Program of NSFC-Guangdong Joint Funds (Grant No.U0735002).

References

1. Watts, D.J., Strogatz, S.H.: Collective dynamics of small-world networks. Nature 393(6684), 440–442 (1998)
2. Barabási, A.-L., Albert, R.: Emergence of Scaling in Random Networks. Science 286(5439), 509–512 (1999)
3. Girvan, M., Newman, M.E.J.: Community structure in social and biological networks. PNAS 99(12), 7812–7826 (2002)
4. Willinger, W., Doyle, J.: Robustness.: The Internet: Design and evolution, http://netlab.caltech.edu/Internet/
5. Boccaletti, S., Latora, V., Moreno, Y., Chavez, M., Hwang, D.-U.: Complex networks: Structure and dynamics. Physics Reports 424(4,5), 175–308 (2006)
6. Wasserman, S., Faust, K.: Social Networks Analysis. Cambridge University Press, Cambrigdes (1994)
7. Pastor-Satorras, R., Vespignani, A.: Evolution and Structure of the Internet: A Statistical Physics Approach. Cambridge University press, Cambridge (2004)
8. Harwell, L.H., Hopfield, J.J., Leibler, S., Murray, A.W.: From molecular to modular cell biology. Nature 402, C47–C52 (1999)

9. Aiello, W., Kalmanek, C., McDaniel, P., Sen, S., Spatscheck, O., Merwe, J.: Analysis of communities of Interest in Data Networks. In: Dovrolis, C. (ed.) PAM 2005. LNCS, vol. 3431, pp. 83–96. Springer, Heidelberg (2005)
10. MAWI Working Group Traffic Archive, http://mawi.wide.ad.jp/mawi/
11. http://www.wireshark.org/
12. Moore, A., Papagiannaki, K.: Toward the accurate identification of network applications. In: Dovrolis, C. (ed.) PAM 2005. LNCS, vol. 3431, pp. 41–54. Springer, Heidelberg (2005)
13. Horvath, S., Dong, J.: Geometric Interpretation of Gene Coexpression Network Analysis. PLoS Comput. Biol. 4(8), e1000117

Design and Implementation of Mobile Learning System Based on Mobile Phone

Qianzhu Shi

College of Information Science and Engineering,
Yanshan University, Qin Huangdao, Hebei, China
shi_qianzhu@126.com

Abstract. Recent years, mobile learning has stepped into the stage of practical application from the academic research rapidly. This paper presents the design of a set of phone-based mobile learning system that focuses on the overall design of the system, the function of each module and concrete implementation methods, as well as the technique used in the process of development. We use WML language to display the foreground page, and use the M3Gate software to simulate real mobile phone interface.

Keywords: Mobile-learning, Mobile Phone, WML.

1 Introduction

Mobile learning means learning at any time, any place with the help of mobile devices. With the support of wireless network technology, learning environment, learning resources and learners could all be movable. As for learning tools, laptops, PDAs, intelligent mobile phones, and other mobile terminal equipments have taken the place of personal computers and wired networks which are inconvenient to move.

This article presents the design and realization of a novel mobile learning system which can display the study content effectively, and provide an interactive teaching method between the teachers and learners.

2 System Analysis and Design

2.1 System Framework

The system takes studying resources as the center. Considering the characteristics of information and the sequence of browsing via mobile phones, we realize the study and interaction functions through organization and management of these resource and information. This platform is B(WEB)-B(WAP) /S (Web Server) pattern, which is shown in Fig. 1, consisting of a WEB version and a WAP version. The application and organization of the information, the resources and the simple flow may be realized through the WAP service completely, but the management and more complex flow have to be carried out through the WEB service, for the sake of the limited computing ability of mobile devices.

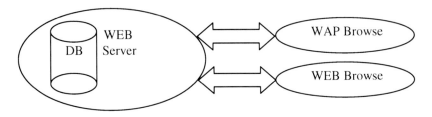

Fig. 1. System Framework

2.2 System Requirements Analysis

The roles in this system involve administrator, teacher and learner. The tasks that the roles should concern could be summarized as follows:

1. The system administrator is the super user who can add and remove teachers or learners, and also can manage the course materials the teachers uploaded. The administrator initializes the system platform and sets legal teacher users with the ability to release and manage the course materials and dispatches learners into the course which they want to participate.
2. Learners access WAP sites via mobile phones. First, they can visit the home page to browse real-time information such as news, pictures and courses, etc., and then register into the study section. After login, learners can modify their personal information and choose what they wish to learn. They can ask questions about specific knowledge points, and they also can discuss and exchange ideas with teachers and other students.
3. Through mobile phones, teachers can communicate with learners and perform only simple managements. The works involving adding or deleting data and managing course materials could only be done through the WEB service.

2.3 System Module Design

1. Resource Sharing Section
Through the WAP access, you can browse with no require for registering, the shared resources are divided into news pages, video pages, pictures pages and ring tones .All these resources are added to an appropriate database through background management service. News forum can be browsed directly, and videos, pictures and ring tones need to be downloaded to mobile phones to watch or play.

2. Learning Module

This section is divided into two subsections of personal management and knowledge study. Users can modify their own personal information. Knowledge browsing is in the way similar to BBS.

3. Background Management Module

Background management is an extraordinary important part to the mobile learning platform. The managers may carry on the simple backstage management through the handset, but complex operations must to be handled through the web service.

Background management involves the management for videos, pictures, ring tones, users and forums. Users are composed of system administrators and general managers, they have different permissions and can conduct different operations. System administrators have privilege over the system; the one with these privileges could be regarded as a super-user. An administrator has all the operation permissions for each section, including adding or removing general managers. The general managers have operation authorities for only a few parts of the news, videos, and other sections, which controlling the operations involving uploading, adding, deleting and other operations over the corresponding resources. In addition, there are some auxiliary sections, such as the link sections, message boards and interactive exchanges. Clicking on the corresponding WAP site will lead you into the link sections, in which the entrance URLs with significant names are presented; Message blocks can be used to put on messages, the user can make comments and suggestions in it through internet; Interaction section is divided into multiple categories, users can enter different class to exchange messages. Furthermore, the cell phone screen is small, so we provide search function on the homepage. Users can search study resources of different sections without page jumping.

3 Implementation of Mobile Learning Platform

3.1 WAP Site Configuration

The Microsoft Internet Information Services (IIS) does not support the development of WAP applications by its default configuration. Because files with the suffix of ".wml" are not recognized by the system. We need to call the MIME type setting screen and input the related suffix and content type into it, as shown in Fig 2. After this step, new document types in IIS can be used directly. These documents types are:

.wml text/vnd.wap.wml
.wmlc application/vnd.wap.wmlc
.wmls text/vnd.wap.wmlscript
.wmlsc application/vnd.wap.wmlscriptc

3.2 Development of Public Documents

In the development process, public documents refer to the files which need to be used in the entire website repeatedly, just as pages linked to database, the website background framework documents and certain public functions, etc.

In the process of the development of this platform, the public page files are conn.asp, function.asp and so on. The development of public files could greatly reduce the codes' repetition, and be easy to read and maintained.

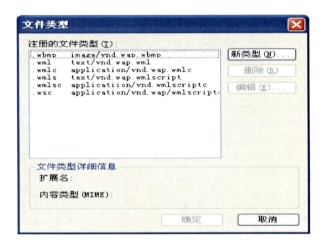

Fig. 2. Add Documents Type

3.3 The Development of Platform Homepage

Because the handset screen is small, number of characters for demonstration is limited and the homepage of the mobile study platform requires not only rich but also succinct links. So we can take the platform homepage as a navigator. In order to acquire detailed contents, users enter the next-level page through the navigator, by which troubles such as jumping into illegal pages may be avoided.

M3Gate is a WAP browser which can translate WML and WML Script marker language. This software is essential for WAP development, the main feature of which is to simulate the handset behavior on computers. Home page of the mobile phone platform is shown in Fig.3.

3.4 The Design and Development of Web Management Platform

The web management platform is realized by frames, providing the following management features, which are the basic establishment, the news management, video management, interaction management, picture management, the ting management, the user management, forum management, results management and other functions mutually.

4 Conclusion

Mobile phone-based mobile learning is a new trend. Its convenient and real-time learning style is obviously different with the traditional ones. It has a broad application prospect and its unique charm will certainly make the world different, and it also will be a new hot research area for the educational technology related researchers.

Fig. 3. Mobile Learning Platform Homepage

References

1. Lin, Y.C., Xu, F.Y.: Mobile Learning and its Theoretical Basis. Open Education Research, 23–26 (2004) (in Chinese)
2. Wu, L.: WAP Developer Guide. People's Posts & Telecom Press, Beijing (2001) (in Chinese)
3. Bao, J., Lu, J.: Dreamweaver MX 2004 the Perfect Example of Web Page Design. China Youth Press, Beijing (2005) (in Chinese)
4. Zeng, L.: Short News Service Learning System Realization based on mobile phone. Modern Educational Technology, 55–57 (2005) (in Chinese)
5. Wang, X.D., Li, Y.M.: Mobile Learning Base on Short News. Chinese Audiovisual Education, 114–117 (2007)
6. Fu, Q.: Motion Education Application New Pattern. Chinese Audiovisual Education, 111–113 (2007) (in Chinese)

Extract Backbones of Large-Scale Networks Using Data Field Theory

Zhang Shuqing[1], Li Deyi[1], Han Yanni[1], and Xing Ru[2]

[1] State Key Labrotory of Software Development Environment,
Beihang University, Beijing, China
[2] Shandong Business Institute, Yantai, China
{zhangshuqing,lideyi,hyn}@nlsde.buaa.edu.cn

Abstract. The last decade has witnessed a thriving development of the World Wide Web and internet networks, people are involved into different communication networks. With the increasing scale of these networks, it has left a challenge to understand their holistic features, to know which members and the relationships among them, that is, backbone, play vital roles. Based on data field theory, this paper proposes an algorithm that offers a practical procedure to extract backbone of large-scale networks. For dealing with the uncertainty of granularities of different backbones, we present a measurement to validate which is the optimal backbone with the most suitable granularity. We apply our method to the identical real-world networks and compare the results with several alternative backbone extracting methods. The experimental results illustrate that this backbone extraction algorithm can get a 0.8 score on precision ratios, and the recall ratios approximate reaches 0.5; the results are superior to the compared methods.

Keywords: backbone, compression, data field, topological potential, granularity.

1 Overview

Network topology is everywhere. It appears in social networks, the World Wide Web, or even the internet. A variety of important patterns has been hidden in these networks. Such as communities or cliques cover in social networks, and internet topologies contain a large number of tree-like structures with a clear hierarchy feather. Traditional backbone-extracting technologies often rely on the patterns or motifs construction, but there exist some limitations. For example, the method detecting backbone in social networks is not applicable for networks with a large number of tree structures, or the results will be ineffective.

Data field theory is applied to measure the interaction force among nodes in a network. For a certain node, the accumulation of the force from other nodes can reflect its importance. The measurement is also called topological potential [1]. Based on the measurement, in this paper, we proposed a backbone-extracting method, which can not only make up a deficiency for the limitations of method based on patterns, but also extract multi-granularity backbones according users' preferences. We also proposed a method to measure the connectivity performance of backbones, which can judge the optimal backbone with the best extraction granularity. It is worth of mentioning that the experimental results not only gain high precision ratios but also high recall ratios.

Data field theory is simply introduced in section 3, after discussing related work in the section 2. In section 4, based on the data field theory, we describes the algorithm extracting backbone of large-scale networks, As applying the algorithm can extract backbones of different granularities, but it's unsure who is the best one, in section 5 we put forward an evaluation method to decide which backbone has the best quality. In section 6 we experimentally verify our results on the algorithm and nine real-internet data sets compare our results with the method proposed in arc [2]. Finally, we conclude the paper in section 7.

2 Related Work

We discuss the related work as the following three aspects.

Extraction algorithm. Gilbert C. and Levchenko K. defined several compression schemes in arc[2], including vertex similarity measures and vertex ranking and presented nine compression methods based on the two schemes. In fact, due to the established extraction granularity, the backbones obtained were generally small-scale, so the precisions remain high, but the recall ratios extremely low.

Nan Du, Bin Wu and Bai Wang [3] propose an algorithm to discovery backbones in social networks, which needed to identify all cliques in the network. The complexity is quite high on large-scale networks. The procedure throughout satisfy the condition that backbone is a tree structure to insure the backbone to be enough thin, so the final obtained backbone is a minimum spanning tree of all kernel nodes.

Extracting application. Scellato S., Cardillo A. and Latora V., in arc[4], devise a method to show how it is possible to extract the backbone of a city by deriving spanning trees based on edge betweenness and edge information. Parter M., Kashtan N. and Alon U. [5] consider a reduced backbone should maintain a similar modularity as the original modules, so they describe a procedure for metabolic network reduction for visualization.

M. Angeles Serrano, MariAin Bogu and Alessandro Vespignani [6] propose a filtering method that offers a practical procedure to extract the relevant connection backbone in complex multi-scale networks. To use the filtering method, the weight of edges should be known, so it's not applied in networks without weight information.

Network data. Spring N., Mahajan R. and Wetherall D. [7] implemented a novel technique named roketfuel to derive approximate real-world IP topologies. The backbone data detected by rocketfuel can be used to verify extraction effectiveness of an algorithm or technology.

In arc[8], Winick J. and Jamin S. proposed a simulation internet network topology tool, inet-3.0, which was used to generate synthetic network data sets.

Newman [9] collected data sets of scientists-cooperation networks, and other social networks for the analysis and verification of community structures in complex networks. In this paper, we use several of the publicly available data sets in [9-10] for experiment.

3 Data Field Theory

Deyi L. and Yi D. [1] proposed data field theory to solve data mining and knowledge discovery problems, such as classification and clustering. According to the field theory in physics, potential in a conservative field is a function of position, which is inversely proportional to distance and directly proportional to magnitude of particle's mass or charge. Inspired from physical idea, they introduced field into network topological structure to describe the relationship among nodes being linked by edges and to reveal the general characteristic of underlying importance distribution.

Topology potential. Given the network G= (V, E), V is the set of nodes, E is the set of edges and |E|=m. A network can be viewed as a whole system constructed with nodes and interactions between each pair of them. The interactions is generated and transformed by the edges. Surrounding each node there is a virtual strength field, any node in the network will be subject to the combined effects of the other nodes. Thus, on the entire network topology a data field was established, called the topology potential field [11]. Generally the Gaussian function, which could respect the role of short-range field and have good mathematical properties, is used to describe the interaction between the nodes. According to the data field potential function definition, topology potential score of any node in the network can be expressed as the below formation

$$\varphi(v_i) = \sum_{i=1}^{n} m_j \cdot e^{-\left(\frac{d_{ij}}{\sigma}\right)^2} \tag{1}$$

where d_{ij} is the distance between node v_i and v_j, parameter σ is used to control the influence region of each node, called influence factor, $m_i \geq 0$ is the mass of the node $v_i (i = 1, \cdots, N)$. This article assumes that the mass of each node are equal and meet the normalization $\sum_{i \in N} m_i = 1$. The topology potential formula may be simplified description as below.

$$\varphi(v_i) = \frac{1}{n}\sum_{i=1}^{n} e^{-\left(\frac{d_{ij}}{\sigma}\right)^2} \tag{2}$$

4 Backbone Extracting

We suppose that backbone is constituted by kernel members, including kernel nodes and important edges. Kernel nodes are influential nodes whose topological potential score are often relatively high. Generally the edges between kernel nodes are important. To extract the backbone of networks, our algorithm iterates the step of searching backbone ties between the kernel nodes until the whole backbone becomes connected.

Kernel nodes. We sort the topological potential score of each node in a network. For a given a parameter $\alpha(0 \le \alpha \le 1)$, which is established to be a threshold to control the scale of the obtained backbone. We define the nodes ranking top α as the original kernel nodes. The procedure of extracting the whole backbone is divided into the following two steps:

1. Finding the kernel nodes as the original backbone members, denoted as *source*. As this step is completed, each isolated node in *source* is an island subnet;

2. Finding the bridge ties to connect those island subnets, and joining the ties to *source*, loop the two operations until *source* is connective. We define the distance between two island subnets as the equation (3) below,

$$dist(subg1, subg2) = \min_{v1 \in subg1, v2 \in subg2} |shortestpath(v1, v2)| \quad (3)$$

where subg1 and subg2 are two island subnets in *source*, v1 and v2 are respectively arbitrary nodes of the two subnets. While searching the backbone ties, in each step we choose a pair of island subnets with the shortest distance between them. If the distance equals 1, directly select the tie contain one hop between the two subnets. After this the two subnets are merged into one. Otherwise, we find node in each of the two subnets, as one new kernel node, which ranks the top topological potential score in all neighbors of the subnet. Intuitively the distance between the two island subnets is very likely to be reduced. The algorithm continuously iterates the above step until the backbone has no island subnets.

We describe the algorithm as below.

```
Procedure backbone extraction:
  Input: network G, α(0≤α≤1)
  Output: backbone B(α)
  Matrix Sp: compute shortest path length of all pairs of
  nodes; var i: = 1;
  Evaluate hops: = avg(Sp); evaluate factor: = √2 /3*avg(Sp);
  Begin:
  repeat:
    i: = i + 1;
    for each node v ∈ G, compute φⁱ(v) :topological potential
      within i hops;
    sort(φⁱ(v),v∈G); source: = ∅ ;
    for each node v ∈ G,
      if φⁱ(v),v∈G rank Top α, source:= source∪{v};
  repeat:
    for each pair of island subnets subg1,subg2∈ source,
      if distance between subg1 and subg2 is the shortest,
        if distance(subg1, subg2) = 1
          merge(subg1, subg2);
        else
```

```
         find  neig1∈ subg1, φ'(neig1) = max{φ'(v∈ subg1)} , tie1 link neig1 and subg1
               neig2∈ subg2, φ'(neig2) = max{φ'(v∈ subg2)} , tie2 link neig2 and subg2
         source:= source∪{neig1, neig2, tie1, tie2};
           end if
         end if
     until network generated from source is connected
     B(α) := B(α)∪source ;
   until i ≥ hops
End
```

Performance analysis: While searching the shortest path length between all pairs of nodes in a given network, we use the breadth-first searching strategy. In an undirected and weightless network, this will cost $O(|V||E|)$. To calculate topology potential score for each node, we need to cost $O(|V||E|)$, where $|V|$ denotes the number of nodes, and $|E|$ denotes the number of edges. Searching all backbone ties costs $O(\alpha |V|^2 avg(Sp))$, because if we assume the number of the original island subnets is $\alpha |V|$, then to make them connective, it at most needs to seek $O(\alpha |V| \cdot |V| avg(Sp))$ ties. Therefore, the overall will cost $O(max\{\alpha |V|^2 avg(Sp), |V||E|\})$ in the worst case.

5 Performance Measurement

As our algorithm should accepts a parameter $\alpha (0 \leq \alpha \leq 1)$, the backbone scale will be uncertain when the value of $\alpha (0 \leq \alpha \leq 1)$ changes. This seems to be able to obtain backbones with multi-granularities, but which backbone is the best? If we remove backbone from the network G, it will generate a new network, which we denote it as G'. Now we care the network connectivity changes.

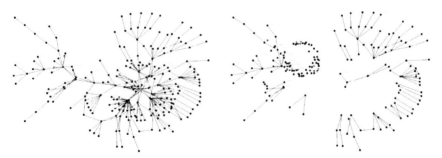

Fig. 1. After removing backbone from the original network as4755 (such as the left), it generates a number of island subnets (such as the right)

After the removing operation, network G' may contain a number of island subnets. Fig.1 presents the situation after remove backbone from a network named as 4755. Here we denote the number of island subnets as $cnt\_subgs(G')$. Intuitively the larger *cnt_subgs(G')* is, the worse the connectivity of network G' is. The reduction of connectivity is caused by the missing of the backbone. The worse connectivity of the network G' indicates that the backbone make greater contribution to the performance of network connectivity.

On the other hand, the size of the island subnets also reflects some important information. For instance, if the largest island subnet is large enough to achieve half of the original network, the connectivity loss of network G' is weak. It still remains kernel nodes or important ties in the island subnet. On the contrary, if the largest island subnet is small, the possibility that it contains backbone nodes will be low, theoretically the possibility of other island subnets contains backbone nodes will also be low. Here we denote the size of the largest island subnet of network G' as *lar_subgs_size(G')*. The smaller *lar_subgs_size(G')* is, the more possible network G' lost kernel nodes and important ties. Thus, we consider that the connectivity loss of network G' can be measured by the size of the largest island subnet *lar_subgs_size(G')*.

We use the publicly available real-world networks data sets offered in [10] as the test-bed. These networks also will be used to carry out validation analysis in the next section (Sec.6). We choose four of them, which respectively are named as as3356, as4755, as2914, and as7018.

How can we optimize the backbone within the scale of effective compression ratio? If the size of island subnets in network G' are small enough, then the probability that these subnets contain kernel nodes is extremely low. Under the circumstance, it can be considered that while removing backbone from the original network, the connectivity of network G' is collapsed. We want to discover the changes of the size of the largest island subnet. In order to unify measurement, *lar_subgs_size(G')* is normalized as equation (1). According to BA model and Eppstein Power law model proposed in [12-13], we use computer to randomly simulate complex networks. Using these synthetic networks and the four real-world networks described above, we observe changes of *r_lar_subgs_size(G')* on each network shown in Fig.2.

$$r\_lar\_subgs\_size(G') = \frac{lar\_subg\_size(G)}{size(G)} \qquad (4)$$

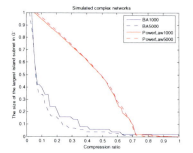 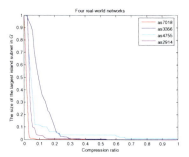

Fig. 2. When the compression ratio increases to an enough high value, the size of the largest island subnet *r_lar_subgs_size(G')* changes little

We discuss the granularity of backbones, which we measure with the number of backbone nodes or the compression ratio. As Fig.2 shown, when the backbone granularity reaches a certain scale, the further expansion of its granularity would lower the value of $r\_lar\_subgs\_size(G')$ extremely little, but reduce the extraction precision. So the further expansion is little gain but great loss. In fact, as Fig.2 shown, the value of $r\_lar\_subgs\_size(G')$ monotonically decrease while the compression ratio increases. When the compression ratio increases to a certain point, the value of $r\_lar\_subgs\_size(G')$ remains almost unchanged or decreases extremely low. Therefore, on the fitted curve exists a critical point. Summarizing the above analysis, we consider that the algorithm needs to obtain the optimal backbone with the compression ratio at the critical point. In order to judge which point is the critical point, we define an elasticity coefficient as equation (2) describes.

$$\text{Elasticity} = \frac{d(r\_lar\_subgs\_size(G'))}{d(compress\_ratio)} \tag{5}$$

It's assured that the critical point could tradeoff between the scale of backbones and the extraction precision best. Using the definition of elasticity on equation (2), we define the critical point as the point on the curve while elasticity coefficient equals 1. As the fitted curve is monotonically decreasing, using the binary searching strategy to detect the critical point, satisfying enough precision, the optimal compression ratio can be searched rapidly.

6 Experiment

Measuring the performance of backbone network is to extract backbone with high-performance. According the analysis of Sec.5, we first need to search the optimal compression ratio. As the networks describes the topology of different autonomous systems and the real-world backbone nodes were detected and filled into the provided datasets, which can be used to verify the extraction quality of the algorithm. The related data on the optimal backbone and the results that compare it with the real backbone is shown in table 1 below.

Table 1. Related data of the extracting backbones and the results that compare it with the real backbone in nine tested networks

Network name	Original network scale	Optimal backbone scale	Compression ratio	Size of the largest island subnet	Real backbone scale	Precision ratio	Recall ratio
as4755	226	16	0.071	13	43	81.3%	30.2%
as3356	1786	385	0.216	94	454	86.2%	73.1%
as2914	6253	295	0.058	193	794	76.3%	28.3%
as7018	11745	590	0.043	119	531	61.0%	57.6%
as1239	10180	355	0.035	120	574	76.6%	47.4%
as1755	300	34	0.113	40	117	91.2%	26.5%
as3257	506	64	0.126	27	174	98.4%	36.2%
as3967	424	77	0.182	25	115	66.2%	44.3%
as6461	654	88	0.135	23	152	86.4%	50.0%

In arc[2], the authors proposed a variety of compression methods. For the four real-world networks named as1239, as2914, as3356 and as7018, table 2 presents the precision ratio and recall ratio of the methods in arc[2] and the method based on data field theory in this paper. Although the former can obtain results with higher precision ratios, most of which can reach above 0.9, but the recall ratios below 0.2. However, the latter can maintain the backbone precision ratios to about 0.8, and the recall ratios to about 0.5. An excellent extraction method requires the results maintain high recall ratio. From this point of view, the data-field method is superior to the methods proposed in arc[2]. Some other relevant extraction algorithms provide no effective validation, so the extraction quality was uncertain.

The optimal backbones extracted using the data field method is shown in Fig.3-4. The nodes with diamond shape are the real backbone nodes extracted correctly, while the nodes with square shape are the nodes mistaken as backbone nodes.

Table 2. Compare the precision ratio and recall ratio of this method with the method introduced in arc[2]

Compressing method	as1239 precision ratio	as1239 recall ratio	as2914 precision ratio	as2914 recall ratio	as3356 precision ratio	as3356 recall ratio	as7018 precision ratio	as7018 recall ratio
Deg/One	0.96	0.066	0.957	0.049	1	0.046	0.895	0.088
Beta/One	0.97	0.097	0.774	0.091	1	0.06	0.909	0.104
Path/One	1	0.055	1	0.044	1	0.046	1	0.052
Deg/All	0.916	0.27	0.966	0.188	0.927	0.176	0.911	0.213
Beta/All	0.94	0.348	0.89	0.267	0.969	0.216	0.912	0.242
Path/all	0.953	0.169	1	0.143	0.971	0.158	0.956	0.112
RVE+Deg/One	1	0.058	0.909	0.044	1	0.046	0.941	0.083
RVE+Beta/All	0.913	0.262	0.847	0.221	0.935	0.167	0.89	0.231
RVE+Path/One	1	0.055	1	0.046	0.85	0.039	1	0.054
Data Field	**0.766**	**0.474**	**0.763**	**0.283**	**0.862**	**0.731**	**0.61**	**0.576**

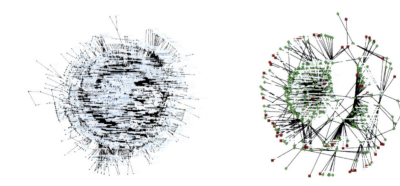

Fig. 3. The as3356 network and its optimal backbone extracted using the method proposed in this paper

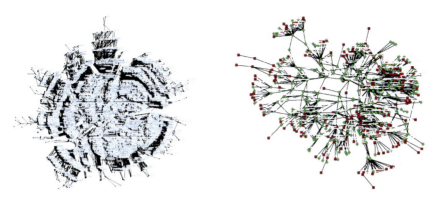

Fig. 4. The as7018 network and its optimal backbone extracted using the method proposed in this paper

7 Conclusion

This paper presents an algorithm based on data field theory to extract backbones of the large-scale networks. Moreover, a novel backbone performance measurement was designed according the changes of connectivity of network while removing backbone from the original network, with which the optimal extraction granularity could be obtained. Verifying the method on several large-scale real-world networks, the extracted backbones have both high precision ratio and high recall ratio.

Acknowledgements. This work was supported by national basic science research program of China under grant no.2007CB310803 and the China natural science foundation under grant no.60496323 and no.60803095.

References

1. Deyi, L., Yi, D.: Artificial Intelligence with Uncertainty, pp. 153–200. M. Chapman & Hall/CRC (2007)
2. Gilbert, C., Levchenko, K.: Compressing Network Graphs. In: Proceedings of the LinkKDD Workshop at the 10th ACM Conference on KDD (2004)
3. Nan, D., Bin, W., Bai, W.: Backbone Discovery in Social Networks. In: IEEE/WIC/ACM International Conference on Web Intelligence, pp. 100–103. IEEE Computer Society, USA (2007)
4. Scellato, S., Cardillo, A., Latora, V.: The Backbone of a City. J. The European Physical Journal B 50(1), 221–225 (2006)
5. Parter, M., Kashtan, N., Alon, U.: Environmental Variability and Modularity of Bacterial Metabolic Networks. J. BMC Evolutionary Biology 196 (2007)
6. Angeles Serrano, M., Bogu, M., Vespignani, A.: Extracting the Multi-scale Backbone of Complex Weighted Networks. J. PNAS 106(16), 6483–6488 (2009)

7. Spring, N., Mahajan, R., Wetherall, D.: Measuring ISP Topologies with Rocketfuel. In: ACM SIGCOMM 2002 Proceedings, pp. 133–145 (2002)
8. Winick, J., Jamin, S.: Inet-3.0: Internet Topology Generator. Technical Report CSE-TR-456-02,University of Michigan (2002)
9. University of Michigan, http://www-personal.umich.edu/%7Emejn/netdata/
10. University of Washington, Rocketfuel: an ISP Topology Mapping Engine, http://www.cs.washington.edu/research/networking/rocketfuel/
11. Yanni, H., Jun, H., Deyi, L.: A Novel Measurement of Structure Properties in Complex Networks. In: Complex 2009 (2009)
12. Albert, L., Barabási, et al.: Emergence of Scaling in Random Networks. J. Science 286(5439), 509–512 (2002)
13. Eppstein, D., Joseph, W.: A Steady State Model for Graph Power Laws. In: International Workshop on Web Dynamics (2002)
14. Nan, H., Wenyan, G., Deyi, L.: Evaluate Nodes Importance in the Network using Data Field Theory. In: ICCIT, pp. 1225–1230. IEEE Computer Society, Los Alamitos (2007)
15. Han, J., Kamber, M.: Data Mining: Concepts and Technologies, 2nd edn. M. Morgan Kaufmann, San Francisco (2005)

An Efficient Collaborative Recommendation Algorithm Based on Item Clustering

Songjie Gong

Zhejiang Business Technology Institute, Ningbo 315012, P.R. China
eigsj@sina.com

Abstract. To help people to find useful information efficiently and effectively, information filtering technique emerges as the times require. Collaborative recommendation is becoming a popular one, but traditional collaborative recommendation algorithm has the problem of sparsity, which will influence the efficiency of prediction. Unfortunately, with the tremendous growth in the amount of items and users, the lack of original rating poses some key challenges for recommendation quality. Aiming at the problem of data sparsity for recommender systems, an efficient collaborative recommendation algorithm based on item clustering is presented. This method uses the item clustering technology to fill the vacant ratings where necessary at first, then uses collaborative recommendation to form nearest neighborhood, and lastly generates recommendations. The collaborative recommendation based on item clustering smoothing can alleviate the sparsity issue in collaborative recommendation algorithms.

Keywords: recommender system; collaborative recommendation; item clustering; sparsity.

1 Introduction

With the rapid growth and wide application of the Networks, the amount of information is increasing more quickly than people's ability to process it. All of us have known the feeling of being overwhelmed by the number of new things coming out each year. Now technology has dramatically reduced the barriers to publishing and distributing information. It is time to create the technologies that can help people sift through all the available information to find that which is most valuable to us and we can use it [1,2].

Collaborative recommendation is becoming a popular technology. The task in collaborative recommendation is to predict the use of items to a particular user the active user based on user item ratings database. Memory based algorithms operate over the entire user database to make predictions and model based algorithms in contrast uses the user database to estimate or learn a model which is then used for predictions [3,4,5].

When dealing with collaborative recommendation, two fundamental problems of recommender systems have to be taken into account. The first is the sparsity of the data and the second is the scalability problem. The sparsity problem, which we encounter when rating is missing many values, arises from the fact that with the growth of the number of users and the number of items, the basic nearest neighbors algorithm fails to scale up its computation [6,7,8].

In order to help people to find useful information efficiently and effectively, information filtering technique emerges as the times require. Collaborative recommendation is becoming a popular one, but traditional collaborative recommendation algorithm has the problem of sparsity, which will influence the efficiency of prediction. Unfortunately, with the tremendous growth in the amount of items and users, the lack of original rating poses some key challenges for recommendation quality. Aiming at the problem of data sparsity for recommender systems, in this paper, an efficient collaborative recommendation algorithm based on item clustering is presented. This method uses the item clustering technology to fill the vacant ratings where necessary at first, then uses collaborative recommendation to form nearest neighborhood, and lastly generates recommendations. The collaborative recommendation based on item clustering smoothing can alleviate the sparsity issue in collaborative recommendation algorithms.

2 Employing the Item Clustering to Alleviate Sparsity Issue

2.1 Clustering Items to Form Centers

Item clustering techniques work by identifying groups of items who appear to have similar ratings. Once the clusters are created, predictions for a target item can be made by averaging the opinions of the other items in that cluster. Some clustering techniques represent each item with partial participation in several clusters. The prediction is then an average across the clusters, weighted by degree of participation. Once the item clustering is complete, however, performance can be very good, since the size of the group that must be analyzed is much smaller [9,10,11].

The idea is to divide the items of a collaborative recommendation system using item clustering algorithm and use the divide as neighborhoods, as Figure 1 show. The clustering algorithm may generate fixed sized partitions, or based on some similarity threshold it may generate a requested number of partitions of varying size.

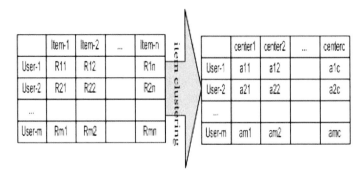

Fig. 1. Item clustering to form centers

An Efficient Collaborative Recommendation Algorithm Based on Item Clustering

Where Rij is the rating of the user i to the item j, aij is the average rating of the user i to the item center j, m is the number of all users, n is the number of all items, and c is the number of item centers.

2.2 Smoothing Vacant Ratings Where Necessary

One of the challenges of the collaborative recommender is the data sparsity problem. To prediction the vacant values in user-item rating dataset where necessary, we make explicit use of item clusters as prediction mechanisms.

Based on the item clustering results, we apply the prediction strategies to the vacant rating data as follows:

$$R_{ij} = \begin{cases} R_{ij} & \text{if user } i \text{ rate the item } j \\ c_j & \text{else} \end{cases}$$

where cj denotes the item center.

3 Producing Recommendation

Through the calculating the vacant user's rating by item clustering, we gained the dense user item rating database. Then, to generate prediction of a user's rating, we use the collaborative recommendation algorithms.

3.1 The Dense User-Item Matrix

After we used the item clustering, we gained the dense ratings of the users to the items. So, the original sparse user-item rating matrix is now becoming the dense user-item matrix.

3.2 Measuring the User Rating Similarity

There are several similarity algorithms that have been used in the collaborative filtering recommendation algorithm.

Pearson's correlation, as following formula, measures the linear correlation between two vectors of ratings.

$$sim(i, j) = \frac{\sum_{c \in I_{ij}} (R_{i,c} - A_i)(R_{j,c} - A_j)}{\sqrt{\sum_{c \in I_{ij}} (R_{i,c} - A_i)^2 \sum_{c \in I_{ij}} (R_{j,c} - A_j)^2}}$$

Where Ri,c is the rating of the item c by user i, Ai is the average rating of user i for all the co-rated items, and Iij is the items set both rating by user i and user j.

The cosine measure, as following formula, looks at the angle between two vectors of ratings where a smaller angle is regarded as implying greater similarity.

$$sim(i,j) = \frac{\sum_{k=1}^{n} R_{ik} R_{jk}}{\sqrt{\sum_{k=1}^{n} R_{ik}^2 \sum_{k=1}^{n} R_{jk}^2}}$$

Where Ri,k is the rating of the item k by user i and n is the number of items co-rated by both users.

The adjusted cosine, as following formula, is used in some collaborative filtering methods for similarity among users where the difference in each user's use of the rating scale is taken into account.

$$sim(i,j) = \frac{\sum_{c \in I_{ij}} (R_{ic} - A_c)(R_{jc} - A_c)}{\sqrt{\sum_{c \in I_{ij}} (R_{ic} - A_c)^2 * \sum_{c \in I_{ij}} (R_{jc} - A_c)^2}}$$

Where Ri,c is the rating of the item c by user i, Ac is the average rating of user i for all the co-rated items, and Ii,j is the items set both rating by user i and user j.

In this paper, we use the cosine measure to calculate the similarities of users.

3.3 Selecting the Target User Neighbors

In this step, we select of the neighbors who will serve as recommenders. We employ the top-n technique in which a predefined number of n-best neighbors selected.

3.4 Recommender Using Collaborative Recommendation Algorithm

Since we have got the membership of user, we can calculate the weighted average of neighbors' ratings, weighted by their similarity to the target user.

The rating of the target user u to the target item t is as following:

$$P_{ut} = A_u + \frac{\sum_{i=1}^{c} (R_{it} - A_i) * sim(u,i)}{\sum_{i=1}^{c} sim(u,i)}$$

Where A_u is the average rating of the target user u to the items, Rit is the rating of the neighbour user i to the target item t, Am is the average rating of the neighbour user i to the items, sim(u, i) is the similarity of the target user u and the neighbour user i, and c is the number of the neighbours.

4 Dataset and Measurement

4.1 Data Set

The MovieLens site continues to be used by people every week. MovieLens data sets were collected by the GroupLens Research Project at the University of Minnesota. Ratings data from the MovieLens systems have been responsible for many publications. The historical dataset consists of 100,000 ratings from 943 users on 1682 movies with every user having at least 20 ratings and simple demographic information for the users is included [12,13]. Therefore the lowest level of sparsity for the tests is defined as 1 − 100000/943*1682=0.937. The ratings are on a numeric five-point scale with 1 and 2 representing negative ratings, 4 and 5 representing positive ratings, and 3 indicating ambivalence.

4.2 Evaluation Measurement

Several metrics have been proposed for assessing the accuracy of collaborative recommender system algorithms. They are divided into two main categories. The first one is the statistical accuracy metrics and the other one is the decision-support accuracy metrics [14, 15].

Statistical accuracy metrics evaluate the accuracy of a prediction algorithm by comparing the numerical deviation of the predicted ratings from the respective actual user ratings. Some of them frequently used are mean absolute error (MAE), root mean squared error (RMSE) and correlation between ratings and predictions. All of the above metrics were computed on result data and generally provided the same conclusions [14].

Formally, if n is the number of actual ratings in an item set, then MAE is defined as the average absolute difference between the n pairs. Assume that p1, p2, p3, ..., pn is the prediction of users' ratings, and the corresponding real ratings data set of users is q1, q2, q3, ..., qn. See the MAE definition as following:

$$MAE = \frac{\sum_{i=1}^{n} |p_i - q_i|}{n}$$

Decision support accuracy metrics evaluate how effective a prediction engine is at helping a user select high-quality items from the set of all items. The receiver operating characteristic (ROC) sensitivity is an example of the decision support accuracy metric. The metric indicates how effectively the system can steer users towards highly-rated items and away from low-rated ones [15].

Assume that p1, p2, p3, ..., pn is the prediction of users' ratings, and the corresponding real ratings data set of users is q1, q2, q3, ..., qn. See the ROC-4 definition as following:

$$ROC\text{-}4 = \frac{\sum_{i=1}^{n} u_i}{\sum_{i=1}^{n} v_i}$$

$$u_i = \begin{cases} 1, & p_i \geq 4 \text{ and } q_i \geq 4 \\ 0, & otherwise \end{cases}$$

$$v_i = \begin{cases} 1, & p_i \geq 4 \\ 0, & otherwise \end{cases}$$

The larger the ROC-4, the more accurate the predictions would be, allowing for better recommendations to be formulated.

5 Conclusions

In order to help people to find useful information efficiently and effectively, information filtering technique emerges as the times require. Collaborative recommendation is becoming a popular one, but traditional collaborative recommendation algorithm has the problem of sparsity, which will influence the efficiency of prediction. Unfortunately, with the tremendous growth in the amount of items and users, the lack of original rating poses some key challenges for recommendation quality.

Aiming at the problem of data sparsity for recommender systems, in this paper, an efficient collaborative recommendation algorithm based on item clustering is presented. This method uses the item clustering technology to fill the vacant ratings where necessary at first, then uses collaborative recommendation to form nearest neighborhood, and lastly generates recommendations. The collaborative recommendation based on item clustering smoothing can alleviate the sparsity issue in collaborative recommendation algorithms.

References

1. Sarwar, B., Karypis, G., Konstan, J., Riedl, J.: Item-Based collaborative filtering recommendation algorithms. In: Proceedings of the 10th International World Wide Web Conference, pp. 285–295 (2001)
2. Chee, S.H.S., Han, J., Wang, K.: Rectree: An efficient collaborative filtering method. In: Kambayashi, Y., Winiwarter, W., Arikawa, M. (eds.) DaWaK 2001. LNCS, vol. 2114, p. 141. Springer, Heidelberg (2001)
3. Breese, J., Hecherman, D., Kadie, C.: Empirical analysis of predictive algorithms for collaborative filtering. In: Proceedings of the 14th Conference on Uncertainty in Artificial Intelligence (UAI 1998), pp. 43–52 (1998)

4. Bridge, D., Kelleher, J.: Experiments in sparsity reduction: Using clustering in collaborative recommenders. In: O'Neill, M., Sutcliffe, R.F.E., Ryan, C., Eaton, M., Griffith, N.J.L. (eds.) AICS 2002. LNCS (LNAI), vol. 2464, pp. 144–149. Springer, Heidelberg (2002)
5. Kelleher, J., Bridge, D.: Rectree centroid: An accurate, scalable collaborative recommender. In: Procs. of the Fourteenth Irish Conference on Artificial Intelligence and Cognitive Science, pp. 89–94 (2003)
6. Grcar, M., Mladenic, D., Fortuna, B., Grobelnik, M.: Data Sparsity Issues in the Collaborative Filtering Framework. In: Nasraoui, O., Zaïane, O.R., Spiliopoulou, M., Mobasher, B., Masand, B., Yu, P.S. (eds.) WebKDD 2005. LNCS (LNAI), vol. 4198, pp. 58–76. Springer, Heidelberg (2006)
7. George, T., Merugu, S.: A scalable collaborative filtering framework based on coclustering. In: Proceedings of the IEEE ICDM Conference (2005)
8. Rashid, A.M., Lam, S.K., Karypis, G., Riedl, J.: ClustKNN: A Highly Scalable Hybrid Model- & Memory-Based CF Algorithm. In: Nasraoui, O., Spiliopoulou, M., Srivastava, J., Mobasher, B., Masand, B. (eds.) WebKDD 2006. LNCS (LNAI), vol. 4811, pp. 147–166. Springer, Heidelberg (2006)
9. Xue, G., Lin, C., Yang, Q., et al.: Scalable collaborative filtering using cluster-based smoothing. In: Proceedings of the ACM SIGIR Conference 2005, pp. 114–121 (2005)
10. Cantador, I., Castells, P.: Multilayered Semantic Social Networks Modelling by Ontologybased User Profiles Clustering: Application to Collaborative Filtering. In: Staab, S., Svátek, V. (eds.) EKAW 2006. LNCS (LNAI), vol. 4248, pp. 334–349. Springer, Heidelberg (2006)
11. Symeonidis, P., Nanopoulos, A., Papadopoulos, A., Manolopoulos, Y.: Nearest-Biclusters Collaborative Filtering. In: Nasraoui, O., Spiliopoulou, M., Srivastava, J., Mobasher, B., Masand, B. (eds.) WebKDD 2006. LNCS (LNAI), vol. 4811, pp. 36–55. Springer, Heidelberg (2006)
12. Herlocker, J.: Understanding and Improving Automated Collaborative Filtering Systems. Ph.D. Thesis, Computer Science Dept., University of Minnesota (2000)
13. Sarwar, B., Karypis, G., Konstan, J., Riedl, J.: Recommender systems for large-scale e-commerce: Scalableneighborhood formation using clustering. In: Proceedings of the Fifth International Conference on Computer and Information Technology (2002)
14. Qin-hua, H., Wei-min, O.: Fuzzy collaborative filtering with multiple agents. Journal of Shanghai University (English Edition) 11(3), 290–295 (2007)
15. Fengrong, G., Chunxiao, X., Xiaoyong, D., Shan, W.: Personalized Service System Based on Hybrid Filtering for Digital Library. Tsinghua Science and Technology 12(1), 1–8 (2007)

An Intelligent Solution for Open Vehicle Routing Problem in Grain Logistics

Hongyi Ge[1], Tong Zhen[1], Yuying Jiang[1], and Yi Che[2]

[1] College of Information Science and Engineering, Henan University of Technology,
Zhengzhou 450001, China
[2] Academic Administration, Anhui Audit Vocational College, Hefei 230601, China
gehongyi2004@163.com

Abstract. This paper studies the grain logistics open vehicle routing problem (OVRP), in which the vehicles are not required to return to starting depot after completing service, or if they are required, they must return by traveling the same route back. The grain logistics OVRP is a well known hard combinatorial optimization problem, the objective is to minimize the fleet size following routes of minimum distance. We present a hybrid intelligent algorithm for solving the open-version of the well-known open vehicle routing problem (OVRP). Computational results are given for several standard test instances, which show that the proposed procedure obtains better solutions than those found in the open literature, and also indicate that the proposed hybrid method is capable of obtaining optimal solutions very efficiently.

Keywords: grain logistics, open vehicle route problem, simulating annealing, hybrid particle swarm optimization, optimization.

1 Introduction

Grain logistics is an important component of the logistics system for agricultural products, which has the characteristics of great quantity, wide range and multipoint, and is different from other physical distribution of goods. Grain circulation plays a major role in the economy of the State, especially in rural. The efficiency of the grain logistics is a key factor in the economic prosperity, as the cost of moving grain directly impacts on grain growers' incomes. In order to reduce the cost of grain logistics, which will enhance peasants' income, increase enterprise benefits and guarantee nation grain security. However, there is a large span about grain logistics in the time and space, and will cause higher logistics cost. Accordingly, the solution to the minimization the cost of grain logistics is to optimize grain logistics vehicle routing problem, arrange the grain circulation route scientifically and reasonably, it has the great strategic and realistic significance.

The vehicle routing problem (VRP), which was first proposed by Dantzig and Ramser in 1959, is a well-known combinatorial optimization problem in the field of service operations management and logistics [1]. The open vehicle routing problem (OVRP) is a special variant of the standard vehicle routing problem (VRP). The most important feature consists in that the route of VRP is Hamiltonian cycle, whereas the OVRP is Hamiltonian path. So the vehicles in the OVRP are not required to return to the central depot, or if they are required to do so, they must return along the same route in the opposite order. The OVRP is a basic distribution management problem that can be used to model many real-life problems, such as, the third party logistics company without its vehicle fleet contracts its logistics to the hired vehicles. In such case, the third party logistics company is not concerned with whether the vehicles return the depot and does not pay any traveling cost between the last required depot and the central depot. It can be modeled as an OVRP. Other applications include the newspaper home delivery problem [1], school bus problem and emergent scheduler problem etc.

In the past several years, a lot of new algorithm had been proposed to solve the VRP, such as current heuristic algorithms, genetic algorithms, simulated annealing algorithms and so on, which also have made some preferably effect [2]. Only very few people has studied the OVRP. So far as we know, the first author to mention the OVRP was Schrage [3] in a paper dedicated to the description of realistic routing problems, bringing attention to some of its applications. Sariklis and Powell [4] use the "Cluster First, Route Second" method, in the second phase, they generate open routes by solving a minimum spanning tree problem. Their method is rapid, but doesn't get so good solution. Brandao et al. [5] apply the hybrid tabu Search algorithm for this problem. They generate the initial solution using a variety of methods including nearest neighbor heuristic and K-tree method.

Particle Swarm Optimization (PSO) [2] is a population based stochastic optimization technique, inspired by social behavior of bird flocking or fish schooling. Which has many advantages, such as less number of individuals, simple calculation, and good robustness, and can get better results in the various multidimensional continuous space optimization problems [3]. However, basic PSO algorithm suffers a serious of problem that all particles are easy to be trapped into the local minimum in the later phase of convergence. Aiming at the feature of the grain logistics and the shortage of current situation, we proposed a hybrid algorithm which combines PSO with the simulated annealing algorithm (SA) for solving OVRP problem, and obtained good results.

2 Problem Definition and Mathematical Formulation

In the classical version of Vehicle Routing Problems (VRP), the vehicles are required to return to the depot after completing service. In grain logistics OVRP, however, the vehicles need not do so. The results show that, the vehicle routes are not closed paths but open ones, starting at the central grain depot and ending at one of the required depots. Fig .1, which shows the optimal solutions to both the open and closed route, all required depots have demands and the vehicle capacity is to a certain value, in general, which also describes the optimal solution for the open version of a VRP can be quite different from that for the closed version.

An Intelligent Solution for Open Vehicle Routing Problem in Grain Logistics

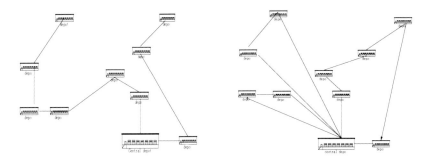

Fig. 1. Open routes and closed routes with different required depot, the optimal solutions to both the open and closed route for the same input data. The central depot is represented by big house and the required depot by small houses.

The Open Vehicle Routing Problem (OVRP) concerns the determination of routes for a fleet of vehicles to satisfy the demands of a set of customers. Grains are to be delivered to a set of required depot by a fleet of vehicles from a central depot. The locations of the depot and the customers are given. The objective is to determine a suitable route which minimizes the distance or the total cost under the following constraints: each depot, is served exactly once by exactly one vehicle to service its demand; each vehicle starts its route at the central depot, the total length of each route must not exceed the constraint; the total demand of any route must not exceed the capacity of the vehicle.

The Open Vehicle Routing Problem (OVRP) can be described in the following way: A complete undirected graph $G = (V,E)$ is given, with $V=\{0, \ldots, n\}$. Vertex 0 represents the grain central depot, the other vertices represent required depot. The cost of travel from vertex i to vertex j is denoted by c_{ij}, and we assume costs are symmetric, so $c_{ij} = c_{ji}$. A fleet of K identical vehicles, each of capacity $Q>0$, is given. Each required depot i has a demand q_i, with $0 < q_i <= Q$. Each required depot must be serviced by a single vehicle and no vehicle may serve a set of required depot whose total demand exceeds its capacity. Each vehicle route must start at the depot and end at the last customer it serves, the vehicles are not required to return to starting depot, the objective is to define the set of vehicle routes that minimizes the total costs. The grain logistics OVRP can be formulated as follows:

$$\min Z = \sum_I \sum_J \sum_K c_{ij} x_{ijk} + P_E \sum_{i=1}^{n} \max(ET_i - s_i, 0) + P_L \sum_{i=1}^{n} \max(s_i - LT_i, 0) \quad (1)$$

$$s.t. \sum_i d_i j_{ik} \leq q_k \quad (2)$$

$$\sum_k y_{ki} = 1 \quad (3)$$

$$\sum_i x_{ijk} = y_{kj} \qquad (4)$$

$$\sum_j x_{ijk} = y_{ki} \qquad (5)$$

$$x_{ijk} = 0 \, or \, 1, \; i,j = 0, 1, \cdots, n; \; \forall k \qquad (6)$$

$$y_{ik} = 0 \, or \, 1, \; i,j = 0, 1, \cdots, n; \; \forall k \qquad (7)$$

$$ET_i \leq t_i \leq LT_i \qquad (8)$$

Where objective function (1) minimizes the total cost; Equation (2) ensure that the demand of every required depot is at most a vehicle of capacity Q; Equation (3) ensure that each required depot is served exactly once; Equation (4),(5)ensure that every required depot is serviced only once; Equation (6),(7)ensure that the variable only takes the integer 0 or 1; Equation (8) shows that time windows are limited.

3 Solution Method

In this section, the concepts of the SA and PSO are briefly discussed. At the beginning, the particle swarm optimization is described.

3.1 Particle Swarm Optimization

Particle swarm optimization (PSO) algorithm [6] is based on a social-psychological metaphor, was originally proposed by J.Kennedywhich, which shares many similarities with evolutionary computation techniques such as Genetic Algorithms (GA). However, PSO has no evolution operators such as crossover mutation and duplication. Particles update themselves with the internal velocity. The mathematical expression of Particle Swarm Optimization (PSO) algorithm as follows [7]:

Each particle is regarded as a point in a D dimensional space. The position of a particle corresponds to a candidate solution of the considered optimization problem. At any time, Pi has a position, which is represented as $X_{it} = (x_{i1}, x_{i2}, \ldots x_{iD})$ and a velocity $V_{it} = (v_{i1}, v_{i2}, \ldots v_{iD})$ associated to it, these particles fly through hyperspace and have two essential reasoning capabilities: their memory of their own best position and knowledge of the global or their neighborhood's best.. The particle position and velocity update equations in the simplest form are given by:

$$V_{id}(t+1) = w*V_{id}(t) + c_1 * rand()[P_{id} - X_{id}(t)] + c_2 * rand()[g_{id} - X_{id}(t)] \qquad (9)$$

$$X_{id}(t+1) = X_{id}(t) + V_{id}(t+1) \quad 1 \leqslant i \leqslant n \quad 1 \leqslant d \leqslant D \quad (10)$$

Where w is the inertia weight; V_{id} is the velocity of particle i, X_{id} is the partical position, p_{id} is the current position of particle; c_1, c_2 are the two positive constants, called acceleration parameters; rand () is the random functions in the range[0,1]; The position of particles are restricted in interval[-Xmaxd, Xmaxd], while the velocity of particles are restricted in interval[-Vmaxd, Vmaxd] in the d (1≤d≤D) dimensional space; t is one iteration. The initial position and velocity are randomly generated, and then used to iterate by formula (9) and (10), until to meet the end of conditional expression of iteration.

3.2 Simulated Annealing Algorithm

The simulated annealing (SA) algorithm was derived from statistical mechanics, was first introduced by Metropolis et al. (1953) [8], Kirkpatrick et al. [9] were the first to apply SA to solve combinatorial optimization problems. SA is a stochastic hill-climbing search algorithm. It consists of a sequence of iterations, each changes the current solution to a new solution in the neighborhood of the current solution. The neighborhood is defined by the chosen generation mechanism. Once a neighborhood solution is created, the corresponding change in the cost function is computed to determine whether the neighborhood solution is accepted as the current solution. If the cost is improved, the neighborhood solution is directly taken as the current solution. Otherwise, it is accepted according to Metropolis's criterion [10]. Based on Metropolis's criterion, the neighborhood solution is accepted as the current solution only if r ≤exp (-ΔE/T), in which r is generated from U [0, 1], ΔE is the difference between the costs of the two solutions, and T is the current temperature. Three parameters need to be specified in designing the cooling schedule: an initial temperature; a temperature update rule; and the number of iterations to be performed at each temperature stage.

3.3 Hybrid PSO for Grain Logistics OVRP

3.3.1 Particle Representation

In general, how to find a suitable particle representation is the key problem of implementation algorithm, which the location of particle corresponds with the solutions to the problem. In this paper, using the coding idea of document [11] for reference, supposing the total of particle, is denoted as service nodes n (required depot), we use 2n-dimensional particle representation, each service node, which is corresponded to the vehicle and the order of the vehicle route. In this way, the dimension size of the particle's position equals to the number of required depot. When the particle position is decoded, it is convenient for readjusting the particle position when updated. For L required depot, each particle is encoded as a real number vector with L dimensions. The integer part of each dimension or element in the vector represents the vehicle.

3.3.2 The Hybrid Algorithm Procedure

The SA algorithm is introduced into PSO to solve the OVRP of grain logistics, in order to avoid trapping into local minimum. Simulated annealing mechanism is introduced into the speed and location renewal processes of each particle to accept the optimal solution of fitness after the particle swarm evolutionary via the Metropolis criterion and the deterioration solution by a certain probability, and then jump out of the local extremism and convergence to global optimal solution. The hybrid PSO algorithm is represented by the flowchart in Fig. 2.

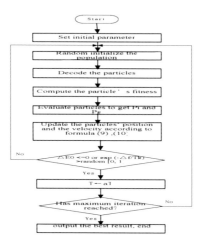

Fig. 2. Procedure of hybrid PSO algorithm for grain logistics open vehicle route problem

4 Experimental Results

All algorithms in the study are developed in C# language, and all computational experiments are tested on a 2.83 GHz Pentium 4 PC with 2 GB RAM running in windows environment.

In the experiment, the Solomon benchmark [12] is used. In the benchmark, there are three types data: heap distribution (C), the random distribution (R), semi-heap distribution (RC). Solomon instances are selected in every test type.

In order to choose the population of algorithm, the C101, R101, RC101 are tested. Experiments have been done with the following parameter settings, the number of particle, respectively 5, 10, 20...., the first column shows the size of particle, the iteration is 200, the algorithm run 50 times randomly. And the results of the mean and standard deviation are shown in the following table 1. According to the table 1, the populations are smaller; the results of optimization are poor. This is because a small population of particles, which are easy to jump into local optimum. As the population size increases, when population sizes is more than 80, the results of the algorithm have no obvious influence, the algorithm is more stable.

Table 1. Result of with different populations

populations	C101 S.D.	C101 Mean	R101 S.D.	R101 Mean	RC101 S.D.	RC101 Mean
5	76.5	378.3	71.1	636.9	65.4	483.7
10	52.2	272.8	58.2	583.6	57.8	436.8
20	18.5	192.9	35.4	532.1	30.4	385.1
40	10.1	166.6	22.3	476.4	17.3	334.8
80	8.5	157.6	16.7	474.2	15.6	328.6
160	9.1	157.2	15.3	473.6	14.8	327.5

For comparison, standard genetic, standard PSO and hybrid PSO algorithm are used to solve the Solomon instance. In the experiments, the parameters are set as following: population size n= 80, crossover probability p_c=0.8 and mutation probability p_m=0.05; $c_1=c_2$=2.0, w=0.75, the number of iterations is 100; the initial temperature is set at 100, the cooling rate is set at 0.9. The results of algorithm are shown in Table 2. Compared with the PSO, GA algorithms, hybrid PSO algorithm is performed much better than the results of PSO and GA algorithms.

Table 2. Hybrid PSO with other algorithms

	Hybrid PSO	PSO	GA
C101	152.6	184.1	289.3
C102	189.5	215.7	270.1
C103	169.7	189.1	263.3
C104	198.2	226.2	275.2
C105	151.4	182.3	246.5
R101	463.1	481.6	572.3
R102	448.3	465.9	520.6
R103	448.1	453.2	553.7
R104	432.6	470.2	571.9
R105	443.4	497	542.1
RC101	314.5	345.2	500.6
RC102	312.2	343.1	492.1
RC103	268.4	321.7	431.3
RC104	269.7	324.3	435.8
RC105	307.5	335.2	402.5

5 Conclusions

This paper studies the hybrid PSO for solving the grain logistics Open Vehicle Routing Problem, although Open Vehicle Routing Problem has been developed for more than two decades, there are only a few solutions available. In the paper, based on PSO, a solution-solving scheme for grain logistics OVRP is proposed. The performance of the hybrid PSO algorithm is evaluated in comparison with some other algorithm. The results showed that the hybrid PSO algorithm is efficient for grain logistics OVRP.

Acknowledgments. This work is partially supported by The 11th Five Years Key Programs for Science and Technology Development of China (No.2008BADA8B03, No. 2006BAD08B01), Henan Program for New Century Excellent Talents in University (No. 2006HANCET-15).

References

1. Laporte, G., Gendreau, M., et al.: Classical and modern heuristics for the vehicle routing problem. International Transactions in Operational Research 7, 285–300 (2000)
2. Golden, B., Wasil, E.: The open vehicle routing problem: Algorithms, large-scale test problems, and computational results. Computers and Operations Research
3. Schrage, L.: Formulation and Structure of More Complex/realistic Routing and Scheduling Problems. Networks 11, 229–232 (1981)
4. Sariklis, D., Powell, S.: A Heuristic Method for the Open Vehicle Routing Problem. Journal of the Operational Research Society 51(5), 564–573 (2000)
5. Brandao, J.: A Tabu Search Algorithm for the Open Vehicle Routing Problem. European Journal of Operational Research 157(8), 552–564 (2004)
6. Kennedy, J., Eberhart, R.C.: Particle Swarm Optimization. In: Proceedings of IEEE International Conference on Neural Networks, Piscataway, NJ, pp. 1942–1948 (1995)
7. Parsopoulos, K.E., Vrahatis, M.N.: Particle Swarm Optimization Method in Multiobjective Problems. In: Proceedings of the 2002 ACM symposium on applied computing, pp. 603–607 (2002)
8. Kirkpatrick, S., Gelatt, C., Vecchi, M.: Optimization by simulated annealing. Science 220, 671–680 (1983)
9. Kirkpatrick, S., Gelatt, C.C., Vecchi, M.P.: Optimization by Simulated Annealing. Science 220, 671–679 (1983)
10. Metropolis, N., Rosenbluth, A., Teller, A.: Equation of state calculations by fast computing machines. The Journal of Chemical Physics 21, 1087–1092 (1953)
11. Salmen, A., Ahmad, I., Al-Madani, S.: Particle swarm optimization for task assignment problem. Microprocessors and Microsystems 26, 363–371 (2002)
12. Solomon, M.M.: Algorithms for the Vehicle Routing and Scheduling Problems with Time Window Constraints. Operations Research 35, 254–265 (1987)

E-Commerce Comparison-Shopping Model of Neural Network Based on Ant Colony Optimization[*]

Kang Shao and Ye Cheng

School of Economics and Management,
Anhui University of Science and Technology,
Huainan, China
e-mail: sboy@ah163.com

Abstract. A new model of comparison-shopping and key contents is discussed in the paper to solve the problem of the user bias' filtering and learning. On the basis of traditional comparison shopping method, trains the BP neural networks by ant colony optimization algorithm to obtain the users' preference information. It also adopts the growth-oriented method of network structure to decrease the learning error. And the sequence of search results is reorganized based on the information, to provide users with the personalized shopping guide service to meet their needs. Besides, the application of Web 2.0 can be optimized by using the knowledge of preference to build a better comparison-shopping e-commerce website.

Keywords: comparison-shopping; user preferences; ant colony algorithm; neural network; search engine.

1 Introduction

The rapid development of Internet Technology and E-commerce makes shopping sites increased rapidly, and the user's shopping options increasing, comparison-shopping came into being as a new way online shopping [1]. Users can find all the goods and prices provided by online shopping sites by simply visit a particular comparison-shopping Web site, it had raised the scope and efficiency of online shopping [2, 3]. All these sites, however, only pay attention to the requests of consumers about commodity prices, while ignoring the other aspects of consumers' requests about the commodity, such as character, warranty, delivery dates, etc. Therefore, what the customers searched will be not always the very commodity that suited them.

People-oriented, user-centered in the web2.0 era has become important [4]. The individual demand of customers has been also raised increasingly. How to

---

[*] This work is supported by 2008' scientific and research program of Huainan City Science and Technology Bureau to Shao Kang.

improve the results showed by comparison-shopping search engine and by use of the model of users stored in the knowledge base to guide them making a decision itself, to provide users with personalized shopping guide service to meet their needs [5], will be one of the important technologies to improve the performance of comparison-shopping search engine.

For this reason, this paper presents a comparison-shopping model, to learn the personalized preferences of customers automatically on Neural Network based on Ant Colony Optimization [6, 7, 8]. At the same time, the information will be searched and filtered for customers by the personalized preferences [9], and sort the requirements of goods according to personalized preferences of customers when searches finished [10], so that customers will be able to buy well-content commodity without need to visit a large number of sites.

2 Neural Network Training Based on Ant Colony Algorithm

Artificial Neural Network has a complex nonlinear mapping ability, function approximation and large-scale parallel distributed processing capabilities. BP Network is very extensive in the application of Artificial Neural Network, but the existence of BP Network has the defects of very slow training speed and problems that local minimum point to escape. Ant Colony Algorithm [11, 12, 13] is a heuristic global optimization algorithm, so it can be used to train the network to compensate for deficiencies in BP Network [14, 15].

The basic idea of the algorithm is that assumed all the weight value and threshold in Neural Network that needed training have a total of parameters m, are denoted as $P_1, P_2, \ldots P_m$, Neural Network parameters P_i ($1 \leq i \leq m$) are set for the possible values of N within the random non-zero value, to form a set, denoted as I_p. Defines the number of ants is S, each ant starting from the set I_p, choose a weight value, select the weights of a group of Neural Network in all of the sets. When the ants get through with the choices of elements in all the sets, it has reached the food source. Afterwards regulate all the elements of the pheromone in set. The process will be repeated until the optimal solution has been found.

The steps of Neural Network based on Ant Colony Optimization Algorithm are as follows:

1) Initialization: Let the time t zero, the pheromone of elements j in the collection I_{p_i} is $\tau_j(I_{p_i}) = C$, the number of ants is S, set the maximum number of iterations.

2) Start all ants, each ant starting from the set I_{p_i}, ant k ($k = 1, 2, \ldots S$), select its elements j in random according to the following formula for the probability.

$$\Pr(\tau_j^k(I_{P_i})) = (\tau_j^k(I_{P_i})) / \sum_{g=1}^{N} \tau_g(I_{P_i}) \qquad (1)$$

3) Calculate the output error of the value of the selected parameters for Neural Network training samples through all the ants, record the optimal solution in

currently selected parameters. After *m* units of time to complete the element of choice, all the path of the amount of information to be updated by pressing style:

$$\tau_j(I_{P_i})(t+m) = \rho \tau_j(I_{P_i})(t) + \Delta \tau_j(I_{P_i}) \qquad (2)$$

Among them, parameter ρ ($0 \leq \rho \leq 1$) denotes the persistence of pheromone, then 1-ρ denotes the vanish degree of pheromone.

$$\Delta \tau_j(I_{P_i}) = \sum_{k=1}^{S} \Delta \tau_j^k(I_{P_i}) \qquad (3)$$

$$\Delta \tau_j^k(I_{P_i}) = \begin{cases} Q/e^k, \text{The ant k select element } P_j(I_{P_j}) \text{ in this iteration} \\ 0, \text{or else} \end{cases} \qquad (4)$$

Equation (4) denotes pheromone that the ant k left in this cycle in the element j of set I_p. Among them, Q is a constant, denotes the speed of adjustment; e^k is the maximum output errors of training samples when the element that ant k selected to be Neural Network, define:

$$e = \sum_{1}^{D} \frac{1}{2}(d-o)^2, \qquad (5)$$

D is the quantity of the sample; d and o denote respectively the actual output and the desired output of Neural Network, the smaller the margin errors, the more the pheromone increase correspondingly.

4) If all the ants converge to one path or more cycles, then end the circle and output the calculate results, or else repeat 2), 3).

3 E-Commerce Comparison-Shopping Search Model of Neural Network Based on Ant Colony Optimization

The model consists of two modules class, concrete structure shown in Figure 1.

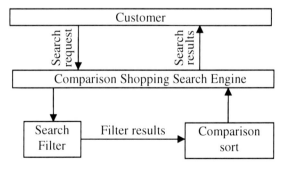

Fig. 1. Basic Search Model of Comparison-Shopping

1) Search filter module: To filter out the higher value of information for the users from the eligible information. According to user's needs, filter algorithm, filter the web pages and then put the search results into a commodity database. There are many methods of information filter, such as collections of literature-based approach, vector method based on the keyword, evolutionary information filtering method, multi-agent filtering method, etc. Here is a brief introduction of information filter that based on key vector. It is a similar filtering method with operability. Its main principle is that the key word of characteristic information unit and the key word of searching will be seen as two vectors (v_1, v_2), and the similarity between the two vectors will be gained by making cosine $cos(v_1, v_2)$ folder angle calculations. The greater its value is the more similar to vectors. The information unit will be retained when the similarity up to a certain degree. Search filter module will array search back page in descending order according to similarity and pass the top side of the page to the comparison-sorting module.

2) Compare-sorting module: To compare and sort the information that provided by search filtering module. According to the information searched by search filter module, it can predict user's overall satisfaction of the approximation in commodity prices, character, warranty, delivery time, etc. and other indicators according to the information in the knowledge base preferences, and present them to customers in descending order. The model is shown in Figure 2.

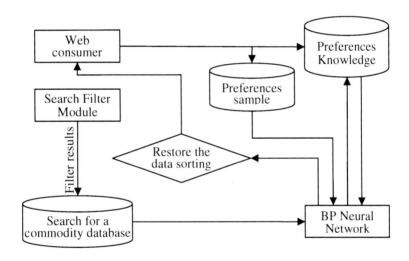

Fig. 2. Basic model of comparison sorting

Specific process of comparison-sorting module is as follows:

1) The BP Neural Network will be trained before the users sending a search request. There are two training ways:
   a) Give BP Neural Network weights and threshold assignment by read the history of knowledge base of preferences directly.

b) New users will be guided to select a sample approach to them from preference sample database to train BP Neural Network. The training results will be deposited in knowledge base of preferences in order to be used directly next time.

2) The search filter module will be started after the training of the BP Neural Network to search the commodity information automatically according to the requests of customers.

3) The commodities data obtained by the search filter module will be processed by BP Neural Network; the results are the degree of satisfaction with the indicators of commodities for customers. Comparison-sorting module will present the outputs to customers by the degree of satisfaction in descending order.

4) Customers select what commodities they want to buy according to the returned information.

4 Experiments and Discussion

The text improves the Neural Networks by use of Ant Colony Optimization algorithm. The determination of the numbers of hidden layer nodes is the key to success for BP Networks. Too small quantity of the numbers to obtain the enough information to solve the problem. Too large quantity of the numbers will let to the new problem, too. It not only increases the training time, but also result in appearing of so-called "transition fit" problem, known as increased test errors will led to the ability of generalization reduced. Hence, choosing a reasonable numbers of nodes in hidden layer is very important. The method for choosing the numbers of the hidden layer and the nodes is quite complex, so the general principle is adopted—based on reflecting the input-output relationship correctly, the fewer numbers of the hidden layer nodes should be used to allow the network structure as simple as possible. This paper adopts the growth-oriented method of network structure to decrease the learning error. In the method, the fewer nodes is set firstly, then train the network and test the learning errors, and then gradually increase the number of nodes, until the learning errors are no longer decreased significantly.

Based on the previous analysis of Comparison-shopping site, the following sample format to train the BP Network has been adopted, formats as follows:

[X, Y], here $X=(X_1, X_2, X_3)$, X_1, X_2, X_3 denotes the indicators of customer's preferences respectively for the goods about the price, warranty and delivery time, their values from 0 to 1. Y denotes the set of the degree of customer's satisfaction to the indicators on X.

To train a BP Network of three layers with a single hidden layer, input layer is three nodes, the output layer is one node, and options hidden layer nodes based on the following empirical formula: $n_1 = \sqrt{n+m} + a$. Where, n is the number of input nodes, m is the number of output nodes, a is a constant between 1 and 10. In the present instance, n_1 ranges from 3 to 12; here n_1 takes 12 according to the above-mentioned principles. In order to reduce errors, instead of having all of input-output variables to be normalized, only output variables would be normalized according to the input-output range of sigmoid transfer function. Parameter ρ, which denotes the levels of pheromone residue, takes 0.6, Q takes 50, the number of ants is 100, and N is 100.

Adopt the sample data in Table 1 to train the Neural Network. Some user's preferences are embodied in the sample data.

a) Customers usually regard as very important things are price, next are delivery time, finally are the character and warranty;
b) As the prices raise high, the customer's satisfaction will decline, and when the price is close to 1500, the customer's satisfaction decline sharply.

Table 1. Training Samples

Commodity prices (Yuan)	Warranty period (months)	Delivery time (days)	Satisfaction
1260	8	4	0.81
1260	12	10	0.88
1260	24	10	0.89
1320	8	7	0.73
1320	8	10	0.76
1320	12	10	0.77
1380	4	7	0.63
1380	8	7	0.64
1380	12	14	0.71
1440	12	7	0.55
1440	12	10	0.58
1440	24	10	0.59
1500	8	7	0.12
1500	12	14	0.19
1500	24	14	0.21

To test the data in Table 2 with the training neural network:

Table 2. Test Samples

No.	Commodity prices (Yuan)	Warranty period (months)	Delivery time (days)
1	1250	4	10
2	1300	4	4
3	1350	8	4
4	1350	8	12
5	1350	24	4
6	1450	8	7
7	1485	12	14

Results in the following table:

Table 3. Test results

No.	Satisfaction	Sort Satisfaction
1	0.887425	1
2	0.710253	3
3	0.647591	5
4	0.780306	2
5	0.673248	4
6	0.348719	6
7	0.281532	7

Contrasted with the three tables will show that the Neural Network based on Ant Colony Optimization have good ability for learning. Through the learning and training in Table 1, the user's preference habits can be grasped perfectly. Test results from comparing the group 1 and group 4 as well as the group 3, group 4 and group 5 will also embody that the prices are most valued for users of preferences, followed by the delivery time, and finally the character and warranty. With the commodity prices increasing, the customer's satisfaction is declining notably. When the price rose to 1,450 Yuan, the satisfaction of customers began to decline sharply. This is just matched to the character of customers' preferences described above.

5 Conclusions

The rapid development of E-commerce makes the competition between Comparison-shopping sites fierce increasingly, how to maintain undefeated and continue to expand our market share in such a competitive environment became the point that must be taken into account by comparison-shopping site. This paper presents a Comparison-shopping search model of Neural Network based on Ant Colony Optimization; take their preferences information to improve the results showed by Comparison-shopping search engine, in order to provide users with personalized shopping guide services to meet their needs. The experiment proved that Neural Network based on Ant Colony Optimization could predict the consumers' preferences accurately; Comparison-shopping search model of Neural Network based on Ant Colony Optimization is effective.

References

1. Wangyuan, X.: Analyze the development of comparison shopping in China. Northern Economy and Trade (2), 62–63 (2008)
2. Lee, H.K., Yu, Y.H., Ghose, S., Jo, G.: Comparison Shopping Systems Based on Semantic Web – A Case Study of Purchasing Cameras. In: Li, M., Sun, X.-H., Deng, Q.-n., Ni, J. (eds.) GCC 2003. LNCS, vol. 3032, pp. 139–146. Springer, Heidelberg (2004)
3. dos Santos, S.C., Angelim, S., Meira, S.R.L.: Building Comparison-Shopping Brokers on the Web. In: Fiege, L., Mühl, G., Wilhelm, U.G. (eds.) WELCOM 2001. LNCS, vol. 2232, p. 26. Springer, Heidelberg (2001)
4. Yao, C.: Research on the Innovative Mode of E-business Based on Web2.0. Information Science 25, 1559–1562 (2007)
5. Qun, J.: The investigation and suggestion on the development of comparative shopping websites in China. Journal of BUPT (Social Sciences Edition) 9, 6–10 (2007)
6. Hong, M., Yong, W., Rongqi, Z.: Study on the Feedforward Neural Network Based on Ant Colony Optimization. Journal of Wuhan University of Technology (Transportation Science & Engineering) 33, 531–533 (2009)
7. Kang, L., Ling, Y.: Analyse of Continues Problem about Ant Colony Algorithm. J. Sichuan Institute of LICT 17, 42–46 (2004)
8. Chen, L., Shen, J., Qin, L., Chenp, H.: An improved ant colony algorithm in continuous optimization. Journal of Systems Science and Systems Engineering 12, 224–235 (2003)
9. Yanyan, S., Peigang, L., Yong, L.: A Personalized Search System Fitted User Preference. Information Science 26, 1248–1251 (2008)
10. Cheng, Y., Shao, K.: The applications of user preference to comparison shopping. Enterprise science and technology & development (22), 209–211 (2009)
11. Haibin, D.: Principle and application of ant colony algorithm. Science Press, Beijing (2005)
12. Liu, B., Li, H., Wu, T., Zhang, Q.: Hybrid Ant Colony Algorithm and Its Application on Function Optimization. In: Kang, L., Cai, Z., Yan, X., Liu, Y. (eds.) ISICA 2008. LNCS, vol. 5370, pp. 769–777. Springer, Heidelberg (2008)
13. Socha, K., Blum, C.: An ant colony optimization algorithm for continuous optimization: application to feed-forward neural network training. Neural Computing & Applications 16, 235–247 (2007)
14. Chongzhi, S., Lu, W., Nenggang, X.: Research on Neural Networks Training Based on Ant Colony Optimization. Automation & Instrumentation (5), 10–12 (2006)
15. Blum, C., Sampels, M.: An Ant Colony Optimization Algorithm for Shop Scheduling Problems. Journal of Mathematical Modelling and Algorithms 3, 285–308 (2004)

PPC Model Based on ACO

Li Yancang and Hou Zhenguo

College of Civil Engineering, Hebei University of Engineering,
Handan056038, China
liyancang@163.com

Abstract. In order to find a new method for solving the shortcoming of the projection pursuit comprehension assessment algorithm, an improved ACO based on information entropy was introduced. The proved ant colony optimization algorithm was used to optimize the function of the projected indexes in the PP. Application results show that the method can complete the assessment objectively and rationally.

Keywords: comprehensive assessment; Projection Pursuit; ant colony optimization algorithm; information entropy.

1 Introduction

PP techniques were originally proposed and experimented with by Friedman, J. H., and Tukey, J. W [1]. Project pursuit comprehensive assessment algorithm is an effective method to deal with the high-dimension-data assessment. Yet, this method has one serious drawback: defect to optimize the function of the projected indexes. To overcome this shortage, an Ant Colony Optimization (ACO) algorithm was introduced. ACO is a group of novel population-based algorithms that combine distributed computation, autocatalysis (positive feedback) and constructive greedy heuristic in finding optimal solutions for NP-hard combinatorial optimization problems. ACO has been successfully applied to most combinatorial optimization problems, e.g. TSP (Traveling Salesman Problem), JSP (Job-shop Scheduling Problem), QAP (Quadratic Assignment Problem), SOP (Sequential Ordering Problem) and so on [2]. Here, the information entropy based ACO proposed in [3] was employed to solve the problem of the projected indexes optimization.

The paper is organized as follows. In the following part, attention was paid to the generalities of the PPC and ACO. Then, the improved PPC model based on the improved PP was set up. Finally, its application in practice was introduced and the advantages of the method we proposed were pointed out.

2 Basic Knowledge of PPC and ACO

The most exciting feature of PP is that it is one of the very few multivariate methods able to bypass the "curse of dimensionality" caused by the fact that high-dimensional space is mostly empty. In addition, the more interesting PP methods are able to ignore irrelevant (i.e. noisy and information-poor) variables. This is a distinct advantage over methods based on interposing distances like minimal spanning trees, multidimensional scaling and most clustering techniques. PP emerges as the most powerful method to lift one-dimensional statistical techniques to higher dimensions. The PP methods have been successfully applied in many fields [4-5].

ACO is a family of meta-heuristics stochastic explorative algorithms inspired by the natural optimization mechanism conducted by real ants. ACO algorithms can find the best solution by using the evolutionary procedure. As shown in [3], ACO is based on the following ideas. (1) From a starting point to an ending point, each path is associated with a candidate solution to a given problem. (2) The amount of pheromone deposited on each edge of the path followed by one ant is proportional to the quality of the corresponding candidate solution. (3) The edge with a larger amount of pheromone is chosen with higher probability. As a result, the ants eventually converge to a short path, hopefully the optimum or a near-optimum solution to the target problem.

Since the establishment of the first ACO system, called Ant System, several refined versions including the information entropy based ACO we mentioned have been proposed to to solve the premature convergence problem of the basic Ant Colony Optimization algorithm. The main idea is to evaluate stability of the current space of represented solutions using information entropy, which is then applied to turning of the algorithm's parameters. The path selection and evolutional strategy are controlled by the information entropy self-adaptively. Simulation study and performance comparison with other Ant Colony Optimization algorithms and other meta-heuristics on Traveling Salesman Problem show that the improved algorithm, with high efficiency and robustness, appears self -adaptive and can converge at the global optimum with a high probability [3].

3   Improved Model Based on ACO

The ACO is an effective method for the global optimization. So, we can use the ACO to deal with the indexes function. Here, the ACO proposed in [3] was employed.

3.1 Procedure of Modification

The improved PP can be shown as follows.

(1) Normalization

In order to finish the elimination of all dimensions and uniform changes in the scope of the value of the indicators, the normalization is needed:

For the greater and better indicators:

$$x(i,j) = \frac{x^*(i,j) - x_{min}(j)}{x_{max}(j) - x_{min}(j)} \quad (1)$$

For the smaller and better indicators:

$$x(i,j) = \frac{x_{max}(j) - x^*(i,j)}{x_{max}(j) - x_{min}(j)} \quad (2)$$

Where $x_{max}(j)$ and $x_{min}(j)$ are the Jth maximum and minimum indicators values respectively. $x(i,j)$ is indicators normalized sequences.

(2) Construction of projection function

The key to the PP is to find the optimal projection which can fully show the features of the data. Suppose $a = a\{a(1), a(2), \cdots, a(p)\}$ is the $p-$ dimensional unit vector, PP is to project $x(i,j)$ to $\overline{a}$ and obtain the value of one-dimensional projection $z(i)$.

$$z(i) = \sum_{j=1}^{p} a(j)x(i,j), \quad i = 1,2,\cdots,n \quad (3)$$

Where $a(j)$ $j = 1,2,\cdots,p$ is the projection vector, and it is unit vector.

Projection indicators in the integrated value $z(i)$ require that the projector is characterized by the spread of local projection point intensive, and the projection target function can be expressed as:

$$Q(a) = S_z \cdot D_z \quad (4)$$

Where S_z is the standard deviation of the projection value, and D_z is the local density of $z(i)$. They can be expressed as follows.

$$S_z = \left\{ \sum_{i=1}^{n} [z(i) - \overline{z}]^2 \Big/ n-1 \right\}^{\frac{1}{2}} \quad (5)$$

$$D_z = \sum_{i=1}^{n} \sum_{j=1}^{p} (R - r_{ij}) \cdot I(R - r_{ij}) \quad (6)$$

Where $\bar{z}$ is the mean value of $\{z(i), i = 1 \sim n\}$, R is the radius of windows and $r_{max} + \dfrac{p}{2} \le R \le 2p$. r_{ij} is the distance of the the samples.

(3) Optimization of projection function

When the index value of the program is given, the projection function $Q(a)$ only changes with the projection direction a. Different projection direction reflects different construction of data characteristic. The best projection direction which is most greatly possible to expose the characteristic structure is the high-dimensional data. We can estimate the best projection direction through the solution projection target function maximization question.

$$Max: Q(a) = S_z O_z$$
$$s.t. \sum_{j=1}^{n} A^2(j) = 1 \tag{7}$$

It is a complex misalignment optimization question. Here, we will employ the information entropy ACO in the step acceleration method style. At the early stage of the evolution, the value of $\alpha'_{(t)}$ is small and with the process proceeding, the value becomes bigger in order to explore the solution space in the beginning and reinforce the local search ability at final stage to avoid stagnation. At the same time, $\beta'_{(t)}$ is biggest at early stage in order to make the algorithm find the optimal route as many as possible and later it becomes smaller to reinforce the function of random operation, which can also avoid the stagnation.

3.2 Application

To validate the efficiency of the modification proposed in the paper, we compared its performance with other PP algorithms. All algorithms were benchmarked on the same problem in [6].The data used are shown in Table 1.

Table 1. Comprehensive assessment of marine ecological indexes in a bay

Item	No. 1 spot	No.2 spot	No.3 spot	No.4 spot	No.5 spot	No.6 spot
suspended matter (mg/L)	6	7	7.7	6.4	6.3	6.1
Petroleum (mg/L)	0.020	0.036	0.0205	0.018	0.010	0.012
inorganic nitrogen(mg/L)	0.479	0.465	0.4065	0.046	0.195	0.066
activated phosphate(mg/L)	0.012	0.251	0.025	0.01	0.007	0.002
		9.85	9.76	9.81	9.78	9.92

Table 1. *(Cont.)*

dissolved oxygen(mg/L)	10.2 31.5	31.3	31.3	31.5	31.4	31.6
salinity of sea water(‰)	32.1	45.7	36.2	27.8	22.6	32.3
Cu (ug/L)	39.5	42.1	37.1	39.8	42.5	47.8
Pb (ug/L)	103.5	87.5	92.3	83	121.5	78.5
Zn (ug/L)						
Cd (ug/L)	0.61	0.87	0.96	1.32	0.81	0.77
organic carbon (ug/L)	240	3300	2100	2600	1700	1600
sulphide (ug/L)	0	550	969	1349	448	535
Shannon-Weaver indexes	874 2.46	2.24	1.99	2.42	2.48	1.47
Shannon-Weaver indexes of zooplankton	3.16	2.68	2.42	2.33	2.72	1.98
Shannon-Weaver indexes of zoobenthos	0.99	1.192	2.570	3.348	3.169	3.736
Gross quantity of bacteria in monitoring spots(/L)	23000	1700 0	1400 0	2000 0	1600 0	1300 0
Analysis result of organism samples Cd	0.234	0.205	0.147	0.356	0.127	0.109

The data shown above are 17- dimension data. Then, employ the information entropy ACO to optimize the projection function. And in the ACO, $\alpha = 1.5, \beta = 4.0, \rho = 0.6, Q = 50$, when the information entropy is smaller than a given 0.01, the algorithm terminates. Then, we can obtain a^* =(0.10076, 0.14587, 0.1721, 0.1692, 0.10786, 0.10755, 0.1094, 0.1215, 0.1094, 0.1188, 0.1133, 0.1058, 0.1523, 0.2090, 0.1203, 0.1124, 0.1672), and $z^*(j) = (2.1563, 1.9651, 2.8927, 2.8651, 3.1622, 1.2329)$. We can draw a conclusion that No.5 spot is best, No.6 is the worst. The occlusion is same as the fact and more rational than [6].

4 Conclusion

The development of science and technology has signified the need for the study of higher- dimension data solution methods. We introduce the information entropy

ACO to the PPC which is an effective method for this problem. The ACO was used to optimize the projection function which is the bottleneck of PPC. Engineering practice shows that this method can deal with the high- dimension data effectively and rationally. This study provides a promising method for the clustering problems and comprehensive assessment problems.

References

1. Friedman, J.H., Tukey, J.W.: A projection pursuit algorithm for exploratory data analysis. IEEE Trans. Comput. 23, 881–889 (1974)
2. Dorigo, M., Stutzle, T.: Ant Colony Optimization. MIT Press, Cambridge (2004)
3. Li, Y., Li, W.: Adaptive Ant Colony Optimization Algorithm Based on Information Entropy: Foundation and application. Fundamenta Informaticae 3, 229–242 (2007)
4. Huber, P.J.: Projection Pursuit. The Annals of Statistics 2, 435–475 (1985)
5. Qiang, F., Yonggang, X., Zimin, W.: Application of projection pursuit evaluation model variations in the Sanjiang Plain. China Pedossphere 3, 249–256 (2003)
6. Hu, J.: Research on ecological environment comprehensive assessment methods for coast marine Area. Dalian University of Technology (2007)

Image Algorithm for Watermarking Relational Databases Based on Chaos

Zaihui Cao, Jianhua Sun, and Zhongyan Hu

Department of Art and Design, Zhengzhou Instiute of Aeronautical Industry Management
450015 Zhengzhou, China
czhhn@126.com, jianhuasun@zzia.edu.cn, huzhongyan@zzia.edu.cn

Abstract. Database watermarking is an important branch of information hiding technology, and it is chiefly used to protect copyright. This paper creatively proposes a novel method for watermarking relational database, which uses the encrypted images sequence as watermark embed into database according to the weight of attributes. Experimental results show that the proposed scheme is robust against various forms of attacks. Our approach is more intuitive, and it support easy watermark identification.

Keywords: watermarking relational databases;copyright protection; information hiding.

1 Introduction

Data has become merchandise in this information age. Since valuable data could be copied and distributed throughout networks easily, it faces the problem of rights protection as digital multimedia products. With the urgent need for copyright protection of relational database, watermarking relational database is becoming a hotspot in recent years. A lot of efforts have been made in the field, At 2002, R.Agrawal[1]etal proposed relational database watermarking, and introduced a watermark tagging strategy. At 2003, R.Sion[2]et al. tagged to numerical attribution of relational database to implement watermark and protect copyright of relational database. At the same year, Xiamu Niu[3]et al. did further study on relational database watermark to insert a little of watermark which had practical significance. At 2004, Zhihao Zhang[4]et al. successfully embedded an image into relational database. Watermarking relational databases has become an important topic in both fields of database and information hiding.

On the basis of previous studies, they suggested that ownership watermarks should be embedded first, captioning watermarks should be embedded next, and robust watermarks should be embedded last. In this paper, we study a novel method for watermarking relational database. In our scheme, images which encrypted by chaos as the copyright are embedded into numeric attributes according to different weights. To be effective and convictive the watermark embedding process is imperceptible, secure and reliable, the watermark extraction process is a blind detection process and the watermark scheme is resistant to some malicious attacks.

2 Image Algorithm for Watermarking Relational Databases

We assume that some minor changes of some attributes values can be tolerated. And we will embed copyright information into some float attributes. We use the encrypted character images to generate watermark by watermark generation algorithm, then present the corresponding insertion algorithm (see Fig.1) and detection algorithm (see Fig.2).

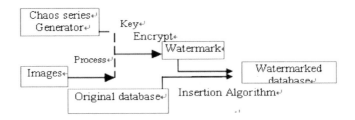

Fig. 1. Database watermarking insertion algorithm

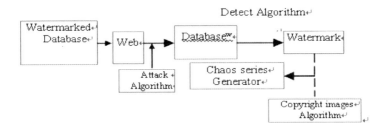

Fig. 2. Database watermarking detect algorithm

In fact, there are usually a good many attributes in a database (especially in large databases), while they are not of the same importance according their intended using purpose. We now present a technique for watermarking an attribute with a probability according to its weight. Then, we suppose that relation R contains primary key P and numeric attributes A0, A1, ... ,Av-1. Assume that it is acceptable to change one of ξ least significant bits (LSB). We assign a weight to each attribute according to the owner of the database, and denote them as: W0, W1,..., Wi-1($\sum$wi=1, i∈[0,i-1]). Assume that it is acceptable to change one of ξ least significant bits (LSB) in a small number of (1/α) numeric values. Relation R contains η tuples, and a fraction ω=1/α of them can be used for fingerprinting. There are also two important parameter ξ and γ, which describes limit of modification to a database. Table I summarizes the important parameters used in our algorithms[5].

Table 1. Notations used in this paper

η	Number of tuples in the relation
ν	Attribute numbers to be marked
γ	Fraction of tuples marked
ξ	Number of markable bits
ω	Number of tuples actually marked
k	The secret key
W_i	Weight of i-th attribute
α	Significance level
P	The primary key in a relation

2.1 Watermark Generation Algorithm

The procedure contains three main parts:

- Coding and processing of the image signal of the copyright holder
- Use images to generate watermark
- Use chaos series to encrypt the watermark

The detailed process of watermark generation algorithm is as follows: 1) compress the image signal. Watermarking relational databases introduces small errors into the relations by inserting watermark into them. The marks must not have a significant impact on the usefulness of the data, so the watermark should be small. Thus, the compression of the image signal is necessary because of its large information capacity. We use wavelet to compress it;2) images signal convert. Convert the signal waveforms to the 8-bit A-law; The pixels of the copyright image are arranged by the order from left to right and then top to bottom and we can get a set S ={S1,S2,...,SN×N},in which each element is the decimal grey value of the pixel. Convert decimal pixel values to binary numbers and we can get a new set B = {B1, B2,..., BN×N};).3) encrypt the watermark. In our scheme, we use the Logistic chaotic system to encrypt the copyright image series. The chaos model possesses some property of one-way cryptographic Hash function. So, by using the chaos model, the security of the watermarking system is strengthened. Acting as a randomized series generator, the Logistic chaos equation is:$X_{n+1} = \mu X_n (1-X_n)$ { $\mu \in [1,4]$ n=0,1,2,...}, We set μ=4 in this computation. We select a threshold Q change the chaos series to the binary sequences P(i), The copyright image is encrypted as E(i) =W(i)⊕P(i),i=1,2,..., n.⊕= XOR. The Logistic's chaotic system is iterated until all the elements in the set B= {B1,B2,...,BN×N} is encrypted. Then every element in the encrypted we set E ={E1,E2,...,EN×N}[6]. Figure 3 are the Paralympic Games emblem (right) and the bicycle emblem (left) which we use them as the watermark.

Fig. 3. Watermark Images

2.2 Insertion Algorithm

A one-way hash function (Ex. MD5) H is critical in our algorithm [2][10]. we use the one-way hash function to decide which tuple is to be marked. The same way is used to decide which bit of is marked. As to the attribute, we choose the candidate to be marked by a weighted selection which makes the heavier the more marked. We use id (id=Hash (k, P, k)) and L (the number of bits of the watermark) to calculate the remainder i for each tuple, then collect tuples with same values of i into the same group. The i-th bit of the watermark will be inserted into the i-th group. The ascending order of i ranging from 0 to L-1 naturally preserves the sequence of the marks. This is helpful in the detection phase. Finally we check the usability with respect to the intended use of the data. If not acceptable, we simply give up watermarking this tuple and roll back. The watermark insertion algorithm was given as follows:

```
   // Weighted watermark algorithm, return marked R
   // Hi is one-way hash functions, L is the length of EMC
   // Parameters k, L, α, ξ and ν are private to the owner.
1  E[L]=H(k concatenate M)      // calculate L-bit EMC
2  foreach tuple r∈R do
3     t= H1( k concatenate r.P)
4     if ( t mod γ equals 0) then     // mark this tuple
5        i = select_attribute()      // mark i-th attribute
6        j = t mod ξ                 // mark j th bit
7        k= t mod L         //use the k-th bit of EMC
8        m=Ek XOR (k mod 2)   // value of marked bit
9        set the j-th LSB of r.Ai to m
10 return R
11 procedure  select_attribute()
12    u= H2( k concatenate r.P)
13    d=∑(1/ Wi)      i∈[0,i-1]
14    if (u mod d)  ∈ (∑(1/ Wi-1),∑(1/ Wi]
15       then return i       // W-1=0
```

2.3 Detection Algorithm

Because of the identical distribution of the hash function when seeded by the same key, the selected tuples in each group is of the same order as in the insertion algorithm .For each marked bit, we count the numbers of its value to be zeroes or ones respectively, and then a majority voting mechanism is to decide the final value of this bit. The detected result is a binary sequence which includes the copyright information of the dataset. Thus, we transform the binary sequence return back to the Beijing 2008 Emblem which are the watermark images. The copyright holder can make use of the Watermark image to prove the copyright. The watermark detection algorithm was given as follows:

```
1   for s=0 to L-1 do
2      DM[s]=''              // initialize detected mark code
3      count[s][0]=0, count[s][1]=0    // initialize counter
4      foreach tuple r∈R do
5         t= H1( k concatenate r.P)
6         if ( t mod γ equals 0) then      // select this tuple
7            i = select_attribute()         // mark i-th attribute
8            j = t mod ξ         // select j-th bit
9            k= t mod L         // mark the k-th bit of EMC
10           m= ( j-th LSB of r.Ai) XOR (k mod 2)
11           count[k][m]=count[k][m]+1    // add the counter
12     for s=0 to L-1                    // get the watermark
13        if (count[s][0]>=count[s][1])   // majority voting
14        then DM[s]=0  else DM[s]=1  //the final bit value
15     return  DM[ ]
```

3 Experiments and Analysis

We perform experiments on a computer running Windows XP with 2.4 GHz CPU and 256MB RAM. We apply our algorithms to Forest CoverType dataset, available at the UCI Machine Learning Repository (http://kdd.ics.uci.edu/databases/covertype/covertype.html). There are 581,012 tuples in this dataset, each with 54 attributes. We select 50,000 tuples and 5 attributes of them. In experiment, use MATLAB to simulate the watermark generation algorithm, and use java and JDBC to achieve the insertion algorithm and detection algorithm. .we use ws0 to denote the detection ratio of watermarks. M is the binary sequence which is extracted from the experiment database. $\overline{E}$ (i,j) is not of the E=(i)(0). We applied our algorithms to those data with 5 attributes which are weighted as 0.4, 0.3, 0.2, 0.05 and 0.05 respectively[7].

$$ws_0 = \frac{\sum_{i=1}^{n}\sum_{j=1}^{m} M(i,j) \oplus \overline{E}(i,j)}{m \times n} \times 100\% \qquad (1)$$

Subset Selection Attack: The attacker attempts to omit parts of the watermark by selecting a subset of tuples that are still valuable from the original marked relation. Table 2 shows the testing results at the stochastic subset selection.

Table 2. Watermark recovered in subset selection attack

Tuples selected	80% 60% 40% 20% 15%
Watermark image of A	
ws_0 of A	99.6% 99.1% 98.0% 89.4% 84.7%
Watermark image of B	
ws_0 of B	99.5% 98.7% 97.6% 87.8% 82.9%

Subset Addition Attack: In this part, the attacker randomly selects out part of the watermarked relation and mixes them with similar tuples probably without watermarks to form a new relation of approximately the same size of the original one. Table 3 shows the results.

Subset Alteration Attack: In this part, the attacker modifies some tuples in a marked relation to erase the watermark. Either reset each attacked bit to the opposite or reset each attacked bit randomly. The detection images look a little bit blurry in this attack model. It can accept alteration of 60% in our algorithm. Table 4 shows the results.

The detection results shown above indicate that our method can endure a series of basal attacks. The examinations of above got the clear visible watermark images. Now, we analyze the performance of the algorithm. First, we chooses images which has greater redundancy information as the watermarks, a certain degree of image data loss does not cause a significant decline in the quality of detection [8]. Furthermore, images themselves have error detection and correction; in consequence, it can effectively restore the image at the time of attack. Second, we use chaos sequence to encrypt the watermark and to avoid the watermark

Table 3. Result of subset addition attack

Tuples addition	50%	100%	150%	200%
Watermark image of A				
ws$_0$ of A	99.7%	98.2%	96.3%	94.6%
Watermark image of B				
ws$_0$ of B	99.2%	97.4%	94.3%	92.8%

Table 4. Watermark recovered in subset alteration attack

Tuples alteration	10%	30%	40%	60%	70%
Watermark image of A					
ws$_0$ of A	99.6%	98.0%	94.7%	87.3%	80.0%
Watermark image of B					
ws$_0$ of B	99.3%	97.6%	93.1%	86.2%	79.1%

binary flow appears a mass of all 0 or all 1 before embedding watermark to database. Third, we present a weighted watermarking scheme that mark attributes of a relation on probabilities associated with their weights. Higher weights are assigned to more significant attributes, so important attributes are more frequently watermarked than other ones. Experimental results prove the superiority of the algorithm.

4 Conclusions

In this paper, we propose a novel method for watermarking relational database, which uses the encrypted images sequence as watermark embed into database according to the weight of attributes. In addition, we have presented security analysis to show the robustness of our technique against various attacks. Through the experiments, it is shown that the proposed method of watermarking relational databases using character image is correct, feasible, and robust.

From the above experiments and the corresponding analysis, we can see that different images have different performances on the attack models. In the future, we will study the different performances of images as the watermark and improve the organizing method for the embedded image.

References

1. Agrawal, R., Kieman, J.: Watermarking Relational Databases. In: Proceedings of the 28th VLDB Conference, Hong Kong, China, pp. 155–166 (2002)
2. Radu, S., Mikhail, A., Sunil, P.: Rights Protection for Relational Data. In: Min, H., Cao, J.H., Peng, Z.Y., Fang, Y. (eds.) Proceedings of ACM SIGMOD. A New Watermark Mechanism for Relational, pp. 98–1094 (2003)
3. Niu, X., Zhao, L., Huang, W.: Watermarking Relational Databases for Ownership Protection. Chinese of Journal Electronics 31(12A), 2050–2053 (2003)
4. Zhang, Z.-H., Jin, X.-M., Wang, J.-M.: Watermarking relational database using image. In: Proceedings of the 3rd International Conference on Machine Learning and Cybernetics, Shanghai, pp. 1739–1744 (2004)
5. Hu, Z., Cao, Z., Sun, J.: An image based algorithm for watermarking relational databases. In: International Conference on Measuring Technology and Mechatronics Automation, zhangjiajie, pp. 425–428 (2009)
6. Sun, J., Cao, Z., Hu, Z.: Multiple Watermarking Relational Databases Using Image. In: 2008 International Conference on MultiMedia and Information Technology, Hong Kong, China, pp. 373–376 (2002)
7. Cui, X., Qin, X., Guan, Z., Shu, Y.: Weighted algorithm for fingerprinting relational database. Transactions of Nanjing University of Aeronautics & Astronautics, 149–152 (2009)
8. Liu, S., Wang, S., Deng, R., Shao, W.: A Block Oriented Fingerprinting Scheme in Relational Database. In: Park, C.-s., Chee, S. (eds.) ICISC 2004. LNCS, vol. 3506, pp. 455–466. Springer, Heidelberg (2005)

The Building of College Management Information System Based on Team Collaboration

Yu-you Dong[1] and Shuang Chen[2]

[1] School of Information Technology of Hebei University of Economics & Business, Shijiazhuang, China
[2] School of Public Administration of Hebei University of Economics & Business, Shijiazhuang, China

Abstract. In the 21st century college management information system development direction and the tendency is at based on the network educational administration teaching management. Based on team collaboration development pattern, but safe, effectively carries on the design and the realization to the college management information system. The practice proved that, it can coordinate the teacher, the student; the classroom three relations conveniently, enable the resources to achieve maximized the use. This article introduced one emphatically based on the team collaboration college management information system engineering research.

Keywords: College Management Information System; Team Collaboration.

1 Introduction

Unstable environments are likely to give rise to important new organizational forms and these emerging organizational forms, from time to time, necessitate new concepts about how information systems in the host organization should be designed and developed in supporting the organization to cope with changes. Whether one's primary interest is in the organizational design or information systems design, one eventually comes to terms with the other. The need and opportunity for design and development both in parallel thus are becoming clear; it would ensure better provision for effective development of information systems in terms of the followings:

  -to allow mutual and ongoing adaptation of organization and its information systems;
  -to avoid possible mutual constraint, if there are any, between one another
  - Hence result in more effective alignment of the information systems development objective and the organizational development objective with the business objective.

However, a co-design effort is unusual as it requires high level of creativity in planning, organizing, controlling and monitoring the change process. The development scope would inevitably include the Organizational, social as well as cultural context; from which the information systems are emerged and also from which the creativity is embedded.

This paper proposes how every change process is made more provocative and creative. The major strategies are (1) to induce the creativity of the IS organization, (2) to replicate it to other user organizations, and (3) to coordinate and collaborate the creative activities with other user organizations with an objective to create a stimulating environment that is conducive to group as well as the Organizational learning. A case study in college management context is used for illustration. The enhanced creativity can be partly attributed to the innovative research design from which some creativity elements were enabled and emerged. Before looking into specific matters of the case study, the background of the research is introduced in the next section.

2 Preview Co-design of Information Systems

The propositions on creativity in this paper stem from an on-going research program on Organizational Decision Support Systems (ODSS) in Computer Integrated. The broad objective of the research program is to promote effective use of Information Technology (IT) in college management. The primary interest in the research program is not technological driven. To a large extent, the need for research is induced by social and economical changes, which in turn shape the research directions. The research demands substantial intellectual developments which can be characterized by two major models. The first is engineering oriented, concerned with moving from concepts (e.g. CIM) to prototypes (e.g. ODSS) to production and operation of systems (e.g. to aid production planning and control). The second is oriented to theory building with the aim to develop explanations for consequences of use of information technology. The collective research interest moves reciprocally from the impacts of the technology to the causes leading to the impacts. This reciprocal analysis is inevitably accompanied with a shift in focus from one on human and behavioral aspects surrounding the introduction to adaptation, use and continuing innovation with the information systems. It is believed that research of this kind demands innovative research design and which, in our case, is attended as early as the initial stage of the research program. It would be useful to have a prior understanding of the innovative elements from our research perspective. Collectively, the major aspects are summarized in the following paragraphs.

3 Case Study

The creativity elements of the research design were brought into the case organization. A market survey was conducted to define the domain of the present case study. The change process in the case organization is documented and unfolded to extend our understanding of the mechanism that offers opportunities and constraints in making use of Information Technology for competitive advantage. The subsequent sections extracted the milestone events that led to the gradual built-up of creativity. After a brief review of the background and other supportive details, the principles for inducing creativity are extracted and summarized.

3.1 Implementation Planning and Monitoring

The management of co-design of IS was not seen as a straightforward, rational process but as a jointly analytical, educational and political process. Power, chance and opportunism are as influential in shaping outcomes as are design, negotiated agreements and master-plans. Seeing the organizational changes as episodes divorced from its social and cultural context would provide limited insights. Better understanding involved continuous reflexive monitoring of the interplay of ideas about context of change, process of change and the content of change. The creative ideas in the case study are organized and summarized around the following organizational interplays:

- Effective Intervention
- Industrial Collaboration
- Learning Laboratory

The attempt to provoke new ideas from Sales, Engineering, Purchasing, Production Planning and Control formed the basis for discussion. The executive team developed a Management Vision, namely: Right Material, Right Quantity at Right Time. This vision aimed to enhance the product quality, reduce its production cost, and at the same time made the organization more responsive to its customer needs; thus providing significant competitive advantage. Therefore, instead of being merely a lip service, the slogan drew management attention to the necessary adjustments in the organization, procedural changes and the need to empower the production staff. Various implementations strategies were applied. The necessary organizational changes, such as re-assignment of work role and revision in procedure, were made before, or in parallel with, the technical system development.

The case organization (WSM) was small in size. WSM had weak Information Technology infrastructure, a limited exposure to and experience in using computers in production. The staff in WSM did not have much expertise to develop necessary computer-based systems. Industrial Sponsored Projects (ISP) in collaboration with the Hong Kong Polytechnic was taken as an intervention strategy to introduce and manage the changes, i.e. technology transfer, organizational development, etc. A new system solution, namely Throughput-Material Requirement Planning was developed out of the collaborative research project.

Advice was sought from the researchers to evaluate the existing technological infrastructure with an objective to make it more flexible. The systems were integrated into an IT infrastructure. Some examples of the integration included: (a) stand-alone personal computers were interconnected through a local area network and were equipped with additional office automation facilities; (b) the overall system architecture was redesigned into a client-server based architecture;

Potential users, managers, staff executive and workers, were given opportunities to become acquainted with the new concepts, systems, and/or technologies through intensive training programmers and hands-on practice sessions. They were encouraged to identify the required skills, comprehend new roles, and suggest new improvements. Their active participation developed a sense of commitment which preempted resistance to change. As learning took place, users began to acquire skills to build their own applications. End user computing was vital to ensure a higher level of integration. These developments provided incentives in motivating all organization members.

3.2 Principles of Creativity Induction

The research study in WSM eventually seeded creative elements and also cultivated a favorable climate for the growth of it. Multitudes of ideas from the reference disciplines can be regarded as impetus to bombard the organization members at various levels, whether manager, researcher, IS staff or worker. Employees were stimulated and were mobilized to think/re-think in a more creative way. The IS staff was enabled to anchor on different perspectives apart from that inherited in traditional model of development. The others became more willing to collaborate with the creative IS staff and to suggest new ideas for improvement, volunteer to take actions. The management team would consciously direct attention to the need of Concurrent Engineering. They were also more willing to allocate organization resources for the forthcoming research and development. A favorable environment was evolved as a consequence. This was found to be vital in sustaining and nourishing the on-going IS development. In the process, several guiding principles were formulated and refined throughout the research. They are identified on reflection as follow:

- Combating Inoculants to Creativity
- Seeding the Creative Elements
- Nourishing creativity
- Cultivating Innovation in Situ

Creativity will not take place naturally. Imagine that if users are de-motivated, or they are hard pressed to meet the schedule, etc., then what kind of creativity could be expected. Sampler and regard these inoculating factors as impediments and identified several, such as lack of basic creative abilities, lack of skills in the task domain, bureaucracy, scheduled constraints, poor management, competition, resistances to change and etc. The presence of any inoculating factor will simply mean creativity is suppressed. Identifying them is one of the earliest steps towards combating them in the creativity inducement process. Identification of inoculants to creativity can be accomplished by group process techniques such as brainstorming or Delphi.

As highlighted in the case study, an organization is likely to be more creative when immersed in an environment characterized by one or more of the following:

- High aspirations
- High needs for achievement, learning and self development
- Concomitant goal shared by participants
- Adequate provision of resources for experiments
- Openness to provoke new ideas;
- Opportunity of reward for innovation

An environment of this kind can be regarded as a creativity incubation chamber. However, building such a chamber usually requires professional process consultation. On the one hand, work study is applied to promote the necessary infrastructure. On the other hand, intervention is required to overcome the organizational defense in order to legitimate the changes and make the changes feasible and desirable. Take the above case as an example, such a creativity chamber takes the form of a several ways. One cost effective way is to collaborate with a Higher Educational Institution.

With a conducive environment in place, the initial seeding of creativity should be put in a proper place to ensure the prosperity of the innovation to be adopted (e.g. ODSS). Seeding as widely as possible is desirable to the extent that it allows more fruitful cross-fertilization of knowledge. In the intricate co-design process, cross fertilization of knowledge over the reference disciplines such as social science, Organizational behavior, information science and etc., is of paramount importance. CO-design of IS implicitly ensures the changes are underpinned in its proper social, organizational as well as technological context. Seeding can be done in various forms such as pilot project, experimental developments, or some experiential training workshops. In the above case study, seeding was primarily done in the form of organizational innovation research projects.

4 Discussion

Intuitively, one of the earliest steps in promoting creativity is "Asking the Right Question". Following a similar logic, "one would never get the answer to a question that was never asked." It is desirable that the IS organization has to work with users, and in which case they mutually enact and ask themselves: "Are We Solving the Right Problems?" In fact, critical introspection is required to choose policies from a mind-boggling array of acronyms, technologies, and methods, e.g. CAD, CAM, TQC, JIT, etc. The answer requires examining diverse areas such as Strategy, Structure, Education, Training, Incentive, etc. Re-thinking is reinforced about real concerns such as need, wish, aim, interest and necessity. These concerns are all action-directed measures that reflect the attitude towards any proposed changes as direct or indirect consequences of the IS developments.

Regarding the research design, the industrial collaborative strategy is found to be well received by local industry. On the industry side, the case organizations can benefit in many ways such as technology transfer, consultation, etc. On the academic side, Higher Educational Institutions (HEIs) may consolidate more novel ideas in professional development of IS people. However, if collaboration with HEIs to study IS phenomenon in its natural setting is to be made more effective, more research must be conducted in such a way that theoretical informed conceptualizations are developed and empirically grounded evidences are solicited.

Contemporary requirements for co-design of IS demand a new breed of IS professionals. Apart from mastering development techniques such system analysis and design, they also have to acquire new level of competence in other non-analytic skills such change process design, client-centered consultation process, etc. Creativity by the IS organization leads to creativity by the client; the reverse is also true. Both the IS organization and the client have to overcome the status quo and acquire new learning habit in order to be more creative in every stage of the IS design and development. This is done in a broader context of organizational, social as well as cultural settings.

5 Conclusion

The study of creative process is embedded in an empirical universe, which makes it subject to observations that make empirical observations possible. Past studies of creativity in IS organizations may lack of an empirical dimension as well as a proper research design. These deficiencies are unlikely to be the consequences of the immutably empirical nature of the discipline but could be regarded as a reflection of the unattended research emphasis in the area; which is correctable by increasing the pace and quality of descriptive, empirical investigations in the area. Contemporary requirements of co-design of IS demand integration of the aspects of social analysis and engineering development of it. Research of such orientation should be conceptually informative and empirically grounded on a broader context. This includes technical, organizational, social, and cultural setting from which the need of IS and other organizational capability are emerged.

References

[1] Chung, W.W.C., Tam, M.M.C., Saxena, K.B.C., Yung, K.L., David, A.K.: Computer Integrated Manufacture (CIM) Research in Hong Kong Polytechnic
[2] Kraemer, K.L., King, J.L.: Social analysis in MIS: TheIrvhe School, 1970-1990. In: Conference Proceeding of International Conference of Information Systems (ICSS 'W), vol. III, pp. 582–590 (2008)
[3] Walton, R.E.: Up and Running: Integrating Information Technology and the Organization. Harvard Business School (2007)
[4] Couger, J.D., Higgins, L.F., McIntyre, S.C.: Differentiating creativity, innovation, entrepreneurship, intra pr-membership, copyright and patenting for I.S. product services. In: Conference Proceedings of Hawaii International Conference on System Sciences, pp. 370–379 (2006)
[5] Elam, J.J., Mead, M.: Designing for creativity: Considerations for DSS development. Information and Management 13, 215–222 (2005)

The Study on Decision Rules in Incomplete Information Management System Based on Rough Sets

Xiu-ju Liu

Dept.of Computer Science and Information Engineering, Heze University,
University street .1, 274000, Heze Shandong, China
exiu0824@126.com

Abstract. In this paper, the concept of non-symmetric similarity relation had been used to formulate a new definition of approximation to an incomplete information management system. By means of the new definition of approximation to an object set and the concept of attribute value pair, the rough-sets-based methodology for certain rule acquisition in an incomplete information management system had been developed. The algorithm could deal with incomplete data directly and do not required changing the size of the original incomplete system. The experiment showed that the algorithm provides precise and simple certain decision rules and does not affect by the missing values.

Keywords: Information systems; Decision tables; Rough sets; Decision rules; optimal certain rules.

1 Introduction

In this paper, we present rough set theory in the incomplete information system and extend the rough set theory in the incomplete information system and the multi valued information system. Some concepts and properties of rough sets based on incomplete information systems are given. We discuss the describing forms of decision of the incomplete information decision system and give some definitions and properties of the certain rules and optimal certain rules. It is easy to transform the describing forms of decision of the incomplete information decision system into the Skolem's standard form and apply it directly to resolution inference in artificial intelligence field.

The paper is organized as follows. In section 2, we present definitions of the incomplete information system and discuss the relationships between the incomplete information system with the incomplete information system and multi valued information system. Section 3 provides the concepts and properties of approximation sets based on the incomplete information system. In section 4 and section 5, we discuss the describing forms of decision of the incomplete information decision system and give some definitions and properties of the certain rules and optimal certain rules.

2 Incomplete Information Systems

Definition 2.1. incomplete information system is a triplet $\varphi = (\theta, AT, f)$, where θ is a non-empty finite set of objects, AT is a non-empty finite set of attributes and f is a mapping such that $f_a : \theta \to \Re(V_a)$ for any $a \in AT$, where V_a is called value domain of an attribute a. Any attribute value domain V_a may contain special symbol "*" which indicates that the value of an attribute is unknown and the real value must be one of the set $V_a - \{*\}$, $\Re(V_a)$ be a power set of V_a. $\inf(x) = \{(a, f_a(x)) | a \in AT\}$ is said to be an information vector set of x. The incomplete information systems φ is called normal if $f_a(x) \neq \phi$ and $* \notin f_a(x)$ for any $x \in \theta$ and $a \in AT$. If $* \notin f_a(x)$ for any $x \in \theta$ and $a \in AT$, the incomplete information system φ is said to be definite, otherwise, it is called singular.

In practical, the value of the attribute is always missing, namely, $f_a(x) = \phi$ for some objects. Here, we assume there exists $a \in AT$ such that $f_a(x) \neq \phi$ for any $x \in \theta$. Otherwise, it will be no meaning for the element, namely, $f_a(x) = \phi$ for any and $x \in \theta$. $a \in AT$.

Proposition 2.1. Let $\varphi = (\theta, AT, f)$ be a incomplete information system, if it satisfies the following conditions:

(1) $f_a(x)$ is a set contained only one element for any $x \in \theta$ and $a \in AT$;

(2) $* \notin f_a(x)$ for any $x \in \theta$ and $a \in AT$;

(3) $f_a(x) \neq \phi$ for any $x \in \theta$ and $a \in AT$;

then $\varphi = (\theta, AT, f)$ is a complete information system.

Proposition 2.2. Let $\varphi = (\theta, AT, f)$ be a incomplete information systems, if it satisfies the following conditions:

(1) $f_a(x)$ is a set contained only one element for any $x \in \theta$ and any $a \in AT$

(2) there exist $x \in \theta$ and $a \in AT$ such that $f_a(x) = *$;

(3) $f_a(x) \neq \phi$ for any $x \in \theta$ and $a \in AT$;

then $\varphi = (\theta, AT, f)$ is an incomplete information system.

Proposition 2.3. Let $\varphi = (\theta, AT, f)$ be a incomplete information system, if it satisfies:

(1) $* \notin f_a(x)$ for any $x \in \theta$ and $a \in AT$;
(2) $f_a(x) \neq \phi$ for any $x \in \theta$ and $a \in AT$;
then $\varphi = (\theta, AT, f)$ is a multi valued information system.

According to the definitions of the complete information system, incomplete information system, multi valued information system and the incomplete information system, the proofs of Proposition 2.1, Proposition 2.2 and Proposition 2.3 are easily completed. Obviously, there is an incomplete information system (see table 1) which is not the complete information system, incomplete information system and multi valued information system. We can obtain the following theorem.

Theorem 2.1. The incomplete information system is a more extensive uncertain information system. The complete information system, incomplete information system and multi valued information system are special cases of the incomplete information system.

Definition 2.2. Let $\varphi = (\theta, AT, f)$ and $\varphi' = (\theta', AT', f')$ be incomplete information systems, the system φ' is referred to as a extension of φ, if it satisfies

(1) $\theta' = \theta$ and $AT' = AT$;
(2) if $f_a(x) \neq *$ then for any $f_a'(x) = f_a(x)$ for any $x \in \theta$ and $a \in AT$.

The set of all extensions of φ will be denoted by $ETN(\varphi)$. The system φ' is called the x-extension, if $\inf(y) = \inf '(y)$ for any $y \in \theta - \{x\}$ and $* \notin f_a'(x)$ for any $a \in AT$. The set of all x-extensions of φ will be denoted by $ETN(\varphi, x)$.

Definition 2.3 Let $\varphi = (\theta, AT, f)$ be a incomplete information system, a certain incomplete information system $\varphi' = (\theta', AT', f')$ is referred to as a example of the system φ, if φ' is an extension of the incomplete information system φ (namely φ' satisfies $* \notin V_{a'}$ for any $a' \in AT'$). The set of all examples of φ will be denoted by $EAM(\varphi)$.

Example 2.1. Table 1 and Table 2 indicate four incomplete information systems, respectively, where "*" expresses that the value of an attribute is unknown, {1,2} two values of the attribute of an element, the blank space indicates that the value of an attribute does not exist. φ' in Table 2 is an extension of the system φ Table 1; φ'' is the extension of the object 2 in the system φ; φ''' is an example of the system φ, namely, $\varphi' \in ETN(\varphi)$ and $\varphi''' \in EAM(\varphi, 2)$.

Any attribute-value pair (a, v) $a \in A \subseteq AT, v \in V_a$ is called an atomic property. Any atomic property or its conjunction is said to be a descriptor. Conjunction of atomic properties (a, v), where $a \in A \subseteq AT$, is referred to as A-descriptor. The set of objects having the atomic property (a, v) is denoted by $\|(a,v)\|$, namely, $\|(a,v)\| = \{x \in \theta \mid f_a(x) \cap v \neq \phi\}$.

Table 1. An incomplete information systems φ

attributes AT / objects θ	a	b	c	d
1	1	2	1	1
2	*	1,2		2
3	1	2	*	1
4	2	1	1,2	1
5	1,2	1	2	2
6	2,3	*	2	1,2
7	3	1	1	1,2

Table 2.

GIS $\varphi' \in ETN(\varphi)$

θ	a	b	c	d
1	1	2	1	1
2	2,3	1,2		2
3	1	2	*	1
4	2	1	1,2	1
5	1,2	1	2	2
6	2,3	2	2	1,2
7	3	1	1	1,2

GIS $\varphi'' \in ETN(\varphi,2)$

θ	a	b	c	d
1	1	2	1	1
2	2	1,2		2
3	1	2	*	1
4	2	1	1,2	1
5	1,2	1	2	2
6	2,3	*	2	1,2
7	3	1	1	1,2

GIS $\varphi''' \in EAM(\varphi,2)$

θ	a	b	c	d
1	1	2	1	1
2	2,3	1,2		2
3	1	2	1	1
4	2	1	1,2	1
5	1,2	1	2	2
6	2,3	2	2	1,2
7	3	1	1	1,2

3 Approximation of Set in Incomplete Information Systems

Definition 3.1. A similarity relation $SR(a)$ (or $SR(A)$) is defined for an attribute $a \in AT$ (or a attribute set $A \in AT$) in the incomplete information system $\varphi = (\theta, AT, f)$ as follows:

$SR(a) = \{(x,y) \in \theta \times \theta \mid, f_a(x) \cap f_a(y) \neq \phi$ or $* \in f_a(x)$ or $f_a(x) \cup f_a(y) = \phi\})$ or $f_a(x) \cup f_a(y) = \phi\})$

Proposition 3.1. The similar relation SR(a) is reflexive and symmetric, but may not be transitive, so it is a tolerance relation.

Proof. According to the definition of the similar relation, we obtain the result of the Proposition 3.1.

Proposition 3.2. The relation $SR(A)$ is a tolerance relation such that $SR(A) = \bigcap_{a \in A} S(a)$ for any $A \in AT$ in the incomplete information system $\varphi = (\theta, AT, f)$.

Proof. For any $(x, y) \in SR(A)$ and $a = A$, (x, y) satisfies

$$f_a(x) \cap f_a(y) \neq \theta \text{ or } * \in f_a(x) \text{ or } * \in f_a(y)$$
$$\Leftrightarrow \forall a \in A, (x, y) \in SR(a)$$
$$\Leftrightarrow \bigcap_{a \in A} SR(a)$$

We denote the set of possibly indiscernible objects $\{y \in \theta \mid (x, y) \in SR(A)\}$ by $SR_A(x)$.

By the definition of the similarity relation $SR(A)$ in the incomplete information system $\varphi = (\theta, AT, f)$, it is obvious to obtain the following results:

The similarity relation $SR(A)$ degenerates a tolerance relation $SIM(A)$ in the incomplete information system. It degenerates an equivalent relation in complete information system.

Definition 3.2. Let $\varphi = (\theta, AT, f)$ be an incomplete information system, $X \subseteq \theta$, $A \subseteq AT$

$$\underline{A}X = \{x \in \theta \mid SR_A(x) \subseteq x\} = \{x \in X \mid SR_A(x) \subseteq X\}$$
$$\overline{A}X = \{x \in \theta \mid SR_A(x) \cap x \neq \theta\} = \cup\{SR_A(x) \mid x \in X\}$$

$\underline{A}X$ and $\overline{A}X$ are called lower approximation and upper approximation of X, respectively, $\underline{A}X$ is a set of objects that belong to X with certainty, while $\overline{A}X$ is a set of objects that possibly belong to X. If $\underline{A}X = \overline{A}X$, X is said to be a precise set, otherwise, X is said to be a rough set.

Proposition 3.3. The relations of sets $\underline{A}X \subseteq X \subseteq \overline{A}X$ hold for any $X \subseteq \theta$ and $A \subseteq AT$; $A \subset B \Rightarrow \underline{A}X \subseteq \underline{B}X$ for any $X \subseteq \theta$ and $A, B \subseteq AT$; $A \subseteq B \Rightarrow \overline{A}X \supseteq \overline{B}X$ for any $X \subseteq \theta$ and $A, B \subseteq AT$

The definitions of the approximation accuracy and the degree of the dependency are similar to that in the complete and incomplete information system. Here, we do not state }

4 Decision Tables and Decision Rules

(Complete or Incomplete) decision table is a (complete or incomplete) information system DT= $(\theta, AT \cup d, f)$, where d such that $d \notin AT$ and $* \notin V_d$ is called a decision attribute and the elements of AT are called condition attribute.

Let us define the function $\partial_A : \theta \to \Re(V_d)$, $A \subseteq AT$, as follows:

$$\partial_A(x) = \{f_d(x) \mid y \in SR_A(x)\}$$

∂_A is referred to as the incomplete decision in DT. $\partial_{AT}(x), x \in \theta$, is a set which contains all decision values based on the available information on x.

Proposition 4.1 Let DT= $(\theta, AT \cup d, f)$, $x \in \theta$, $A \subseteq AT$,

$$\text{card } (\partial_A(x)) = 1 \text{ iff } SR_A(x) \subseteq [x]_d$$

where $[x]_d$ indicates the equivalent class of the element x with decision attribute d.

A decision rule can express the form: $t \to s$ in incomplete information system, namely,

$$t = \bigwedge_{a \in C \subseteq AT} (\bigvee_{v \in B_a \subseteq V_a} (a, v)) \to s = \vee(d, w),$$

where C is a set contained all condition part attributes of the rule, (a, v) atomic property, $w \in V_d \cdot$, and are called the condition and decision part of the rule respectively. A rule with a single decision value in its decision part is called definite; otherwise, it is called indefinite. A rule $t \to s$ in φ is supported by an object x, $x \in \theta$, iff has both attributes t and s in φ.

Let X be a set having attribute $t = \bigwedge_{a \in C \subseteq AT} (\bigvee_{v \in B_a \subseteq V_a} (a, v))$, Y a set having attribute $s = \vee(d, w)$, the rule $t \to s$ in decision table is true, if and only if $\overline{CX} \subseteq Y$, where C is a set of all attributes occurring in the condition part of the rule.

Example 4.1. Table 3 indicates an incomplete decision table about healthy condition of human bodies. The professional types, labor strengths and sport exercises are the condition part attributes (where sport exercises exceeds twice a week, one time surpasses half an hour), the physical condition is the decision part attributes. The domains of attributes are $V_{profession}$={teacher, researcher, office worker, worker}, $V_{professional\ type}$ ={mental work, manual labor}, $V_{labor\ strength}$ ={strong, general, weak}, $V_{sport\ exercise}$ ={ball games, track and field} and $V_{physical\ condition}$ ={good, general, bad}, respectively, the special signal "*" indicates the unknown value of attribute, "#" indicates that the value of attribute does not exist. Table 3 contains a extended column ∂_A which indicates the general decision values of attributes in incomplete information system.

The rules in Table 3 are as follow:

Table 3. The generalized decision table about healthy condition of human bodies

Attribute Universe (people)	profession	professional type	labor strength	sport exercise	physical condition	∂_A
u₁	teacher	mental work	strong	#	bad	{ bad }
u₂	researcher	mental work	strong	#	bad	{ bad }
u₃	researcher	mental work	general	ball games track and field	good	{ good }
u₄	*	*	*	ball games	general	{ good,general }
u₅	worker	manual labor	weak	*	general	{ good,general }
u₆	worker	manual labor	general	ball games	good	{ good,general }
u₇	worker	manual labor	weak	track and field	good	{ good,general }

r₁: (profession, teacher) ∧ (professional type, mental work) ∧ (labor strength, strong) ∧ (sport exercise, #) → (physical condition, bad);

r₂: (profession, researcher) ∧ (professional type, mental work) ∧ (labor strength, strong) ∧ (sport exercise, #) → (physical condition, bad);

r₃: (profession, researcher) ∧ (professional type, mental work) ∧ (labor strength, general) ∧ ((sport exercise, ball games) ∨ (sport exercise, track and field)) → (physical condition, general);

r₄: (sport exercise, ball games) → (physical condition, general); →

r₅: (profession, worker) ∧ (professional type, manual labor) ∧ (labor strength, weak) → (physical condition, general);

r₆: (profession, worker) ∧ (professional type, manual labor) ∧ (labor strength, general) ∧ (sport exercise, ball games) → (physical condition, good);

r₇: (profession, worker) ∧ (professional type, manual labor) ∧ (labor strength, weak) ∧ (sport exercise, track and field) → (physical condition, good);

The definitions of reduction and core of attributes in the generalized information systems and the generalized decision tables are the same as that in the complete (or incomplete) information system and the complete (or incomplete) decision table. Here, we do not give them.

5 Certain Rules and Optimal Rules

Definition 5.1. A rule r: t→s is said to be certain, if the rule r is definite and true in the definite generalized decision table $\varphi = (\theta, AT \cup d, f)$ (namely, $* \in f_a(x)$ for any $x \in \theta$ and $a \in AT$).

Definition 5.2. Let the rule be r: $t = \bigwedge_{a \in C \subseteq AT} (\bigvee_{v \in B_a \subseteq V_a} (a,v)) \rightarrow s = \vee(d,w)$ in the generalized decision table $\varphi = (\theta, AT \cup d, f)$, the rule r is said to be optimal in the φ, if it satisfies:

(1) the rule r is certain;
(2) if there is a rule $r': t' \rightarrow s'$, as $s' = s$, $t = \bigwedge_{a \in C \subseteq AT} (\bigvee_{v' \in B'_a \subseteq V_a} (a,v))$ and $B_a \subset B'_a$ or $s' = s$, $t' = \bigwedge_{a \in C' \subseteq AT} (\bigvee_{v \in B_a \subseteq V_a} (a,v))$ and $C' \subset C$ then the rule r' is uncertain.

Proposition 5.1. If the generalized decision table $\varphi = (\theta, AT \cup d, f)$ is complete (or incomplete), then certain rules and optimal rules in $\varphi = (\theta, AT \cup d, f)$ are certain rules and optimal rules defined in complete (or incomplete) decision table.

Proof. By the definitions of the certain rules and optimal rules in the complete (or incomplete) decision table. Also according to Proposition 2.1 and Proposition 2.2, it is clearly to get the result.

Definition 5.3. A rule r: $t \rightarrow s$ is called the certain rule (or the optimal rule) in the generalized decision table $\varphi = (\theta, AT \cup d, f)$, if each example of the rule r is certain (or optimal) in φ.

By the example 4.1 and the ∂_A column in table 3, the rules r_1, r_2 and r_3 are certain rule, the rest rules are uncertain rules. The objects u_1 and u_2 support not only the rules r_1 and r_2 but also the following rules:

r': (professional type, mental work) (labor strength, strong) (sport exercise, #) (physical condition, bad). $\wedge \rightarrow$

It is clearly that the rule r' is real and definite in the decision table. Therefore, it is certain. We also know that the new produced rules which reduce the condition part attributes and increase new values of attributes in rule r' are false. Hence, the rule r' is a optimal rule in table 3.

6 Conclusions

We extend the rough set theory in incomplete and complete information system. The rough set theory in generalized information system will be applied in more extensive scope. We also present the method and technology dealing with information in generalized information system. The general descriptor of the decision rule is presented in the generalized decision table. The main difference from descriptor of the (incomplete or complete) decision rule to that of the generalized decision rule is that descriptor in the condition part is the conjunctive normal form of the value of the atomic property from the conjunction of the value of the atomic property. This descriptor of the decision rule in the generalized decision table is much more useful for reasoning and computing of the optimal rule. The application of the generalized information system in many fields, for example, machine learning, decision analysis, knowledge acquisition, pattern recognition and data mining, and so on, will be an important task.

References

1. Pawlak, Z.: Rough sets: Theoretical Aspects of Reasoning about Data, vol. 9. Kluwer Academic Publishers, Dordrecht (1991)
2. Pawlak, Z.: Rough sets and intelligent data analysis. Information Sciences 147, 1–12 (2008)
3. Chan, C.-C.: A rough sets approach to attribute generalization in data mining. Journal of Information Sciences 107, 169–176 (1998)
4. Shen, L., Loh, H.T.: Applying rough sets to market timing decisions. Decision Support Systems 37, 583–597 (2008)
5. Huang, C.-C., Tseng, T.-L(B.).: Rough set approach to case-based reasoning application. Expert Systems with Applications 26, 369–385 (2008)
6. Bonikowki, Z., et al.: Extensions and intentions in the rough set theory. Journal of Information Sciences 107, 149–167 (1998)

Collaboration CAD Design Based Virtual Reality Modeling Language in Heterogeneity Assembly Environment

Jian Yingxia[1] and Huang Nan[2]

[1] Scientific Research Department, Xinxiang University,
453003, Xinxiang, P.R. China
zbem1@sina.com
[2] School of Computer and Information Engineering,
Xinxiang University, 453003, Xinxiang, P.R. China
xxxyhn@163.com

Abstract. Collaboration designs now often involve engineers working around the world while current CAD-based virtual assembly system s often only supports single computer operation. This paper presents a VRML (virtual reality modeling language) based virtual assembly system developed to provide a heterogeneity CAD model collaboration assembly environment. The system uses an assembly feature parameter extraction algorithm to realize the assembly constraints among the VRML part models. Furthermore a C-P (command-parameter) and operation token mechanism s are used to handle low speed Internet traffic and avoid operational conflicts. The system also includes a collision detection module based on the V-collide library to dynamically detect collisions between VRML parts during assembly. The system is integrated with the product data management system to create a product life cycle system, including design, demonstration, and maintenance, with effective communication efficiency during the product design stage.

Keywords: Virtual Reality; VRML; Visualization.

1 Introduction

These days, companies are facing drastically changed competition conditions. The situation in vehicle, aircraft and trains is characterized by market globalization, individualized customer demands, increasingly complex products and ever shorter innovation cycles. To cope with these challenges, a company must be able to rapidly launch high-quality, customized products at prices which are readily accepted by the customers. To do so, a company must heavily rely on its development engineers who contribute decisively to its competitiveness. There is indeed an increasing need for virtual product engineering and for tools, systems and methods to realize such a challenge process.

Virtual Reality (VR) provides complex product design a virtual environment for offline programming and simulation. With new user computer interfaces, researchers are starting to incorporate new techniques, which enable the performance of functionality tests and verifications in the advance of interactive media that CAD systems cannot provide. Multimedia techniques are supporting the user computer interaction, as well as dialogues, among the users. Considerable research is in progress to overcome the lack of information in the conceptual part of the design process, introducing functional features as referred to in [1] or with VR techniques as in [2]. VR systems and distributed VR systems have been developed in many fields, such as education [3], architecture [4], medicine [5] and chemistry [6], to mention just a few. Many early applications were developed for the automotive and aerospace industries for the assembly and maintenance process [7] or for manufacturing [8].

VRML (Virtual Reality Modeling Language) is a descriptive language that can be not only connected to the World Wide Web but also used to describe the complex 3Dscene in the world of virtual reality. It employs Scene Graph to establish 3D scene in order to design the special virtual world. Moreover, its files are character-based text ones, and the text file according with the VRML syntax can be interpreted and browsed in VRML browser. It has three main characteristics. (1)User can interact with virtual world; (2) The objects in virtual world can move, thus animation may be made; (3)configured web browser, Cosmo Player and World View are often used in general, is a plug-in of web browser. When VRML files (wrl) are called, Cosmo Player is able to automatically recognize them and set up 3D operational environment in its window. It is this paper that utilizes these characteristics to convert the design or simulation results created in CAD/CAE software into the scene files of VRML2.0 by data interface automatically. Furthermore, 3D visualization is implemented in the browser.

2 System Architecture

This paper realized a new architecture to divide the system into two sub-ones: Programming and Simulation. In Programming, user can make offline programming, and then analyze motion in his own computer to preview program results. In simulation part, user can carry on analysis based on whit the help of CAE software(such as I-DEAS NX,HyperMesh,Ansys).Thus system attained high efficiency in the design and manufacture of complex product well as increasing productivity in industrial applications. System architecture is based on 3 Layers: User Interface, Software layer, Hardware layer.

2.1 User Interface

System provides a friendly interface where user can initialize environment such as numbers and positions of users. Then, user makes task planning and offline program in programming panel as an easy graphical programming environment. With the program user can further control the model. In Viewing Platform, user observes model assembly motion in the screen, analyze and predict potential problems.

2.2 Software Layer

This layer is to build a variety of models-solid models, framework models, surface models, which can be established directly in the software, such as UG, 3D MAX, I-DEAS NX,SolidWorks, Pro/E, and so on. Cosmo is used to develop VRML program, motion animation and environment construction, as well as rendering, lighting and action-event response. EditPlus is a assistant tool for optimizing VRML program; CAE software (such as HyperMesh,Ansys) will give great help for product simulation. In addition, we have developed a VR system based on VRML in this layer. With the help of it, we can establish a virtual world where the behavior and physics attributes can also be embodied.

2.3 Hardware Layer

The third layer is hardware layer, including input and output devices such as data gloves, three dimension space balls, stereo scopic glasses, and graph accelerate cards, etc. The top of this layer is used as interface for connecting hardware devices and its main mission is to obtain data from data gloves or to set up parameter about the vision space of both eyes of stereo scopes glasses.

3 VR System Based On VRML

VRML is the standard for Internet 3D browsing; it can be used as a fundamental base for product design checking etc. Now VRML becomes mature and practical, it can play a versatile role for remote 3D design. In the complex product design system, partners can share or contribute their design capabilities to complete a common product design and assembly via Internet/Intranet. Certainly, some application development is needed to fit specific purpose. In our system, we integrated several modules: design task management, checking and evaluating, design information exchange, virtual assembly.

3.1 System Framework

The system framework is shown in Fig.1.The module of design task management is to handle the existed design schemes to shorten design process. As matter of fact, the most design is more or less related to the existed ones. The system checking and evaluating is the core of the system. Here, by combining with other multimedia information, we use 3D model of a product to be checked and evaluated. Design information exchange is the import and export for the people joined together working for the same product. Virtual assembly is an assistant tool to further checking.

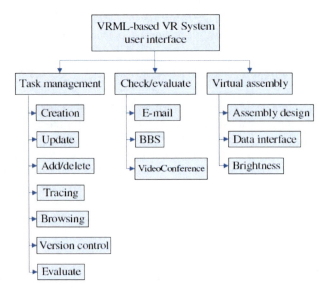

Fig. 1. System Framework

3.2 The Workflow of System

The interactive character of the 3D in VRML makes it much easier for the communication of information. It will help the designer to find errors hidden in the design. So, this interactive 3D form of detail drawing is practical to some degree. The following example will describe the whole process to create it. The main tool used in this example is the VRML, and the aiding applications are Cosmo Worlds and Catia. The final result can be browsed by web-browser. The drawing used in this example is the detail of cover of Car engine. First, finish the 2D drawing in AutoCAD. Then build the 3D model with Catia, and take hyper mesh as the repair tool. Some 3D symbols are added to represent the link. In Cosmo Worlds, setup the camera and the light. Because it is too difficult to build the model directly in VRML editor, the VRML file will be exported from Catia. The file exported by Catia V5R16 is in the form of VRML 97. Using the text editor EditPlus2.0, add the anchor to link the 2D drawing and enlarged detail with the main VRML file (see the code below). As a result, users can browse it by IE or other web browsers. The system workflow is shown in Fig.2.

```
DEFUnnamedTransformTranslationInterp_2Kfa
PositionInterpolator {
Key [0, 0.2 ]
keyTypes [ 2, 2 ]
authorKey[ 0, 0.2 ]
keyValue [ 0 0 0,0 0 -0.5 ]
authorKeyValue [ 0 0 0,0 0 -0.5 ]
}
]
}
```

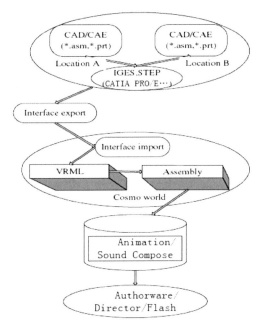

Fig. 2. The workflow of system

4 Visualization of Simulation Results

We realize 3D visualization of simulation results in Cosmo Worlds and Cosmo Player, at the bottom of which, there is a control panel for navigation. By this panel, 3D objects can be moved and rotated so that interactive operation can be realized very conveniently. When the VRML's file is called, Cosmo Player can automatically set up the 3D operational environment. In this system, the shape of Car body is presented in Cosmo Worlds (Fig.3), the model of the engine cover is transformed by Hyper mesh (Fig.4).

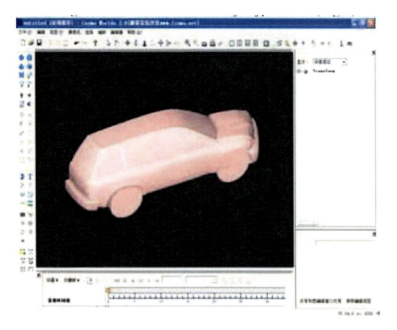

Fig. 3. Moving an object in CosmoWorlds

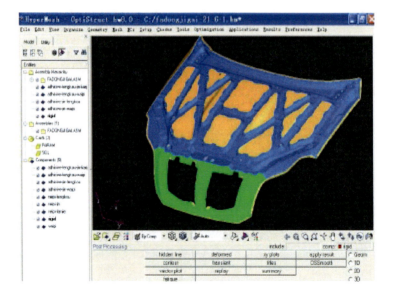

Fig. 4. The model of engine covers

5 Conclusion

VR is recognized as the technology at the moment that can offer to the user the ability to see and explore in a realistic manner new products or concepts before they exist in reality. The costs involved in virtual prototyping are often essentially lower than a similar test on real prototypes. Standard Web technologies are leading to easier, more effective and more generalized applications. In this paper, with the help of special data interface, we used VRML for sharing models over the internet or intranet. The model enables the generation of a virtual environment for a viewing and an evaluation of the functionality in a standard VRML Web browser.

References

1. Brunetti, G., Golob, B.: A feature-based approach towards an integrated product model including conceptual design information. CAD 32, 877–887 (2000)
2. Bimber, O., Encarnacao, L.M., Stork, A.: A multi-layered architecture for sketch-based interaction within virtual environments. Comput. Graph 24, 851–867 (2000)
3. Tate, D., Sibert, L., King, T.: Using virtual environments for firefighter training. IEEE Comput Graph Appl. 17(6), 23–29 (1997)
4. Woo, S.: Shared virtual space for architectural education. In: Proceedings of the 3rd CAADRIA, Osaka, Japan (April 1998)
5. Langrana, N.A., Burdea, G., Ladeji, J., Dinsmore, M.: Human performance using virtual reality tumor palpation simulation. Comput. Graph. Int. J. 21(4), 451–458 (1997)
6. Casher, O., Leach, C., Page, C.S., Rzepa, H.S.: Virtual reality modeling language (VRML) in chemistry. Chem Britain 34, 26 (1998)
7. Gomes de Sa, A., Zachmann, G.: Virtual reality as a tool for verification of assembly and maintenance processes. Comput. Graph 23, 389–403 (1999)
8. Beier, K.P.: Web-based virtual reality in design and manufacturing applications. In: Proceedings of COMPIT 2000, Potsdam, Germany (April 2000)

The Study on the Sharing of Data Sources in CAD Environment Based on XML

Jian Yingxia

Scientific Research Department, Xinxiang University,
453003, Xinxiang, P.R. China
zbeml@sina.com

Abstract. Exchanging and sharing the heterogeneous data sources is the core of the information integration in virtual product development. Using the XML-based method to achieve the product information integration in CAD environment has the merits of clear structure and good interoperability. The article, aiming at the application background of the car mould virtual product development achieves the XML based information integration in CAD environment and provides the quick and effective method of accessing the product information for the other subsystems in car mould victual product development system.

Keywords: Computer application; Apparel CAD; Pattern; XML; File Structure.

1 Introduction

After years of promotion, the application of apparel CAD to domestic apparel enterprise has reached the universal stage and improved the quick response capacity of the apparel enterprise. However, there are also some problems in the application, a prominent one is the system's compatibility systems cannot exchange data with each other successfully, leading many users to a lot of trouble.

However, as DXF format is mainly used to show graph information and that is not specially designed to express the apparel pattern structure information, there are some flaws, therefore, a method of apparel CAD pattern data exchange based on XML is put up in this paper.

XML is an acronym for extensible Markup Language. It was in February 1998 W3C Association (World Wide Web Consortium, known as W3C) formally launched the met language used to create a markup language, which can be used to describe data file format for online data exchange as a linguistic standard. XML is a universal standard of data interface, with simple, easy-to-understand, self-describing, extensible, interactive features, allowing users to create their own tags for different types of data and open data does not rely on the platform and language. Compared with the STEP standard of fixed data model, XML data model is more flexible, and can be modified or added entity information at any time, is more suitable for the information exchange and share in the whole product life cycle [5, 6].

2 Pattern Information Model

For the pattern of a style, it is composed of size, unit, grade rule and sample size piece, of which pattern piece is the basic, and pattern piece is composed of a variety of lines and points. Table 1 gives each of the specific composition of pattern piece [3, 4, 7].

Table 1. Specific composition of pattern piece

No.	Name	Description
1	Sew line	The rough and actual line constituting the outline of piece
2	Cut line	Sew line adding seam allowance to get the outline for cutting.
3	Grain line	Mark of Warp direction as a baseline
4	Notch	Mark for position matching of sewn
5	Internal line	Such as: darts_line, divider line and so on
6	Internal point	Such as: drilling and so on
7	Process mark	Such as: contraction joint, button hole and so on
8	Annotion Text	Pattern information, such as size, name, number and so on

Sew line and cut line are closing lines connected by some points in an orderly manner; these points are of different properties, such as grade point, curve point and so on. Fig. 1 shows the structure of pattern with data tree.

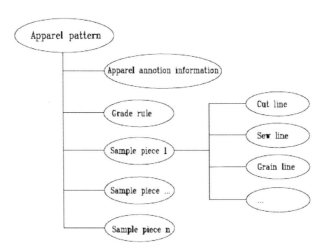

Fig. 1. Pattern structure treeï

3 Description of Pattern by XML

3.1 The Creation and Structure of XML Document [6]

XML describes various types of data in a structured way. The structure of XML document can be divided into element, attribute and value. XML element begins with a label and ends with the corresponding label. Label is defined by the angle brackets- begins with less than (<) and ends with more than (>). The label may contain attribute and can be assigned to it.

The first line of XML document is the XML declaration.

<?xml version="1.0"?>

Each XML document has only one root element, other elements need to be included in the root element. Element can also have attributes, attributes attached to the elements to exist. Such as:

<message text="English"length="200"></ message>

3.2 The DTD and Schema of XML Data

Before creating XML, firstly it is needed to establish its elements (tags) and the structure, and then, According to the definition of the structure, the actual content is filled in it to form a XML document. The structure documents of XML are defined in two ways, that is, the document type definition (DTD) and the model definition (Schema).

3.3 The Design of Pattern XML Document

(1) The elements of pattern. First of all, XML document needs root element. according to Fig.1, Pattern is named root element, pattern annotation information is named element Header, grade rule is named element Grade Rule, piece is named element Piece, and other information are respectively named in Table 2, Table 3, Table 4.

Table 2. Pattern annotation information corresponding elements

No.	Element	Element description
1	Version	Version number of pattern XML
2	Auther	Supplier name of application of creating pattern file
3	CreateDate	Create data
4	CreateTime	Create time
5	Style	Style
6	Units	Units
7	NumberOfSizes	Number of size
8	SizeList	Size list
9	SampleSize	Sample size

Table 3. Pattern grade rule corresponding elements

No.	Element	Element description	Attribute
1	Rule	Grade rule	Num(Number of grade point)
2	Delta	X,Y grade rules values	

Table 4. Sample piece corresponding elements

No.	Element	Element description	Attribute
1	PieceName	Piece name	
2	Quantity	Piece quantity	
3	PolyLine	Composite line	Type
4	Line	Line	Type
5	Point	Point	Type NotchAngle (only notch point) NotchDepth (only notch point) Num(Number of grade point, only grade point)

The Study on the Sharing of Data Sources in CAD Environment Based on XML 447

(2) Pattern structure tree. Table 2, table 3 and table 4 correspond to the structure tree of Fig.2, Fig.3 and Fig.4.

Fig.5 is available after the combination of Fig.2, Fig.3 and Fig.4, together with the root element.

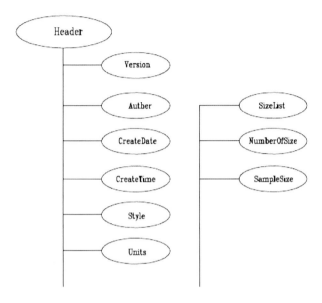

Fig. 2. Corresponding structure tree of table 2

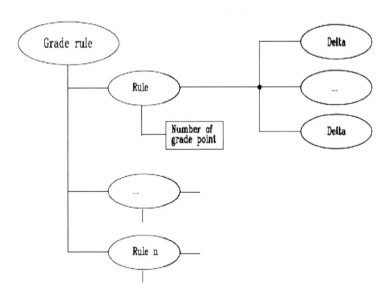

Fig. 3. Corresponding structure tree of table 3

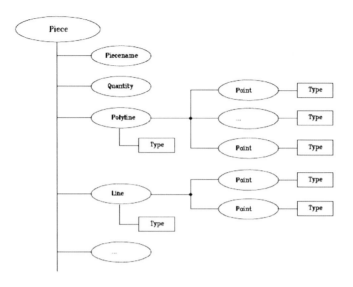

Fig. 4. Corresponding structure tree of table 4

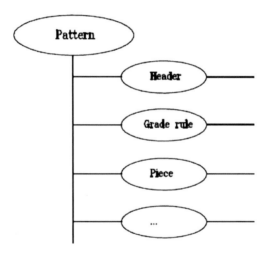

Fig. 5. Main structure tree

(3) XML Schema framework. According to the structure tree above, combining with XML Schema syntax, the framework of XML Schema can be created, assuming named Pattern.xsd, part of the code is as follows:

```
<?xml version="1.0" encoding="UTF-8"?>
<xs:schema
xmls:xs="http://www.w3.org/2001/XMLSchema"
elementFormDefault="qualified"
```

```
attributeFormDefault="unqualified">
<xs:element name=Pattern>
<xs:CpmplexType>
<xs:sequence>
<xs:element name=header>
......
</xs:element>
<xs:element name=GradeRule>
......
</xs:element>
<xs:element name=Piece>
......
</xs:element>
<xs:element name=Piece>
......
</xs:element>
......
</xs:sequence>
</xs:CpmplexType>
</xs:element>
```

4 Conclusions

In order to achieve a successful exchange of pattern data among various apparel CAD systems, this paper puts up using XML to describe the information of apparel pattern, due to the limited amount of information, description is easy to achieve, so using XML to exchange pattern data is totally feasible in the area of apparel CAD. Pattern exchange file based on XML is specialized, high-precision and easy to switch read and write, and it can store graphic and no graphic information, fully meet the needs of pattern data exchange, thus avoiding a lot of repetitive work. With XML technology to improve step by step, not only can be used to describe pattern information, but also can bring a good solution to the knowledge representation of apparel CAPP [8]. With the gradual maturity of XML description and display of the three-dimensional graphics, it will also play a significant role in the field of three-dimensional apparel CAD.

In short, in the information society, to improve the compatibility, openness and standardization of apparel CAD is the trend of the times. Addressing the compatibility issue is conducive to developing China's apparel CAD technology and its spreading and application in China's apparel industry, also conducive to promoting international communication and cooperation, and in favor of the internationalization of the line.

References

1. Li, C.: Theory and Practice of Computer Graphics. Beijing University of Aeronautics & Astronautics Press, Beijing (2004)
2. Su, H., Liu, H.: How to Develop AutoCAD R12. Tsinghua University Press, Beijing (1995)
3. ASTM international, Standard Practice for Sewn Products Pattern Data Interchange—Data Format(D6673-04), ASTM international, West Conshohocken (2004)
4. Japan Industrial Association of Women, CAD Data·Sewn data Exchange system Apparel CAD Exchange to advance Project—Interchange Data Format-V1.00 (Japanese Edition), TIIP,Japan (1997)
5. Lingyi, G., Guangleng, X., Jinching, X., Xudong, C.: Research on the Key Technology in the Integration of Product Information Based on XML. Journal of Computer Aided Design & Computer Graphics 14(2), 105–110 (2002)
6. Jie, W.U.: XML Application Tutorial, 2nd edn. Tsinghua University Press, Beijing (2007)
7. People's Republic of China textile industry standards, Pattern-making(FZ/T 80009-2004). China Standard Press, Beijing (2005)
8. Zhou, N., Liao, W., Shen, J., Li, F., Yang, H.: Integration of CAD/CAPP/CAM Based on XML. Mini-Micro Systems 25(7), 1359–1363 (2004)

Semantic Mapping Approach for Logistics Services Integration in 4PL

Qifeng Wang

Business School of Zhejiang WanLi University,
315100, Ningbo, P.R. China
lhywqf@163.com

Abstract. The forth-party logistics (4PL) is a modern logistics operation mode for logistics industry. The kernel of 4PL operation is the integration and collaboration among the 4PL operator, logistics services providers and customers. Based on the analysis of integration requirements for the logistics services, which including dynamic integration, unified integration mode, integration standard supporting, semantic supporting and loosely coupled etc, this paper puts forward a semantic mapping approach for logistics services integration in 4PL. Firstly, a logistics services semantic mapping framework for 4PL based on the shared logistics ontology is provided. On the basis, the logistics services semantic mapping process based on shared logistics ontology for 4PL is formal described in detail. Finally, the semantic mapping based logistics services integration process is realized, which provides a semantic integration approach for 4PL logistics services integration.

Keywords: The Forth-Party Logistics, Logistics Services, Semantic Mapping, Logistics Ontology, Semantic Integration.

1 Introduction

The 4PL is a modern logistics operation mode for logistics industry. In 4PL, customers outsourcing the logistics business in the whole supply chain to a fourth party, who has the ability of integration management, information technologies, 3PL operation etc. The fourth party develops a set of supply chain integration solutions, and is responsible for the solution to monitor and evaluate the implementation process in order to improve the overall operation of the supply chain performance. The 4PL operator faces global, large number logistics services providers with heterogeneous of information format and systems when setting up the 4PL. How to choose, collaborate and manage the different logistics services provided by different logistics providers is the key of 4PL operation. Most of the logistics services providers have different information systems, applications which developed with heterogeneous system framework and technologies. It is hard to integrate the systems and applications from different system venders and in different technology standard. With the economy globalization, the 4PL faces more and more intense international competition. The IT infrastructure of 4PL are required to be agile and adapted on demand, on the same time, the IT infrastructure of 4PL should integrate the logistics services provided by 3PL and other logistics services providers on demand. The integration of logistics services has become key issue for 4PL operation.

Logistics services integration is the operation basis for 4PL, and the collaboration basis of 4PL, logistics services providers and customers in the internet environment. To support the networked collaboration operation between logistics services providers and customers, and realize the dynamic integration on demand inner the logistics services providers and among different logistics services providers. The integration requirements are as follows, supporting dynamic integration, unified integration mode, integration standard supported, semantic supported services integration and loosely coupled integration mode, etc. The study and application of semantic web services provides a new way to realize the services integration for 4PL, which can satisfy the integration requirements well. The 4PL services integration based on semantic web services involves logistics services semantic modeling, logistics services matching, logistics services semantic mapping, logistics services evaluation, and the construction and execution of logistics services processes, etc. This paper is emphasized on the study of semantic mapping approach for logistics services integration. Firstly, a logistics services semantic mapping framework for 4PL based on shared logistics ontology is provided in section 2. In section 3, the logistics services semantic mapping process based on shared logistics ontology for 4PL is formal described in detail. In section 4, the semantic mapping based logistics services integration process is realized

2 Semantic Mapping Framework for Logistics Services Integration in 4PL

Plenty of domain knowledge is involved in the semantic mapping, which needs domain experts to involve in. Presently, automatic semantic mapping is not reality, while, some attempts of semi-automatic semantic mapping approaches is fulfilled. Firstly, the auxiliary mapping tools mine the mapping relationship and summit to the domain experts, then, the experts decide whether the mapping relationship is reasonable or not. There exists four semi-automatic semantic mapping approach [1], which are ontology concept mapping based on the general ontology base [2], semantic mapping based on the concept similarity of ontology [3], semantic mapping based on the concept hierarchical structure similarity of ontology [4], and semantic mapping integrated with the concept similarity and the concept hierarchical structure similarity [5].

Semantic mapping in the 4PL services operation includes two aspects. One is to solve the mapping of semantic information from different logistics services providers for different semantic description in different logistics services providers. Another is to solve the inner semantic heterogeneous in logistics services providers' information system for different developers and system construction in different times. Semantic mapping is very complex in 4PL logistics services integration. The kernel of semantic mapping in 4PL services integration is to realize the correct mapping with different applications and services in multi logistics services integration with different logistics services providers.

In this paper, the semantic mapping study is based on the shared logistics ontology, the mapping framework is shown in figure 1. If there exists more than one logistics ontology in different logistics services providers, ontology mapping is needed to realize the semantic mapping between integration objects [6-9]. In this paper, the study is limited in one shared logistics ontology.

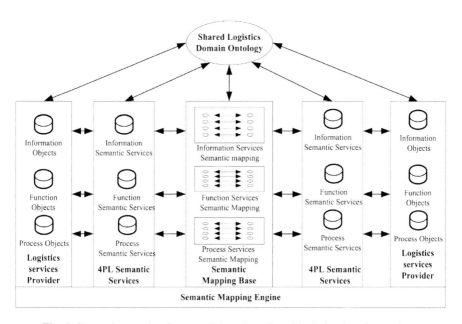

Fig. 1. Semantic mapping framework based on shared logistics domain ontology

The establishing of semantic mapping relationship among logistics services integration objects consist two steps based on the shared logistics ontology.

Step 1. Local schema mapping among logistics services, that is, establishing the mapping relationship between the integration objects of logistics services providers and the integration objects of the 4PL. After the service modeling for the integration objects, the data type and schema of the integration objects should transferred to the local standard based on the shared logistics ontology to keep the uniform with the shared logistics ontology in marking, properties and instances.

Step 2. Global schema mapping, that is, based on the realizing of local schema mapping, establishing the semantic mapping between integration objects of heterogeneous logistics services in marking, properties, instances, etc, by the shared logistics domain ontology.

3 Formal Descriptions for Semantic Mapping Based on Shared Logistics Ontology

Based on the analysis of semantic mapping framework based on shared logistics ontology, the normally formal description of the shared logistics ontology semantic mapping is as follows.

Definition 1. Semantic Mapping, SM=(S, T, LO, LF, RD). The semantic mapping can be expressed by a five-tuple array. S is source integration objects, T is target integration objects, LO is shared logistics ontology, LF is mapping types, and RD is mapping rule base.

Definition 2. Source integration object, S=(SID, Sname,{Sa|Sd},{Sf}). S is the requestor of integration services. SID is the identity marking of source integration object, SName is the name of source integration object, Sa is the data type set of source integration services' properties, Sd is the object type set of source integration services' properties, Sf is the function description set of source integration objects. The Sa and Sd are appeared in pairs, the source integration object can be an information service, function service or process service.

Definition 3. Target integration object, T=(TID, Tname,{Ta|Td},{Tf}). T is the provider of integration services. TID is the identity marking of target integration object, TName is the name of target integration object, Ta is the data type set of target integration services' properties, Td is the object type set of target integration services' properties, Tf is the function description set of target integration objects. The Ta and Td are appeared in pairs, the target integration object can be an information service, function service or process service.

Definition 4. LO= (LC, LR, LF, LA, LI). LO is the shared logistics ontology for 4PL, which is consisted by logistics domain concept (LC), relationship (LR), functions (LF), axiom (LA) and instances (LI).

LC: logistics domain concept. The concept in logistics ontology is a generalized concept, which not only include the general concepts, such as logistics order, human resource and logistics equipments, etc, but also the concept of logistics tasks, function, action, strategy, and reasoning engineering etc. The concepts in the logistics ontology are organized in a hierarchical structure. For instance, with the relationship of subclass, the logistics concepts are organized in a hierarchical structure by category.

LR: the relationship between logistics concepts. LR expresses the relationship between logistics concepts, such as the relationship of subclass, part of etc. Generally, LR: $LC_1 X LC_2 X .....X LC_n$ means that existing N-relationship concepts among LC_1, LC_2,。。。。。LC_n.

LF: function is a special relationship between concepts, function can be expressed in the following forms, LF: $LC_1 X LC_2 X .....X LC_{n-1} ->LCn$.

LA: An axiom which the logistics concepts or the relationship between logistics concepts are satisfied, the LA is a kind of logically true formula. For instance, if concept A includes concept B, and concept B includes concept C, then concept A includes concept C.

LI: logistics instances, which is some basic elements belongs to a logistics concept category. The all instances of the specific logistics domain construct the concept category of logistics domains. The concept category of the instance shows the belonging of the instance.

Definition 5. Mapping types, LF= (LCa,LQa). LF expresses the semantic mapping types between the source integration object and target integration object, which is consisted by two elements, mapping transformation category (LCa) and mapping relationship of mapping transformation (LQa).

The mapping transformation category includes marking mapping, data type mapping, constraint mapping and instance mapping [10].

(1) Marking mapping. Marking mapping expresses the semantic mapping transformation between the name of source integration object and the name of target integration object. The corresponding relationship of the name of source integration object and the name of target integration object is established by the marking mapping.

(2) Data type mapping. Data type mapping expresses the data type properties transformation between the source integration object and target integration object, which includes different data type transformation, transformation between higher case letter and lower case letter, etc.

(3) Constraint mapping. Constraint mapping expresses the constraint mapping transformation between the source integration object and target integration object. For instance, in the source integration object, the constraint of time is the finish time should later than the start time of the logistics tasks, while in target integration object, the time constraint is the construction period of the logistics task should larger than zero. The constraint mapping realizes the unification of the different express type and establishes the mapping relationship.

(4) Instance mapping. Instance mapping expresses the data transformation mapping between integration objects. For instance, in ERP system, the status of logistics task may expressed as follows, 0 presents create, 1 presents release, 2 presents distributing, 3 presents finish, 4 presents settlement, and 5 presents closed. While in the logistics distribution system, the status of logistics tasks may expressed as follows, 10 presents received, 11 presents scheduled, 12 presents distributing, 13 presents finished. To realize the integration of the logistics task status between ERP system and logistics distribution system, it is required to establish the mapping rules for status instances, and realize the semantic integration by the objects instance mapping supported by the mapping rules.

LQa expresses the relationship of mapping category between integration objects. In this paper, six different relationship of mapping category of logistics ontology is used in the establishing of transformation mapping relationship between integration objects [10]. That is, the mapping relationship between integration objects include one to one, one to many, many to one, many to many, null to one and one to null.

Definition 6. Mapping rule base, RD=({R1, R2, ...Rn}). RD is the database which stores the semantic mapping rules. The semantic mapping rules can be created by the domain experts, or improved with the semantic integration practices to support the integration between different integration objects. The semantic mapping rules in RD are described by OWL, which supports reasoning with description logic.

4 The Realization Process of Logistics Services Semantic Integration Based on Semantic Mapping

The realization process of logistics services semantic integration based on semantic mapping includes two sub-processes, that is, process of web services encapsulation for the logistics services integration objects and process of dynamic integration on demand based on semantic mapping.

(1) Process of web services encapsulation for the logistics services integration objects. The process of web services encapsulation for logistics services includes three steps, which is the identification and analysis of the logistics services integration objects, semantic modeling for integration objects, integration objects register and publishing. The identification and analysis of the logistics services integration objects identifies the data, function and process which are required to be integrated and encapsulated them into web services with the requirements of 4PL operation based on the analysis of presently logistics information systems. Integration objects semantic modeling realizes the semantic description for integration objects with semantic modeling language, such as OWL-S, WSDL-S, etc, with the semantic tagging of shared logistics ontology. Integration objects register and publishing realizes the register and publishing of the integration objects in the semantic-riched UDDI based on the process of semantic modeling, and provide an access to the logistics requestors to search.

(2) Process of dynamic integration on demand based on semantic mapping. The Process of dynamic integration on demand based on semantic mapping is an intelligent integration process supported by the semantic integration engine and tools to satisfy the requirements of 4PL dynamic business process execution. The realization of the dynamic integration not only includes the integration of inner heterogeneous logistics information systems of the logistics services providers, but the integration between 3PL and 4PL. The integration realization process includes following four steps.

Step 1. The 4PL establishes the logistics services chain model with the requirements of logistics business process execution. The logistics tasks in the logistics services chain are tagged with semantic information supported by the system integration engine based on the semantic description standard. After this step, the semantic-riched logistics tasks are generated.

Step 2. The semantic-riched integration required tasks mapped with the integration services in the UDDI via different attributes, such as input, output, precondition and results etc, supported by system integration engines and tools. The semantic mapping degree is the multiplication of semantic mapping value of each attribute. If the semantic mapping degree is larger than the given threshold value, the integration service is chosen for the following evaluation and choosing.

Step 3. The integration services are chosen with the constraints of logistics services QoS from semantic services mapping results set by the services choosing engine of system integration engines and corresponding services evaluation algorithms. The chosen services which satisfied with the requirements will be the task executor of the logistics business process node.

Step 4. The integration engines and tools composite the logistics services into logistics services chain on the basis of dynamic business process model and the choosing results of integration services supported by the semantic mapping mechanism, semantic analytic mechanism and semantic routing mechanism, and form them into executable business process with BPEL to realize the logistics services integration and logistics business collaboration.

5 Conclusions

The logistics services integration is the basis for 4PL operation. The 4PL services integration includes many aspects, such as the semantic modeling for logistics services, logistics services mapping, semantic mapping for logistics services, logistics services evaluation, the construction and execution of logistics services process, etc. This paper emphasized on the study of semantic mapping for logistics services. Firstly, a semantic mapping framework for 4PL logistics services based on shared logistics ontology is provided, on the basis, the logistics services semantic mapping process for 4PL is realized by the formal description based on shared logistics ontology. Finally, the semantic mapping based logistics services integration process is realized with two processes, and provided a semantic integration approach for 4PL logistics services integration.

Acknowledgments. This Research is supported by the Natural Science Foundation Project of Ningbo (No.2008A610023).

References

1. He, K.Q., He, Y.F., Wang, C.: Ontology meta-modeling theory, approach and application. Science Press, Peking (2009)
2. Niles, I., Pease, A.: Towards a standard upper ontology. In: 2nd International Conference on Formal Ontology in Information Systems, pp. 221–230. IEEE Press, New York (2001)
3. Ying, D., Schubert, F.: Ontology research and development: a review of ontology mapping and evolving. J. Infor. Sci. 28, 383–396 (2004)
4. Melnik, S., Molina, G.H., Rahm, E.: Similarity flooding: a versatile graph matching algorithm and its application to schema matching. In: The 18th International Conference on Data Engineering, pp. 117–125. IEEE Press, New York (2002)
5. Zheng, L.P.: Research on the ontology mapping. Shandong university, Jinan (2005)

6. Tang, J.L., Bang, Y., Li, J.Z.: Ontology Automatic mapping in Semantic web. Chinese journal of computer science 29, 1956–1976 (2006)
7. Ehrig, M., Sure, Y.: Ontology mapping-an integrated approach. In: Proceedings of ESWS, pp. 76–91. IEEE Press, New York (2004)
8. Helena, S.P., Joao, P.M.: A Methodology for Ontology Integration, http://portal.acm.org/citation.cfm?id=500759
9. Giunchiglia, F., Shvaiko, P.: Semantic matching. The Knowledge Engineering Review Journal 18, 265–280 (2003)
10. Ye, F.B.: Research on knowledge integration of business process of manufacturing enterprise based on ontology. Zhejiang University, Hangzhou (2008)

The Study on Distributed Database Security Strategy

Yan bing

School of Mathematice and Compuer Engineering,
Xihua University, Chengdu 610039, Sichuan, China

Abstract. The rapid development of computer network promotes that of the distributed database management system, but how to ensure the security of the distributed database management system in an open network environment is still a complex question that needs to be analyzed and studied carefully. This paper, in response to this question illustrated the security factors exist in this system, and described the current security strategies, and then gave out the implementation of the security policy based on Oracle database, by which the operation of the distributed database management system can be safeguarded.

Keywords: Distributed database security, distributed database, distributed database management system, distributed database retrieval problems, discretionary security distributed database, query processing.

1 Introduction

"A distributed [1] database is a collection of databases which are distributed and then stored on multiple computers within a network". A distributed database is also a set of databases stored on multiple computers that typically appears to applications as a single database. "Consequently [14], an application can simultaneously access and modify the data in several databases in a network ". A database [2], link connection allows local users to access data on a remote database ". For this connection to occur, each database in the distributed system must have a unique global database name in the network domain. The global database name uniquely identifies a database server in a distributed system. Which mean users have access to the database at their location so that they can access the data relevant to their tasks without interfering with the work of others?

The main difference between centralized & distributed databases is that the distributed databases are typically geographically separated, are separately administered, & have slower interconnection. Also in distributed databases we differentiate between local & global transactions. A local transaction is one that accesses data only from sites where the transaction originated. A global transaction, on the other hand, is one that either accesses data in a site different from the one at which the transaction was initiated, or accessed data in several different sites.

In this paper, we will review all the security features of databases in a general form and distributed databases in particular. We will also investigate the security problems found in both models. Moreover, we will evaluate the security problems unique to each system. Finally, comparing the relative merits of each model with respect to security will be applied as well.

2 Database System Concepts

One of the technology terms that most people have become accustomed to hearing either at work or while surfing the internet is the database. The database used to be an extremely technical term, however with the rise of computer systems and information technology throughout our culture, the database has become a household term.

On a more personal level, your personal computer can have its own database management system. You might have spreadsheets that contain mountains of data. Any time you fill up a spreadsheet with data and run queries to find and analyze data in different ways, you are accessing a database management system. The question is how do you view the data that is the result of a query? The answer is by looking at a report. Most database management systems have a reporting function that is the last step in the data manipulation process.

This functionality also extends to a multi-user database. Such a database management system under this scenario would allow you as one user to operate all functions within the database without having to know what other users are accessing the same database? The user interacts with the database management system in order to utilize the database and transform data into information. Furthermore, a" database [2], offers many advantages compared to a simple file system with regard to speed, accuracy, and accessibility such as: shared access, minimal redundancy, data consistency, data integrity, and controlled access ". All of these aspects are enforced by a database management system. Among these things let's review some of the many different types of databases. The vertical columns are known as the attributes. "Data that is stored on two or more tables establishes a "link" between [3], the tables based on one or more field values common in both tables." A relational database uses a standard user and application program interface called Structure Query Language (SQL). This program language uses statements to access and retrieve queries from the database. Relational databases are the most commonly used due to the reasonable ease of creating and accessing information as well as extending new data categories.

When dealing with intricate data or complex relationships, "object databases [13], are more commonly used "."Object databases in [13], contrast to relational databases store objects rather than data such as integers, strings, or real numbers". Each object consists of attributes, which define the characteristics of an object. Objects also contain methods that define the behavior of an object (also known as procedures and functions). When storing data in an object database there are two main types of methods, one technique labels each object with a unique ID. Every unique ID is defined in a subclass of its own base class, where inheritance is used to determine attributes. A second method is utilizing virtual memory mapping for object storage and management. Advantages of object databases with regard to relational databases allow more concurrency control, a decrease in paging, and easy navigation.

However, there are "some disadvantages [9],of object databases compared to relational databases such as: less effective with simple data and relationships, slow access speed, and the fact that relational databases provide suitable standards oppose to those for object database systems". On the other hand, network databases alleviate some of the problem incorporated with hierarchical databases such as data redundancy. "The network model [7, 10, 14], represents the data in the form of a network of records and sets which are related to each other, forming a network of links."

3 Distributed Databases Design

"The developments [14] ,in computer networking technology and database systems technology resulted in the development of distributed databases in the mid 1970s". It was felt that many applications would be distributed in the future and therefore the databases had to be distributed also. Although man y definitions of a distributed database system have been given, there is no standard definition. A distributed database system includes a distributed database management system (DDBMS), a distributed database and a network for interconnection. The DDBMS manages the distributed database. A distributed database is data that is distributed across multiple databases. In brief term "distributed database [5] is a collection of databases that can be stored at different computer network sites". Each database may involve "different [6], database management systems and different architectures that distribute the execution of transactions". "The objective [10], of a distributed database management system (DDBMS) is to control the management of a distributed database (DDB) in such a way that it appears to the user as a centralized database".

A development of a distributed database structure for centered databases offers scalability and flexibility, allowing participating centers to maintain ownership of their own data, without introducing duplication and data integrity issues.

Moreover, Distributed database system functions include distributed query management, distributed transaction processing, distributed metadata management and enforcing security and integrity across the multiple nodes. "The centralized [7], database system is one of the many objectives of a distributed database system". This system will be accomplished by using the following transparencies: Location Transparency, Performance Transparency, Copy Transparency, Transaction Transparency, Transaction Transparency, Fragment Transparency, Schema Change Transparency, and Local DBMS Transparency. These eight transparencies are believed to incorporate the desired functions of a distributed database system.

The design of responsive distributed database systems is a key concern for information systems. In high bandwidth networks, latency and local processing are the most significant factors in query and update response time. Parallel processing can be used to minimize their effects, particularly if it is considered at design time.

It is the network that enables parallelism to be effectively used. Distributed database design can thus be seen as an optimization problem requiring solutions to various interrelated problems: data fragmentation, data allocation, and local optimization.

Meanwhile, a successful distributed database could include free object naming. "Free object naming means that it allows different users the ability to access the same object with different names, or different objects with the same internal name." Thus, giving the user complete freedom in naming the objects while sharing data without naming conflicts.

Concurrency control (CC) is another issue among database systems. It permits users to access a distributed database in a multi-programmed fashion while preserving the illusion that each user is executing alone on a dedicated system. For this, CC mechanisms are required that interleave the execution of a set of transactions under certain consistency constrains while maximizing concurrency. Two main categories of CC mechanisms are: Optimistic concurrency - Delay the synchronization for transactions until the operations are actually performed. Conflicts are less likely but won't be known until they happen, making rollback operations more expensive. Pessimistic - The potentially concurrent executions of transactions are synchronized early in their execution life cycle. Blocking is thus more likely but will be known earlier avoiding costly rollbacks. Another activity of CC is to "coordinating [8] ,concurrent accesses to a database in a multi-user database management system (DBMS)." There are a number of methods that provide "concurrency control [7], such as: Two phase locking, Time stamping, Multi-version timestamp, and optimistic non-locking mechanisms. Some methods provide better concurrency control than others depending on the system".

4 The Features of Distributed Database System

In this section we will examine the most common features of the distributed database system. One of the main features in a DDBMS is the Database Manager. "A Database Manager [1] is software responsible for processing a segment of the distributed database". A Distributed Database Management System is defined as the software which governs a Distributed Database System. It supplies the user with the illusion of using a centralized database.

Another main component is the User Request Interface, known some times as a customer user interface, which is usually a client program that acts as an interface to the distributed Transaction Manager. A customizable user interface is provided for entering requested parameters related to a database query. The customized parameter user interface provides parameter entry dialogs/windows in correlation to a data view (e.g., form or report) that is produced according to a database query. The parameters entered may provide for modification of the data view. Also, the manager of the database may structure data views of a database to automatically include prompts for parameters before results are returned by the database. "These prompts [12], may be customized by the manager and may be provided according to dialogs such as pop-ups, pull-down menus, fly-outs, or a variety of other user interface components". "A Distributed Transaction [10] , Manager is a program that translates requests from the user and converts them into actionable requests for the database managers, which are typically distributed. A distributed database system is made of both the distributed transaction manager and the database manager". The components of a DDBMS are shown in the "fig. 1" below:

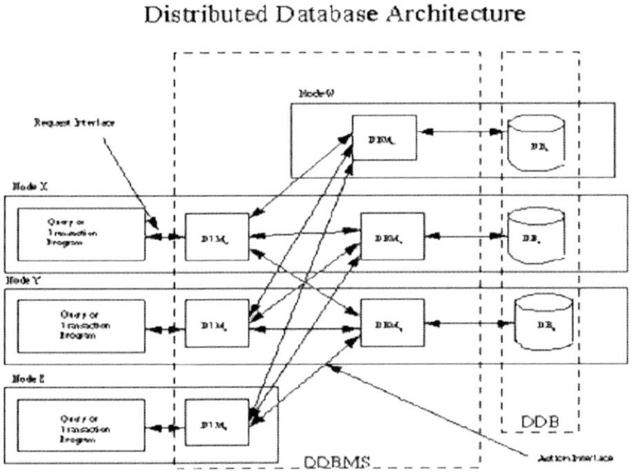

Fig. 1. Shows the architecture of DDBMS

5 Distributed Query Processing

Distributed query processing in a large scale distributed system; it is often difficult to find an optimal plan for a distributed query: distributed systems can become very large, involving thousands of heterogeneous sites. As new databases are added/ removed to/from the system, it becomes harder for a query processor to maintain accurate statistics of the participating relations stored at the different sites and of the selectivity's of the related query operations. Also, as the workload at the various interacting processing servers and the transmission speeds of the links between them fluctuate at runtime, there is the need of distributed query engines that dynamically adapt to large scale distributed environments. The query will be request usually from the user or the client host from a proper user interface.

The user request interface is merely a client program running on one end system that requests and receives a service from a server program running on another end system. Due to the fact that the client and the server run on separate computers, by definition the client/server programs are considered distributed applications. There are two types of client/server applications. One particular client/server application is an implementation of a protocol standard defined in an RFC (request for comments). This type of application forces the client and server programs to abide by certain rules dictated by the RFC. Another sort of client/server application is a proprietary client/server application. "In this case [7], the client and server programs do not necessarily conform to any existing RFC".

In developing a proprietary client/server application the developer must decide whether to run the application over TCP or UDP. TCP connection oriented and ensures reliable byte-stream channel. However "UDP is [4], connectionless and forwards independent packets of data and does not guarantee delivery".

4 Security Aspects in Distributed Database

A proximately all of the early work on secure databases was on discretionary security. But the most important issues in security are authentication, identification and enforcing appropriate access controls. For example, the mechanisms for identifying and authenticating the user, or if a simple password mechanisms suffice? With respect to access control rules, "languages [7], such as SQL have incorporated GRANT and REVOKE statements to grant and revoke access to users". For many applications, simple RANT and REVOKE statements are not sufficient.

There may be more complex authorizations based on database content. Negative authorizations may also be needed. Access to data based on the roles of the user is also being investigated. "Numerous papers [7], have been published on discretionary security in databases ". "Fig. 2", shows a simple security model. DBMS have many of the same security requirements as operating systems, but there are significant differences since the former are particularly susceptible to the threat of improper disclosure, modification of information and also denial of service. Some of "the most [4], important security requirements for database management systems are: Multi-Level Access Control: Confidentiality.

Reliability, Integrity, Recovery". "Security in distributed [11], database systems has focused on multilevel security". Specifically approaches based on distributed data and centralized control architectures were proposed. Prototypes based on these approaches were also developed during the late 1980s and early 1990s. Notable among these approaches are the efforts by Unisys Corporation and the Naval Research Laboratory.

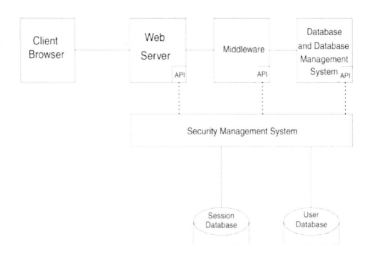

Fig. 2. A Simple Security Model

7 Emerging Security Used in Distributed System Tools

Security impact in most of the distributed database tools became an emerging technology that has evolved in some way from distributed databases and discussions. These include data warehouses and data mining systems, collaborative computing systems, distributed object systems and the web. First, let us consider data warehousing systems. The major issues here are ensuring that security is maintained in building a data warehouse from the backend database systems and also enforcing appropriate access control techniques when retrieving the data from the warehouse. For example, security policies of the different data sources that form the warehouse have to be integrated to form a policy for the warehouse. This is not a straightforward task, as one has to maintain security rules during the transformations. For example, one cannot give access to an entity in the warehouse, while the same person cannot have access to that entity in the data source.

Next, the Warehouse security policy has to be enforced. In addition, the warehouse has to be audited. Finally, the retrieval problem also becomes an issue here. For example, the warehouse may store average salaries. A user can access average salaries and then deduce the individual salaries in the data sources, which may be sensitive and therefore, the inference problem could become an issue for the warehouse. To date, little work has been reported on security for data warehouses as well as the retrieval problem for the warehouse. This is an area that needs much research intention.

Data mining causes serious security problems. For example, consider a user who has the ability to apply data mining tools. This user can pose various queries and infer a sensitive hypothesis. That is, the retrieval problem occurs via data mining. There are various ways to handle this problem. Given a database and a particular data mining tool, one can apply the tool to see if sensitive information can be deduced from legitimately obtained unclassified information. If so, then there is a retrieve problem. There are some issues with this approach. One is that we are applying only one tool. In reality, the user may have several tools available to him or to her. Furthermore, it is impossible to cover all of the ways that the retrieval problem could occur. Another solution to the retrieval problem is to build a retrieval controller that can detect the motives of the user and prevent the retrieval problem from occurring. Such a retrieval controller lies between the data mining tool and the data source or database, possibly managed by a DBMS.

Data mining systems are being extended to function in a distributed environment. These systems are called distributed data mining systems. Security problems may be exacerbated in distributed data mining systems. This area has received very little attention. Other emerging technologies that have evolved in some way from distributed databases are collaborative computing systems, distributed object management systems and the web. Much of the work on securing distributed databases can be applied to securing collaborative computing systems. With respect to distributed object systems security, there is a lot of work by the Object Management Group's Security Special Interest Group.

More recently, there has been much work on securing the web as well. The main issue here is ensuring that the databases, the operating systems, the applications, the web servers, the clients and the network are not only secure, but are also securely integrated.

8 Conclusion

In this paper we introduce distributed database and all common aspects related to distributed database such as database system concepts, features and design of distributed database and distributed query processing as well.

We also indicate some related security issues including multilevel security in distributed database systems. The security aspects lead us to investigate the distributed data and centralized control, distributed data and distributed control and some retrieval problems in distributed databases in since of accessing control and integrity. Moreover, we describe the most common mechanisms of discretionary security and stated the emerging security used in distributed system tools. Finally we believe that as more and more distributed database tools, the impact of secure distributed database systems on these tools will be a significant requirement.

References

1. Bell, D., Grisom, J.: Distributed Database Systems. Addison Wesley, Workinham (1992)
2. Pfleeger, C.P., Pfleeger, S.L.: Security in Computing. Prentice Hall Professional Technical Reference, Upper Saddle River (2003)
3. Haigh, J.T., et al.: The LDV Secure Relational DBMS Model. In: Jajodia, S., Landwehr, C.E. (eds.) Database Security, IV: Status and Prospects, pp. 265–269. Elsevier, North Holland (1991)
4. Köse, İ., Gyte, V.v.A.G.: Distributed Database Security (Spring 2002)
5. Kurose, J.F., Ross, K.W.: Computer Networking: A Top-Down Approach Featuring the Internet. Pearson Education, Inc., New York (2003)
6. Lothian, P., Wenham, P.: Database Security in Web Environment (2001)
7. Pfleeger, C.P.: Security in Computing. Prentice Hall, New Jersey (1989)
8. Wiseman, S.: DERA, Database Security: Retrospective and Way Forward (2001)
9. Ceri, S., Pelagatti, G.: Distributed Databases: Principles and Systems. McGraw-Hill Book Company, New York (1984)
10. Thuraisingham, B.: Security for Distributed Database Systems, Computers & Security (2000)
11. Thuraisingham, B., Ford, W.: Security Constraint Processing In A Multilevel Secure Distributed Database Management System. IEEE Transactions on Knowledge and Data Engineering 7(2), 274–293 (1995)
12. Components of a Distributed Database System (October 24, 2008), http://www.fi/~hhyotyni/latex/Final/node44.html
13. Object Oriented Databases (October 25, 2008), http://www.comptechdoc.org/independent/database/basicdb/dataobject.html
14. Network Databases (October 25, 2008), http://wwwdb.web.cern.ch/wwwdb/aboutdbs/classification/network.html

An Improved Symmetric Key Encryption Algorithm for Digital Signature

Xiuyan Sun

School of Computer Department, Laiwu Vocational and Technical college
Laiwu, China

Abstract. Based on the analysis of the shortage of existing digital signature solution, an improved digital signature solution was proposed according to the advantage of symmetrical key algorithm, asymmetrical key algorithm and unidirectional hashing function. It successfully solved the problem of symmetrical digital signature caused by the difficulty to store key and indispensability of arbitrator. And it also avoided the disadvantage of asymmetrical digital signature that it was too slow to suit for the long file signature.

Keywords: Combined Symmetric Key; Digital Signature; Office Automation.

1 Introduction

We would like to draw your attention to the fact that it is not possible to modify a paper in any way, once it has been published. This applies to both the printed book and the online version of the publication. Every detail, including the order of the names of the authors, should be checked before the paper is sent to the Volume Editors.

With improving information level, more and more governments and companies construct their E-government systems or enterprise e-commerce systems. Ensuring the validity and authenticity of official documents is one focus of the information security for official business work on Internet. Handwritten signature in traditional official document circulation changes into digital signature in the E-government or e-commerce system. This digital signature required that the signing speed must be quick and the system must be simply deployment.

The mainstream technology of digital signature is public key cryptosystems which required constructed the certification authority [1-2]. The maintenance and management of CA is complex and high cost. In additional to, the asymmetric digital signature algorithms, such as RSA that makes the use of large number, have the disadvantages of low-processing speed, which would be restrain official document circulation speed and therefore the work efficiency is not favorable. This paper studied the problems of the digital signature with encryption based on symmetric technology and proposed a method of digital signature with encryption based on combined symmetric key, symmetric technology and hardware technology.

2 Symmetric Key Method

The printing area is 122 mm × 193 mm. The text should be justified to occupy the full line width, so that the right margin is not ragged, with words hyphenated as appropriate. Please fill pages so that the length of the text is no less than 180 mm, if possible.

Combined public key technology was proposed by Nan xianghao[3-6] to solve the problem of large-scale key management in private networks. In public key infrastructure, the user's number must be equal to the public keys number. In combined public key infrastructure, the user public key is not published, but the user public factor key matrix is published and the user public key is created by the user public factor key matrix and relational mapping values. In this paper applied combined method to private-key cryptosystems for solving the difficult key management. Safety secret of the key is an important and difficult problem in private key cryptosystems. Moreover, the cipher system using the same key in each encryption and decryption process is easy to be attacked by the exhaustive key search attack and Trojan horse attack to make the key. Our proposed combined symmetric key can realize to dynamically vary every time by making the key with time information and random numbers, which has the capability to resist replay attack and be traced.

In traditional digital signature system, digital signatures often include timestamps. The date and time of the signature are attached to the message and signed along with the rest of the message. Using timestamps can resist replay attack. Our proposed digital signature method uses timestamps too. Different from the traditional digital signature system, the timestamps is used as a factor of combined symmetric key algorithm in our method.

The main ideas of the combined symmetric key algorithm are on the following points: firstly, Key management center generates key seed matrix for every user. Secondly, timestamps and random number generators get a timestamp and some random numbers. Finally, the combined symmetric key algorithm selects factors form user key seed matrix and combines every encryption and decryption key. The combined symmetric key algorithm has, at the very least, these four requirements.

(a) Assured that different users have different key seed matrixes

(b) The algorithm with the same key seed matrix, timestamp and random numbers always generates the same combined key

(c) The algorithm with different key seed matrix, the same timestamp and the same random numbers always generates different combined key

(d) The limited number of key seed matrix factors can generate almost unlimited number of combined keys. The process of combined symmetric key algorithm is on the following steps:

(a) Constructed user key seed matrix $K_{M \times N}(\forall k_{ij} \neq 0)$ by the factors of key, and then divided $K_{M \times N}$ into N number of child matrixes according to rank.

(b) Selected a rank from the fore six child matrixes by timestamp and added the last N-6 ranks of $K_{M \times N}$ to compose $K'_{N \times N}$

(c) Generated 1-dimensional vector R by random number generator, which $1 \leq r_i \leq N$

(d) Generated unit matrix $E_{N \times N}$ by R, which $e_{i \times r_i} = 1$, the others factors is zero.

(e) Let $K''_{N \times N} = E_{N \times N} \times K'_{N \times N}$

(f) Selected Non-zero factors from $K''_{N \times N}$ and get the combined key by combining them order by rank.

2.1 Generation Method for Key Seed Matrix

Constructed user key seed matrix $K_{M \times N} (\forall k_{ij} \neq 0)$ by the factors of key, and then divided $K_{M \times N}$ into N number of child matrixes according to rank. These child matrixes are on the following name: year child matrix $Y_{L \times N}$ (L is the valid years of the signature system), month child matrix $M_{12 \times N}$, date child matrix $D_{31 \times N}$, hour child matrix $H_{24 \times N}$, minute child matrix $HM_{60 \times N}$, second child matrix $S_{60 \times N}$, the first additional child matrix $A1_{1 \times N}$, the second additional child matrix $A2_{1 \times N}$, ..., the (N-6)the additional child matrix $A(N-6)_{1 \times N}$.

Every child matrix has an n-translation random number and there are a total of N numbers.

2.1 Generation Method for Combined Key

Generating combined key by combined symmetric key algorithm is on the following steps:

(a) Generated timestamp. The timestamp is the current time and its digit is 14. It consists of year, month, date, hour, minute, second. For example: 20080916230156.

(b) Selected a rank from the fore six child matrixes by timestamp and added the last N-6 ranks to compose.

$$K'_{N\times N} = \begin{bmatrix} y_{y_i \times 1} & \cdots & y_{y_i \times n} \\ m_{m_i \times 1} & \cdots & m_{m_i \times n} \\ d_{d_i \times 1} & \cdots & d_{d_i \times n} \\ h_{h_i \times 1} & \cdots & h_{h_i \times n} \\ hm_{hm_i \times 1} & \cdots & hm_{hm_i \times n} \\ s_{s_i \times 1} & \cdots & s_{s_i \times n} \\ a1_{1\times 1} & \cdots & a1_{1\times n} \\ \vdots & \ddots & \vdots \\ a(N-6)_{1\times 1} & \cdots & a(N-6)_{1\times n} \end{bmatrix}$$

in which
y_i = the year value of timestamp
 the year of system deployment
m_i = the month value of timestamp
d_i = the date value of timestamp
h_i = the hour value of timestamp
h_{mi} = the minute value of timestamp
s_i = the second value of timestamp
(c) Generated N-digital N-translation random number. The last combined key is

$$K_{CBD} = \left\{ k'_{1i_1}, k'_{2i_2}, \cdots, k'_{ni_n} \right\}$$

In which: i_j = the ith value of R.

3 Structure of the Digital Signature System

The signature protocol is set up on the client, and the function of digital signature is achieved on the client. The digital signature verification center and the digital signature verification protocol are set up on the web server, and the function of digital signature verification is achieved on web server. The timestamp of this system is generated by web server for avoiding the problem of time synchronization between client and web server.

The client system of our digital signature mainly includes a key card and the client software. The client software is in charge of communicating with key card and the web server, producing a one-way hash of a document and encrypting or decrying a document. Key card stores key seed matrix, symmetric key algorithm, random number generator and combined symmetric key algorithm. Key card is in charge of generating random numbers, computing the key for digital signature and the key for document encryption, encrypting the hash of the document to get the signature of the document and decrypting the result of the digital signature verification.

Web server system of our digital signature mainly includes the digital signature verification center, the digital signature database, key seed matrix database and encrypt card. The digital signature set up the digital signature verification protocol. It is in charge of three tasks.

(a) Accepting the client signature requirement and returning timestamps to the client.
(b) Receiving and storing the document digital signature send by the client
(c) Deal with the requirement of the client digital signature Verify.

The digital signature database stores user's signature information that mainly includes useID, document, the signature of a document, random numbers and timestamps. The key seed matrix database stores the cipher texts of user's seed matrix. The function of encrypt card is similar with key card. In additional, encrypt card stores a pair of symmetric key and use them to encrypt or decrypt user seed matrix. The structure of our digital signature system is as follows.

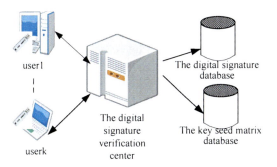

Fig. 1. Structure of the system

4 The Digital Signature Protocol and the Digital Signature Verification Protocol

4.1 The Digital Signature Protocol

Here's how it the digital signature protocol works. Suppose the client is A.

(a) The client A sends web server the signature request. Web server response to the request and return the timestamps TS1 to the client.

(b) The client produces a one-way hash of the document D1.

(c) The client sends the hash and timestamps to key card.

(d) Firstly, key card generates random numbers R1 and random numbers R2.sencondly the combined symmetric key algorithm in key card generates combined symmetric key SK1 by R1 and timestamps TS1 and SK2 by R2 and timestamps TS1. Thirdly, key card encrypt the hash with sky1, thereby the client signing the document D1. Finally, card key sends the signing hash, Sky2, R1andR2 to the client software.

(e) The client encrypts the document with Sk2.

(f) The client A logins The digital signature verification center and sends the data W that includes cipher text of the document D1, the signing hash, userID, Sky2, R1andR2 to web server.

4.2 The Digital Signature Verification Protocol

Here's how it the digital signature verification protocol works. Suppose the client K requests the document D1.

(a) The client K logins the digital signature verification center (web server) and sends the request for the document D1.

(b) Web server gets the data W from the signature database and generates random numbers R3, random numbers R4 and timestamps TS2.

(c) Web server gets the key seed matrix cipher text of client A and the key seed matrix cipher text of client B from user's seed database, and sends these cipher texts, R1, R2, TS1, TS2, R3, R4 and the signed hash of the document D1 to encrypt card.

(d) Encrypt card decrypts these cipher texts of key seed matrixes. The combined symmetric key algorithm in Encrypt card generates combined symmetric key SK2 by R2, timestamps TS1 and A's key seed matrix. Encrypt card sends sk2 to web server software.

(e) Web server decrypts the document cipher text with sk2 and produces the one-way hash of D1.then it sends the hash to encrypt card.

(f) The combined symmetric key algorithm in Encrypt card generates combined symmetric key SK1 by R1 and timestamps TS1. Encrypt card decrypts the signed hash with sk1 .if the signed hash matches the hash produced in step5, the signed is valid.

(g) The combined symmetric key algorithm in Encrypt card generates combined symmetric key SK3 by k's key seed matrix, R3 and timestamps TS2. Encrypt card encrypts the result of the signature verification.

(h) Encrypt card generates combined symmetric key SK4 by k's key seed matrix, R4 and timestamps TS2 and sends the cipher text of the signature verification result and sk4 to web server.

(i) Web server encrypts the document D1 with sk4, and sends R3, R4, timestamps TS2, the cipher text of signature verification result and D1' cipher text to the client K.

(j) The client k' key card generates sk3 and sk4 by R3, R4 and TS2 and decrypts the cipher text of signature verification result with sk3.

(k) If the signed is valid, the client decrypts D1's cipher text with sk4.

5 Advantages of Combined Symmetric Digital

(a) Compared with the public-key signature system, our method uses hardware, Symmetric Cryptosystems and combined symmetric key algorithm to construct the signature system, which has better capability of resisting the attacks from Trojan horse that makes the key.
(b) This method signs and encrypts a document with Symmetric Cryptosystems, which speed increases drastically.
(c) Our signature system assures that every signing key is vary, which improve the security of digital signature. Moreover, the signing key is generated by combined symmetric key algorithm and key seed matrix, so the key needn't update and increase the cost of system management.
(d) The signature key is generated by timestamps, which can resist the attacks from replay attack.
(e) The digital signature verification center is a fair-and square arbitrator and is constructed by web servers and encrypts cards. It has all of user key seed matrix in the same system. This system realizes signing by symmetric key algorithm, thereby does not need the arbitrator verification by certificate, which increases the complex of signature system and management cost.

References

1. Smith, T.F., Waterman, M.S.: Identification of Common Molecular Subsequences. J. Mol. Biol. 147, 195–197 (1981)
2. May, P., Ehrlich, H.C., Steinke, T.: ZIB Structure Prediction Pipeline: Composing a Complex Biological Workflow through Web Services. In: Nagel, W.E., Walter, W.V., Lehner, W. (eds.) Euro-Par 2006. LNCS, vol. 4128, pp. 1148–1158. Springer, Heidelberg (2006)
3. National Center for Biotechnology Information, http://www.ncbi.nlm.nih.gov
4. Schneider, B.: Applied Cryptography Protocols, algorithms, and source code in C, 2nd edn. John Wiley & Sons, Inc., American (1996)
5. Rivest, R., Shamir, A., Adleman, L.: A method for obtaining digital signatures and public-key cryptosystems. Commun. ACM 21, 120–126 (1978)
6. Xianghao, N.: CPK Algorithm and Identification. China Information Security (September 2008)
7. Wen, T., Hao, N.X., Zhong, C.: Elliptic. Curve Cryptography based Combined Public Key Technique. Computer Engineering and Applications, 21 (2003)
8. Xu, Z., Zele, H.: Compared IBE with CPK. Computer Security (June 2008)
9. The Theory of CPK based on identification, http://ehenxen.com/tz4.html

Application of Soft Test Method to Intelligent Service Robots

Wang Hongxing

Department of Mechanical and Power Engineering,
NanChang Institute of Technology, NanChang, China

Abstract. The importance of the soft test technique in developing intelligent service robots is analyzed in this paper. The soft test standards of ISO are of great use to soft test which constitutes one of the critical phases of life cycle of software engineering. The soft test standards of ISO are applied to intelligent service robots for the purpose of unit testing and integration testing. This paper summarized the experience and effective method of soft test in intelligent service robots based on the testing results. Furthermore, the idea of aspect-oriented program was employed to solve problems with regard to intelligent service robots which were difficult to address through the use of traditional method. It is proved that the result is satisfactory.

Keywords: Soft Test, Intelligent Robot, Testing Process, Testing Items.

1 Introduction

Network based Intelligent Service Robot technology has been developed for the past four years in Korea. This technology is also called "Ubiquitous Robot Companion (URC)." The pilot business project of URC has been started as a national project from two years ago. This business model provides new concepts for the robot business industry. During the past and current years, most of countries are developing humanoid robots whose shapes are similar to humans and whose functions are artificially intelligent. This technology is still taking a lot of years to be practically used. Robot experts say that it will need more than 10 years even though they invested more than several ten years in the past. Korean robot experts had strategically interests in another new area. They are to fuse robot technology and IT technology, by which the network based intelligent service robotic project has been created as URC project.

The purpose of the URC is that the robots provide services that the users want anytime and anywhere in low price. At this time, the robot price has been at least more than several USD 10,000s. It is not easy to buy it at home and still the functions of robots are not satisfied. By adopting the server/client systems to the robot technology, the price and the function can be dramatically saved and improved. In URC, the system consists of three parts: computing, sensing and action with a network. If we use servers for computing of robot functions, high quality of service functions can be delivered to client robots using a high speeded network. The general high speeded network for URC robots are more than 10Mbps. During the last year, Korea Telecommunication Company 'KT' leaded the pilot business project and delivered service network based URC robots to 850 apartments. Many URC servers and the high speeded networks were successfully used to provide the various IT services by URC home robots. We estimated the client robot prices as less than USD 2,000. From the first pilot business project, many problems and opinions were given to KT. We are to resolve the problems of malfunctions and problems by restricted quality testing of robot products and also improving the technology, where especially speech recognition service and automatic battery charging were mentioned. During the year 2006, the authentication test daft was made and the most focusing was on complying test to the telecommunication protocol between the robots and the servers. The testing draft should be based on the specifications of robot standards. But this area is still on the beginning stage. This paper is to provide the guide line to be referenced on this purpose.

2 Background of Test

The year 2007 is the second year of URC home robot pilot business project. During the past years, the technology development is the most important but for commercializing the intelligent robots, the strict testing should be applied to the robot products. Otherwise, we may expect many kinds of malfunctions from the serviced robots through the pilot business. The testing is the least filter to prevent these kinds of problems. During the last year, all URC robots produced from six companies failed to be passed in the initial authentication tests of URC robots as shown in Figure 1. Considering the pilot business and encouraging the robot companies, the second test was applied with lower specifications. Many weak points and problems were discovered by the tests. Also the user surveys were carried out after three months of usage for URC home robots. The most important problems of technologies should be improved and resolved. Therefore, we have kept the strict authentication test and also more evaluation items are added in this year.

Fig. 1. URC robots

The testing reference is based on ISO 9126 and ISO 14598 which is the software evaluation rules. About 80 items of ISO standards are too much to the robot tests. We extracted the most important 30 items among the ISO specifications for software test. The software has an important key role in intelligent robots since the intelligence is mostly implemented by computer programs and the software recently is managed in software life method. But robot software is performed with the robot hardware devices. Considering these facts, the performance testing should be more weighted in the testing evaluation of performance. Therefore, the software testing method is suitably combined with the performance tests for authentication tests. This paper suggests that testing methods for intelligent robots are studied and the results are recommended for authentication tests of robots. The testing process and the testing methods are suggested for this purpose in this paper.

3 Soft Test Process

The testing process consists of the following five steps: Test requirement Analysis, Test details, testing design, Processing the tests and Report of Test results.

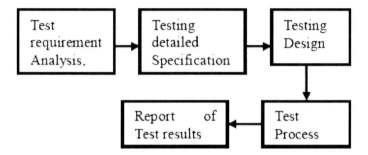

Fig. 2. Testing Process of Robot Software

The testing process of robot software is provided for users of robots. The users can be producers and users of robot products. ISO/IEC 14598-5 defines the above five steps of testing processes and activities for software quality assurance:

• Test requirement analysis: The demanding requirements of users are collected and analyzed to find the demanding functions from the users.
• Testing specifications: The detailed testing specifications are drafted from the user requirements and the product user manuals.
• Testing design: Testing plan is drafted based on testing specifications, target software, testing methods in detail.
• Testing process: Inspection, modeling, measuring and testing are performed based on the testing plan.
• Report of test results: Test results are concluded by acceptance of the test report.

In order to follow the above testing process for the robot software, the evaluation items are defined in the ISO/IEC 9126 as the main characteristics with the sub characteristics for measuring the software.

4 Software Quality Characteristics

The software is measured according to the ISO/IEC 9126 quality characteristics with the sub characteristics. This is called metrics. The software quality characteristics consist of the following six main characteristics with more detailed sub characteristics.

4.1 Functionality

This measures the capability of software products according to the specifications in the user requirement analysis when the software is used under the specified conditions. The following table explains the sub characteristics of the functionality.

Suitability	Capability to provide an appropriate set of functions for specified tasks and user objectives
Accuracy	Capability to provide the right or agreed results of effects with the needed degree of precision
Interoperability	Capability to interact with one or more specified systems
Security	Capability to protect information and data so that only authorized users or systems are accessed to them
Functional Compliance	Capability to adhere to standards related to the functionality

4.2 Reliability

It is the capability to maintain a specified level of performance when it is used under the specified conditions. The following table explains the sub characteristics of the reliability:

Maturity	Capability to avoid failures as a result of faults in the software
Fault tolerance	Capability to maintain a specified level of performance in case of software failure or infringement of its specified interface
Recoverability	Capability to re-establish a specified level of performance and recover the data directly affected in case of failure
Reliability compliance	Capability to adhere to standards related to the reliability

4.3 Usability

It is the capability to be understood, learned, used and attractive to the users under specified conditions.

Understandability	Capability to enable users to understand whether the software is suitable
Learnability	Capability to enable users to learn its application
Operability	Capability to enable users to operate and control it
Attractiveness	Capability to enable users tobe attractive to users
Usability Compliance	Capability to adhere to standards related to the usability

4.4 Efficiency

It is the capability to provide appropriate performance related to the resources used.

Time behavior	Capability to provide appropriate response and processing time and throughput rates
Resource behavior	Capability to provide appropriate amounts and types of resources
Efficiency Compliance	Capability to adhere to standards related to efficiency

4.5 Maintainability and Portability

It is the capability of software maintenance to be modified. Also portability is the capability of software to be transferred from one environment to another. Maintainability consists of analyzability, changeability, stability, testability and compliance. Portability consists of adaptability, instability, co-existence, replace ability and compliance.

5 URC Robot Software Components

The components consisting of the URC robot software is composed of HRI (Human-Robot Interaction), robot smart-action technology, and others.

5.1 HRI Technology

URC robots interact with users for user recognition or some kind of expressions, which is different from other information machinery. Robot technology requires the recognition and response of user's voice, the verbal transmission through voice composition on the information, the recognition to user's facial expression and gesture from the image signal through cameras. Chart 6 contains the contents of the human-robot interaction technology which is under development. This software module is offered by URC service, and it is developed by the aim focused on the reaction to 100 clients within one second with simultaneous contacts. The final aim of voice cognition is to recognize the dialogue style with user's natural utterance in the noise environment. But at present, it maintains a standard of using the cognition apparatus of isolating languages and the cognition

Table 1. Hri Technology

HRI component technology			
	Infra-Image HRI component technology	User Recognition	User's cognition module by an quasi-organism information, facial expression at a long distance in the situation of robot's upward sight
		Gesture Recognition	User's pursuit and gesture cognition module development Caller Identification module
		Expression Recognition	Facial component detection and special feature extraction module
	Infra-Voice HRI component technology	Voice Recognition	Cognitive module of user's voice information transmitted from robot
		Voice Decomposition	Conversion module changing arbitrary sentence with composition voice
		Voice Pursuit	Exclusion module removing noises in the robot and outside noises for tracking the original sound
		Speaker Recognition	Speaker cognition module by the context speaker independent method

lexicon for sentence units considering user's natural utterance. We have developed the voice recognition, aiming at the cognitive rate of 95% as a basis of linking language cognition, within 2 meter, in the limit of 20dB, with only once utterance, perceiving subsidiary lexicon which make up important lexicon and declinable word corresponding to the key words in URC. Voice composition is to convert arbitrary sentence into composition voice similar to human voice. According to the various services, URC can change the output composition voice being suitable to the usage, as a basis of limitless lexicon composition. Tagged Text Processing is used for the purpose of the various outputs. That is, it modulates the speed, the volume, the pitch, chooses the various composition voices of children, men, women, and supplements the background sound or effect sound through the input of XML pattern. User's cognition is a kind of technology that finds out the master of the robot and the master's identity by robot's analyzing the inputted facial image in camera. In order to overcome the limitation of existing, cooperative, user's cognition technology in URC, we have developed the technology that recognizes the natural, acting user with using the information of semi-Biometrics, such as height, dress color, and so forth, in ordinary life, and we have also been developing and studying user's emotion through caller's facial expression and discernment module that make the robot accept the user's order, dialogue or request spontaneously with recognizing the visual information of user's special actions.

5.2 URC Robot's Smart Action Technology

Smart Action Technology can be divided into the intellectual covering technology and the intellectual manipulating technology for realizing the physical action of URC (Table II).

The Intellectual Covering Technology is divide into 'Position-Deciding Technology' that recognize the present position of robot, ' Map-Forming

Table 2. Smart Action Technologies

Smart Action Technology	The Intellectual Covering Technology	Position-Deciding Technology	The correctly searching technology of robot's position
		Map-Forming Technology	Map-forming technology presented by digital data that grasps the robot's covering environment
		Route-Framing Technology	Route-controlling technology that make the robot plan its course, make it move as the original course
	The Intellectual Manipulating Technology	Environment-Object Cognitive Technology	The cognitive technology that makes the robot recognize the manipulating objects and circumference-environment with using camera, distance sensor
		Manipulation-Control Technology	The technology that manipulates the special object with controlling robot's arms articulation

Technology' that grasps the present operating environment, and 'Route-Framing Technology' that can move safely after generating the moving route for the operation accomplishment. The decision of the moving robot is very easy to human, but very difficult to robot. That is why robot can't possess the intelligence that deals with various uncertainties effectively in real life. In order to develop position-deciding technology applicable to the real environment through URC task, the intensive research is in progress. Map-forming technology is to store the estimated environment information of the periphery in the memory unit from all the sensor information and to use its information if necessary. The most difficult area in map-forming technology is to form the map to the environment sustaining the circulation section. So far, its level is in the static map-forming level, and it has been developed in the kinetic environment applicable to the technical development. Route Framing method is to cover its course safely, with generating the optimum course toward the final destination, in case of the given map information among the major points, the present and final positions, the optimum factors---safety, time, distance, energy, etc.---when covering. This also generates the optimum course with the real time, securing the safety in the kinetic environment, and makes the research to execute the given task with success by only the least sensor information.

6 Conclusion

URC robot offers users necessary services anytime and anywhere, using the high speeded network to the robot, expanding the applicable service, keeping all the functions in the servers, it undertakes functions for the causing problems through the network. This paper focuses on the testing technology for quality assurance of URC software components. The test case design using scenario is an effective methodology to cover all user requirements. And we are continuously studying to adapt ISO standards for URC component Software testing based an ISO/IEC 9126. This paper suggests the testing process and evaluation items for software testing of robots. This material could be referenced to establish standard software testing of intelligent robots. Nevertheless the test cases for practical evaluation of robots should be studied and developed further more. ETRI is continuously developing the evaluation methods for evaluation of robots.

References

[1] Kim, H.: Network based intelligent service robot –URC. ETRI CEO Information (15) (2008)
[2] Kim, H., Cho, Y.J., Oh, S.R.: CAMUS: A Middleware Supporting Context-aware Services for Network-based Robots. In: IEEE Workshop on Advanced Robotic and its Social Impacts (2005)
[3] ISO/IEC TR 9126, Software engineering –Product quality - Part 4: Quality in Use Metrics (2003)
[4] Pol, M., Teunissen, R., Veenendaal, E.V.: Software Testing: a guide to the TMap Approach, p. 7 (2002) ISBN 0 201 74571 2

The Research of the Role Information of the Marketing Channel Management Personnel of the Domestic Leisure Garment Brands

Zhang Junying

Fashion Design and Engineering department,
Henan institute of engineering, Zhengzhou, China

Abstract. Based on the garment brands randomly selected in Shanghai, Zhejiang, Guangdong, Fujian, Henan, Jiangsu, and Hubei, this paper learns about the basic information of the garment brand channel manager through the manager himself, store manager, and the company's other staff. Making use of telephone, Email, interviews, networking online contact and other data collection methods, understanding the survey questionnaires about the channel manager's gender, age, education, occupation, marital status, working years, salary, and treatment, sorting out the role information of the channel manager in the current domestic leisure garment market, and representing the data in charts, this paper is aimed to provide some referential information for clothing brand enterprises while they choose their channel manager.

Keywords: Leisure garment brands, channel, manager, role, information.

1 Introduction

"Those who get these channels may win the world." Marketing channel is a mobile carrier for the product to reach the consumer and a bridge to connect the enterprise and the market. Only through this channel, can the products be turned into commodities. Only through this channel, can the company turn its goods into the funds needed to expand its production and reproduction so as to make great benefits.

Since the 1990s, leisure garment gradually enters our country. The development of leisure garment brought a great impact to Chinese garment industry known as the world's clothing kingdom. In 1992, "Giordano" appeared in Guangzhou for the first time, followed by "JeansWest", "Bossini" and so on which came into Guangdong from Hong Kong and quickly engulfed the entire country. Casual clothing stores spread throughout the major cities in China, changing people's sense of dress and aesthetic taste. In the era of "channel being king", a reasonable marketing channel strategy has become the key for the textile and garment enterprises to seize the market and achieve success in competition. With the increasing importance of channels, the role of channel managers is increasingly to the fore.

With the development of economy and the progress of society, leisure is becoming a fashion, and casual wear apparel has become the dominant trend of clothing. At present, the number of professional manufacturer of leisure garment has reached ten thousand in our country, and the domestic casual brands are also up to more than 2000 kinds. Casual wear clothing takes a dominant position of China's garment industry. In 2008, China's casual wear clothing takes up at least 2,200 billion consumer markets. Not only are the domestic casual wear brands ready to flex their muscles, a large number of foreign-funded retail enterprises also compete for the Chinese market, making China's casual wear brand competition in the market heat up. As the market become increasingly competitive, the competition between various manufacturers is essentially the competition of marketing channels. Faced with the increasingly intense competition, many clothing brands make salary adjustments to hire experienced channel management. In the face of competitors from thousands of brands, to understand the channel manager's role information in the current domestic leisure garment market is of great benefits for the clothing brand enterprises to select the channel management personnel.

2 The Research of the Role Information of the Marketing Channel Management Personnel of the Domestic Leisure Garment Brands

2.1 The Selection of Survey Samples and Research Methods

120 apparel brands of Shanghai, Zhejiang, Guangdong, Fujian, Henan, Jiangsu, and Hubei were randomly selected. The basic information of the clothing brand channel managers were acquired through the channel manager himself, store manager, and the company's other staff. In view of the fact that the research subjects are too scattered and the questionnaire are easy to answer, the survey is carried out through telephone, Email, interviews, and networking online contact to collect data. 100 valid questionnaires were recovered and the efficiency rate of the questionnaire is 83%.

2.2 Research Data Analysis

2.2.1 Gender, Age Structure of the Respondents

In the 100 valid samples, 48 are male channel managers and 54 females. Thus, the female managers of the domestic casual brand marketing channel take a major proportion.

The age structure of the respondents is set according to the traditional classification methods: with a 5-year age group to be divided, and divided into five stages, namely less than 30 years of age, 31-35 ,36-40 ,41-45, and over 45 years of age. The age distribution of the sample is represented in Figure 1.

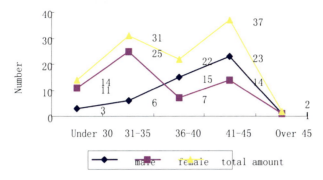

Fig. 1. The age distribution of respondents

From the figure, it can be seen that the age of the respondents mainly distributes among 31-35, and 41-45 years of age, and the samples of these two age groups account for a large part of the whole sample; of these 31-35 age group, males account for 6%, and females account for 25%; while of the 41-45 age group, males account for 23%, women account for 14%; therefore it can be seen that in the 31-35 age group, female channel managers are more, while in the 41-45 age group, male channel managers are more; the respondents under the age of 30, males account for 3%, and females account for 11%; thus it can be seen that female channel managers in this age group are more than males; in 36-40 age group, women account for 7%, while men account for 15%, that is to say, male channel managers in this age group are obviously more than female managers; in the age group of over 45, the proportion of men and women are almost of the same, and in all the age groups, the number of channel managers in this age group is the smallest; it can be concluded that in the domestic leisure garment brands market, the 45-year-old managers only take a minor proportion.

2.2.2 The Marital Status of Respondents

In the valid questionnaires, the sample is divided into three groups, i.e., "unmarried", "married without children" and "married with children ", and then the data in Figure 2 are obtained, which are shown in Figure 2.

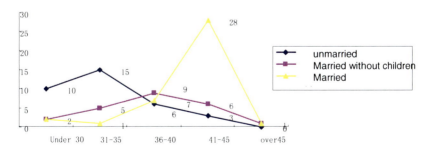

Fig. 2. The Marital Status of Respondents

It can be seen from the figure that among channel managers of the apparel brands, the elderly unmarried are very common.

2.2.3 The Education Profile of the Respondents

In the 100 valid samples, 32 of the respondents are with college degree, 51 are of Bachelor's degree, and 17 are of Master's degree or above. The education profile of respondents is shown in Figure 3. It can be seen that the respondents primarily are undergraduates, accounting for more than half of the total number; the number of people of college and master's degree and above practitioner in this work accounts for 32% and 17% respectively.

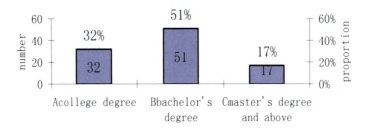

Fig. 3. Distribution of Respondents' Education Profile

The figure shows that in the sample the decision-making power the marketing channel structure of the domestic leisure garment brands is primarily under the control of persons with Bachelor's degree.

2.2.4 Distribution of the Positions of the Respondents

Due to the slight differences of the titles at home and abroad, this research is based on some domestic common titles of the channel managers so as to conduct research on the titles of first-time visitors. The results of the position titles of the respondents in the sample are shown as in Figure 4.

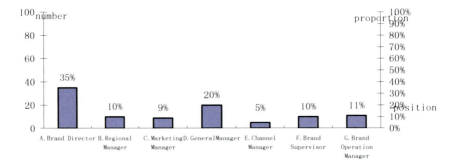

Fig. 4. Distributions of the Positions of the Respondents

As can be seen from the above figure, brand manager takes 35% of the total samples, which is the largest proportion; followed by the general manager, 20%of the total samples; the proportion taken by other duties are relatively balanced; the percentage of the channel manager is the smallest, accounting for only 5%. It indicates that the channel managers in the domestic leisure garments market are mostly titled as brand manager. However, the title of "channel manager" that are commonly used in foreign countries is rarely used in China.

2.2.5 The Practitioner Time Distribution of the Respondents

As the threshold of apparel industry being lower, and staff mobility being relatively large, this paper sets the practitioner time of three years as a certain level, totally set to five levels: 1-3, 4-6, 7-9, 10-12, and above 12 years. And this paper also conducts comparative studies on the homework time of the respondents in the field of leisure garment brands and the practitioner time in the present leisure garment brand. The practitioner time of the respondents and its corresponding number of practitioners are shown in Figure 5.

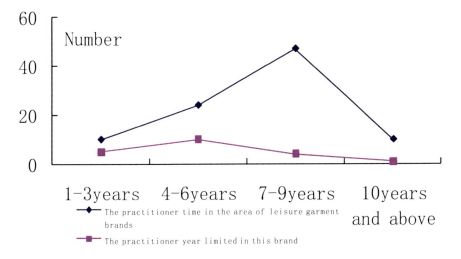

Fig. 5. The practitioner time of the respondents and its corresponding number of practitioners

From Figure 4, we can see that the practitioner time of the respondents in the field of the casual wear brand mainly falls in 7-9 years, with the number as, accounting for 47% of the total number, of them 26 are men, and 11 are women; the total number of persons whose practitioner time falls in a 1-3 years period is 10, accounting for 10% of the total number; the total number of persons whose practitioner time practitioner time is within 4-6 year period is 24, accounting for 24% of the total number; the number of persons whose practitioner time practitioner time is 10 years and above accounts for 10% of the total number. In addition, the proportion for 1-3 years ,4-6 years ,7-9 years and 10years and above of the practitioner time in the current casual brand respectively were 25%, 50%, 20%, 5%.

It is found in the research that 35% of the respondents have been engaged in the research of and making decision for marketing channelsof the same brand, those who have changed jobs but have been engaged in the marketing channels operation of leisure garment brands account for 55%; those who switch from other types of work to marketing channel operations of casual wear brand account for 10%.

2.2.6 Pay Treatment

As to the research on pay treatment in order to protect the privacy of respondents, this salary survey will divide the pay treatment into a classes according in accordance of the annual salary, i.e., an annual salary of 50 thousand yuan or less per year, 50-100 thousand yuan per year,100-150 thousand yuan per year ,150-200 thousand yuan per year, 200 thousand yuan per year. After collating the survey data, it is found that 7 persons' pay treatment (annual salary) is 50 thousand yuan or less, 12 persons' 50-100 thousand yuan, 27 persons' 100-150 thousand yuan, 31 persons' 150-200 thousand yuan, and 23 persons' is 200 thousand or above. The data distribution is shown in the figure below.

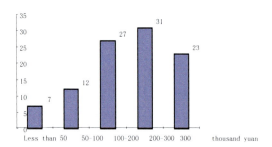

Fig. 6. Pay treatment of the respondents

It can be seen from the above data that the proportion for the salary treatment of 200-300 thousand per year for the marketing channel manager of the current apparel brand is relatively large. On the one hand, from the survey data it is found that the pay treatment for the channel manager in coastal cities such as Shanghai, Zhejiang, Guangdong, Fujian and other two provinces was significantly higher than that of in Henan and Hubei provinces, and Jiangsu is in between of them. On the other hand, there is a direct proportional relationship between the working years and the salary treatment.

3 Conclusions

Through this research and analysis, we can see that the decision-making power of the marketing channel of the domestic leisure garment brands primarily falls under the control of people with undergraduate education, and working years within the period of 7-9 years, and these people mainly are men of 41-45 years old and women of 31-35 years old, whose positions in the company are mainly brand directors and general managers. Among the channel managers of apparel brand, elderly unmarried people are common. In addition, people whose pay treatment is 200-300 thousand per year take a larger proportion.

References

[1] Who detonated the first year of apparel circulation changes? Textile Information Weekly (2), 26 (2007)
[2] Xiaolong, Z.: The Role of Marketing Channels. Economist, 186 (May 2006)
[3] Marketing Channel Strategy Forum held in 2005, China Textile Garment Net (March 30, 2005), http://www.wears.com.cn/glyx/12/index18.html
[4] 2003-2004 Development Report of China's Apparel Industry, China Garment Industry Association, p. 52
[5] 2003-2004 Development Report of China's Apparel Industry, China Garment Industry Association, p. 40
[6] The Analysis of China's Leisure Garment Market Development Situation and the Characteristics of the Market, China Garment Network (April 4, 2006), http://www.efu.com.cn/data/2006/2006-04-04/144410.shtml
[7] Fen, L.: The Course of Eight-year Innovation of Casual Wear in Shaxi. Fashion Times (October 16, 2007)
[8] Casual Clothes into the Era of competition Among Supermarkets, Fashion Times

The Study on Data Mining to CRM Based on Rough Set

Zhang Wei-bo

School of Information Sciense and Technology,
Northwest University, Xi'an, China, 710127

Abstract. The key of competition is a corporation can satisfy the client's requirement, it makes the CRM turn into the focus of the development, research of CRM based on data mining is become a pop in this correlative domain. The concept, character and basal effects of CRM, research of data mining with CRM are introduced. In the Data Mining technology, Rough Set Theory is introduced. At present, knowledge reduction and decision rule mining in decision are the research hotspots. With heuristic information, minimal decision rule acquirement based on Rough Set Theory is researched. Finally, an example is used to verify the effectiveness and feasibility of this method.

Keywords: CRM; Data Mining; Rough Set Theory; Optimization.

1 Introduction

1.1 CRM

1) Conception.
CRM (Customer Relationship Management) is a new management mechanism used to improve the relationship between the enterprise and customers. It is implemented in business marketing, sales, service and technical support fields, which are related to customers. Its core is "customer centric", to improve customers' satisfaction and change the customer relationship [1]. Thereby, the competitiveness of enterprises can be enhanced. Essentially, CRM is to realize the value-added services for customers. This process will carry out specialized sales and services strategies for customers in specific market through an extension of the supply chain. As the resources of enterprises, customers, CRM is the hottest market in the world.
2) Basic function
A complete CRM system should include 3 major modules: the collaboration layer, the operation layer and the decision-making layer. As shown in Figure 1.
 a) The collaboration layer. Its function is to deal with the requirements integration and automation for customer communication means (such as telephone, fax, Internet, Email). It can provide consumers much more convenient modes of interaction.

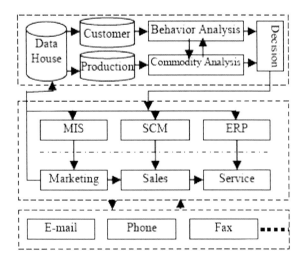

Fig. 1. Structure of CRM system

b) The operation layer. It includes 3 parts: marketing, sales and customer service. These 3 parts are managed by MIS, SCM (Supply Chain Management) and ERP. In this respect, the role of these processes and technologies is to improve the efficiency and accuracy of operation. A good CRM solution can effectively manage the various channels of interaction, whether they are used in marketing, sales or service, the use of channels of interaction will be more efficient.

c) The decision-making layer. This layer's function is to deal with information generated by the two parts above. It can build a customer intelligent system to provide supports for enterprise's strategic decision. It mainly includes: Data Warehousing, customer databases, production databases, customer segmentation systems, reporting and analysis system. These functions can provide capabilities of analyzing the customers' behaviors, and basic principles of decision-making.

Thus, throughout the CRM process, the analysis of data plays a key role. To make full use of data for strategic decision-making, it is necessary to find that valuable information, hide and never been found previously, between consumers and suppliers. Such as the industry trends, direction and other significant factors. The Data Mining is a suitable technology, which is designed to help develop new strategies [2].

1.2 Data Mining

1) Conception.

Data Mining is a new technology which can discover and extract hidden and useful information from large databases or data warehouses, and it is a promising new area of application in database research. Data Mining can help users extract useful business information from a large number of data, in order to support decision-making. Applied in CRM, Data Mining is able to select and purify the data, seemed no relationship, in the massive customer database, and to extract valuable customer relationships. This process also can predict the future development trends and behaviors [3].

2) Data Mining process

The general process of Data Mining is as follows:

a) Definition of problem. Defining the business problem clearly and determining the purpose of Data Mining.

b) Data preparation. It includes: Data Selection --Extracting the target data sets in a large database and data warehouse for Data Mining; Data Preprocessing -- Reprocessing the extracted data, including data integrity, data consistency, noise eliminating, filling and removing data, and so on.

c) Data Mining. In accordance with the types and features of data, appropriate algorithm is selected and used purify and converse data sets. This step is the most important one in whole process.

d) Results analysis. Explaining and evaluating the results, building proper model, and conversing them to become knowledge able to understand by users.

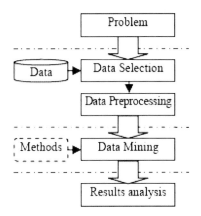

Fig. 2. Process of Data Mining

There are many Algorithms. Rough set theory, in this paper, considering the actual situation, is used to establish a mining model for customer classification. This model can divide customers into different levels accurately, providing decision-making for CRM.

2 Rough Set Theory

Rough set theory combines the classification and knowledge together, and it can portray approximately the uncertain or imprecise knowledge based on the known knowledge in database. Here are a few related basic concepts.

(1) An information system S can be expressed as: S={U, A, V, f}. U: sets of nonempty finite objects, universe U={x_1, x_1 ... x_n}; A: nonempty finite sets of attributes, A={a1,a2...am}; V: range of all attributes; f: a information function U×A→V, $x_i \in U$, $a_j \in A$, $f(x_i, a_j) \in V_a$, if A is represented by condition attributes set C and decision attributes set D, A=C∪D and C∩D=φ, so we may name S decision table[4,5,6].

(2) For any B, B∈A, denoted as ind(B) ={(x, y): f(x, a_j)= f(y, a_j), $a_j \in B$}, and indiscemible relation ind(B) divides U into many parts, U/ind(B)={[x]B: x∈U}, [x]B={y: (x, y)∈ind(B)}.

(3) If X ∈ U, R_(X)={x ∈ U: [x]R ⊆ X} is lower approximation set, posR(X)=R_(X);
R⁻(X)={x∈U: [x]R⊆X∩X≠φ} is upper approximation set, $neg_R(X)$=U-R_(X). And the boundary region is $bn_R(X)$=R⁻(X)-R_(X).

(4) S= {U, A, V, f}, R∈A, R is an equivalence relation. r∈R, if ind(R)=ind(R-|r|), so r is unnecessary in R; On the other hand, r is necessary in R. If every r in R is necessary, R is independent, else R is reliant.

Define Q∈P, if Q is independent, and ind(Q)=ind(P), so Q is a reduction of P on U. core(P) is a set composed of reductions, core(P)=∩red(P).

(5) S= {U, A, V, f}, a(x) is the value of x on attribute a, a(x)=f(x, a), m_{ij} is an element in matrix M.

$$m_{ij} = \begin{cases} \{a \in C : a(x_i) \neq a(x_j)\} & D(x_i) \neq D(x_j) \\ \emptyset & D(x_i) = D(x_j) \end{cases}$$

i, j=1, 2, 3,..., n, and n=|U|.

3 Algorithm Rules

Traditional heuristic method of attribute is simple, but the efficiency of calculating core is not high. When the decision-making table is complex with many conditions, it needs more storage space. It is hard to simply attribute through this method.

In this paper, the improved algorithm can directly simply attribute. The definition of rules is as follows:

Information system S={U, A, V, f}, $A = C \cup D$ and $C \cap D = \Phi$. C is condition attribute set and D is a decision attribute set. If U/C={$X_1, X_2,...,X_n$}, U/D={$Y_1, Y_2,...,Y_n$} [7], so the support degree of D is $k_C(D)$, (0≤$k_C(D)$≤1):

$$k_c(D) = \frac{1}{U} \sum_{i=1}^{m} pos(Y_i |, Y_i \in U/D) \qquad (1)$$

When $k_C(D)$=1, the decision is determined by condition; $k_C(D)$=0, the result is different.

If $k_C(D) \neq 0$, c∈C, the importance of attribute c for U/D is $sig_{C-|c|}^D$, $(0 < sig_{C-|c|}^D \leq 1)$:

$$sig^D_{C-|c|} = k_C(D) - k_{C-|c|}(D) \qquad (2)$$

The steps of decision rules are:

Input: S= {U, A, V, f};
Output: the minimal decision rule set [8, 9]:

(1) According to formula (1), we can get $k_C(D)$, and set R=φ,
(2) For every c, c∈C-R, from formula (2), $sig^D_R(c)$ is obtained. We choose the largest one, R=R ∪ {c}.
(3) If $k_R(D)=k_C(D)$, the process is end, and R is a reduction; else, return to step (2).

After reduction R is got, a data analysis method is used to simply the decision table. At last, we can get the minimal decision rule set.

4 Case Analysis

Here, a tourism enterprise is taken for example. Tourism provides invisible services which reach customers directly, and customers are the most important intangible assets for enterprises. The distinction between the types of customers and the development of potential customers are both very important. Customer classification should be based on the customer value. In this example, Date Mining technology will be used in CRM based on Rough Set Theory.

In Table 1, D= {d}: decision attribute set, it describes the types of customers. 1- Customers with high value; 2-Normal customers; 3- Customers with low value. $C = \{c_1, c_2, c_3, c_4\}$: condition attribute set.

Table 1. Table Type Styles

U	c_1	c_2	c_3	c_4	d
x_1	2	1	1	2	2
x_2	3	3	2	2	2
x_3	1	2	2	1	1
x_4	1	3	1	2	2
x_5	3	1	3	3	3
x_6	2	1	2	2	1
x_7	2	3	3	3	2
x_8	3	2	1	1	3
x_9	1	2	2	3	2
x_{10}	1	1	3	2	1

C1: Scale of enterprise, 1-Large; 2-Medium; 3-Small.
C2: The nature of enterprise, 1-Overseas-invested enterprise; 2-State-owned enterprise; 3-Private enterprise.
C3: Production and service, 1-Good; 2-Ordinary; 3-Bad.
C4: Profit level, 1-High; 2-Normal; 3-Low.

$U/ind(C) = \{\{x_1\},\{x_2\},\{x_3\},\{x_4\}\}$,

$U/ind(D) = \{\{x_3\},\{x_6\},\{x_1,x_2,x_4,x_7,x_9\},\{x_5,x_8\}\}$,

From formula (2), after calculated, $k_C(D)=1$. And the support degrees of $C-|c_i|(i=1,2,3,4)$ are respectively $K_{C-|C1|}(D) = 2/5$, $K_{C-|C2|}(D) = 1$, $K_{C-|C3|}(D) = 3/5$ and $K_{C-|C4|}(D) = 1$

Then the importance of ci (i=1, 2, 3, 4) is calculated as follows:

$sig^D_{C-|C_1|}(c_1) = k_C(D) - k_{C-|c_1|}(D) = 1 - 2/5 = 3/5$

$sig^D_{C-|C_2|}(c_2) = k_C(D) - k_{C-|c_2|}(D) = 1 - 1 = 0$

$sig^D_{C-|C_3|}(c_3) = k_C(D) - k_{C-|c_3|}(D) = 1 - 3/5 = 2/5$

$sig^D_{C-|C_4|}(c_4) = k_C(D) - k_{C-|c_4|}(D) = 1 - 1 = 0$

From the results, we can get the reduction R, $R = \{c_1, c_3\}$. So back to the example, 6 rules are proposed.
If c_1=1 and c_3=1 then d=2;
If c_1=1 and c_3=2 then d=1;
If c_1=1 and c_3=3 then d=1;
If c_1=2 and c_3=2 then d=1;
If c_1=2 and c_3=3 then d=2;
If c_1=3 and c_3=3 then d=3.
In order to understand the result clearly, a tree is introduced. In Figure 3:

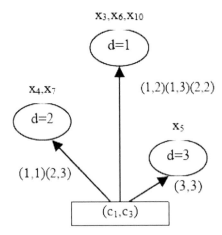

Fig. 3. Result tree

From the example, characteristics of customer value can be identified through the Data Mining using Rough Set Theory. For enterprises, they can separate the ones with high-value from the total customers and concentrate the limited resources to them to provide personal service and enhance their loyalties. In a word, this method can increase their economic benefits.

5 Conclusions

Rough Set Theory can characterize the hierarchy, uncertainty and incompleteness of human studying better. It is in line with the laws of human perception and becoming a promising method of soft computing. The above analysis shows that: Rough Set Theory may provide a powerful means for Data Mining.

Of course, this theory is still in continuous development and improvement, so far there are a number of theoretical and practical problems need to be solved. But the unique idea of this theory in dealing with uncertainty issues makes it to be received wide attention and successfully applied in many fields. It shows that this method has a very promising prospect with vitality.

References

1. Rongqin, H.: Design Practical of CRM Principle. Publishing House of Electronics Industry, Beijing (2003)
2. Yan, F.: Building CRM and Improving the Core Competitiveness of Tourism Enterprise. Statistics and Decision 7, 123–124 (2005)
3. Sin, W., Xuedong, G.: Data Warehousing and Data Mining. Metallurgical Industry Press, Beijing (2008)
4. Zhenxing, H., Qi, Z.: A Survey on Rough Set Theory and Its Application. Journal of Control Theory and Application, 153–157 (1999)
5. Mingchun, W.: Study on Method of Data Mining and Text Mining Based on Rough Set, pp. 23–24. Tianjin University, Tianjin (2005)
6. Ju, L., Jun, W., Xing, W.: Rough Set-based Approach to Feature Selection in Customer Relationship Management. Computer Engineering and Design, 5830–5836 (2008)
7. Zhi, T., Baodong, X.: Study on Knowledge Reduction Algorithms Based on Decision Attribute Support Degree. Journal of Northeastern University (Social Science), 1025–1028 (2008)
8. Yang, Z., Peiyou, C.: Application of Decision Tree Classification Algorithm in Estimating Loan Customer Credit based on Rough Sets. Science Technology and Industry 8(1), 57–60 (2008)
9. Shangzhi, W.: Research on Attribute Reduction Algorithms Based on Rough Sets, pp. 45–47. Northwest Normal University, Lanzhou (2006)

The Data Mining Method Based on Rough Sets in Economic Practice

Luo Shengmin

Qingyuan Polytechnic
Qingyuan City, Guangdong Province, China

Abstract. As the computer technology rapid development, each kind of data grows suddenly. How to extract useful information from these massive data has become a very real and important issue. It is very important that how to get the rule in the theory of rough set, it may obtain some inherent laws from the rule. It is helpful to analyze the decision table. Based on the economic data of Hebei Province in recent years, analyses the system of Hebei economics by the rule got. The result get from the rules is consistent with the reality.

Keywords: Data mining; Rough sets; Fuzzy variable precision rough sets.

1 Introduction

In the study of modern economics, mathematical models are often constructed to simulate phenomena. A case in point is the Black-sholes Equation, which models the equities in stock market. However, because of the effect of vast perturbation, it is difficult to build exact models for all economic phenomena. So, the empirical research methods are introduced. This method utilizes the historical data, instead of theoretical model, as the base of the study. More explicitly speaking, it is by data mining techniques instead of by theoretical analysis that the potential patterns locked up in databases from economic activities are extracted.

Every economic activity yields a great amount of data, such as the commercial sales records, credit history of bank customers, and other statistics. Potentially useful information often underlies them. Bring the information forth and using it as a guide to decision making is of great importance in economic activities. In fact, some famous international companies such as IBM, GM and Microsoft have their data mining research teams. In China, the research center on fictitious economy and data science has been built in China Academy of Science, one of whose focuses is mining the economic data. As a huge economic entity, the city of Baoding is supposed to build its own data mining center. Providing with useful information, the data mining techniques will benefits our city's economy.

Many data mining theories and techniques have been developed, among which are statistical methods such as SVM, Bayesian network, decision tree and so on. But, these researches are based on the assumption that more or less prior knowledge beyond the data such as probability distribution of some random variables is available. Owing to this assumption, these tools are limited in application. The rough set theory, proposed by Z. Pawlak in 1980s, is a mathematical tool dealing with the vague and incomplete data with no use of any information beyond the data. The data mining techniques based on rough set theory are investigated and applied to actual economic practice. However, most of the investigations and applications are conducted within the framework of classical rough set or variable precision rough set theory, in which the data are characterized with the symbolic attributes. When the data of continuous values are to deal with, they must be discredited in advance, which results in the loss of information in the original data. To overcome the drawback, fuzzy rough set theory and other extension of classic rough set theory are developed. The reference [7] investigated the data mining technique based on fuzzy rough set theory. However, these extensions of classic rough set model are sensitive to the noise in the data. As an improvement to them, a new theory called fuzzy variable precision rough sets is proposed in the reference [12].

The first aim of the present research is to review several cases of data mining based on the rough set theory in economic practice and to draw some guidelines from them. The second aim to propose a methodology for the rough economic analysis by combing the guidelines with the model of fuzzy variable precision rough sets. This research as a preparation to the application of rough sets in the economic sectors of Baoding City has the potential to enhance the economy of Hebei Province. The remainder of the paper is organized as follows: section 2 gives a brief introduction to the classic rough set theory; section 3 reviews several cases of data mining in actual economic practice and draws guidelines from them; section 4 introduces the fuzzy variable precision rough set theory, and by combining it with the guidelines, a methodology of rough analysis in economic field is developed; section 5 concludes this research.

2 Basic Notions in Rough Set Theory

The concept of rough sets was first proposed by Z. Pawlak in 1982, aiming at describing a target concept i.e. a subset of a universe of discourse by the use of the incomplete information about the universe. To that end, the equivalent classes induced by equivalent relations on the universe are used to define the lower approximation and upper approximation of the concept. Specifically, the lower approximation of it is the union of the equivalent classes contained in it, and the upper approximation is the union of the equivalent classes having a non empty intersection with it. When the lower approximation of a set differs from its upper approximation, the set is called a rough set. Information about the universe is often depicted by a decision table, in which objects of the universe are described by some conditional attributes and a decision one. The equivalence of two objects consists in that they have the same values on a set of the attributes. So, a concept i.e. a subset of the universe can be described by its lower and upper approximations. One of the most important goals of rough set theory is finding the minimal subsets of the conditional attributes that preserve the classification ability of the set of the conditional attributes, which are called reducts of the conditional attributes. The classification ability of a set of conditional attributes is characterized by its positive field, which is the union of the conditional equivalent classes that are contained in decision classes. Besides the basic method for obtaining the reducts proposed by Z. Pawlak, the discrimination matrix proposed by Skoron is an very useful method [2]. After obtaining reducts of the conditional attributes, the decision rules may be reduced further. Finally, a set of reduced decision rules is acquired. Some software based on rough set theories such as Rough Enough and Rose have been developed and applied in many fields, specially in economic practice.

3 Some Guidelines Drawn from Several Cases of Application of Rough Sets in Economic Practice

Since the rough set theory can deal with the vague and incomplete data, it is applied in data mining in a wide range of economic practice including stock exchanging, insurance marketing and so on. The first case to review is the analysis on the stock market data from the New Zealand Exchange Limited (NZX) in [10]. Its work is comprised of three parts: data preparation, rough set analysis and validation. In the first part, initial data are chosen from a specific period representing the closing price, opening price, highest price reached during the day, and the lowest price reached during the day. Then after data conversion, data cleaning, conditional attributes and decision attribute selection, attributes discretization, a subset of original data is selected to comprise a decision table, with other data left for validation. In the second part, a software called Rosetta Rough Set Toolkit is performed on the decision table to produce reducts and decision rules. Then by filtering out those reducts that are unable generate rules with sufficient support of the data, 18 reducts are obtained. Only those rules with higher support and higher accuracy are reserved. In the third part, matching the rules from part two with the validation subset of the original data, their new support, coverage and accuracy are obtained. Rules of the similar measures are reserved for future use, for example, of predicting.

The second case is the application of rough sets to predicting insolvency of Spanish non-life insurance companies in [11]. In that work, 72 insurances were chosen to comprise the universe; 19 variables were chosen as conditional attributes based on a detailed analysis of the variables, studies on the bankrupted insurance companies and researchers' preferences and knowledge; the decision attribute took the value of 1 or 0 in consistence with the firm's being healthy or not. Data were split into analysis and validation sets. The Rose software was performed on the analysis set, and after validation 25 decision rules containing 5 conditional attributes were obtained. Comparison of the rough set approach with discriminant analysis demonstrated the former's superiority.

In China, rough set theory has been studied to be applied in a wide range of economic practice. The author of [5] gathered a data set from some business bank, then applied to it a reduction method proposed by him based mainly on indiscrimination matrix and entropy, and at last matched reducts to the coming sample with the help of k-nearest neighbor algorithm. Summarizing the above work and reference [4], we can deduce the following guidelines to utilizing rough sets in data mining in economic practice:

(1) Building rough models. The purpose of this stage is to prepare data and establish a decision table. Select a set of objects forming the universe from a specific period of a business and a set of attributes characterizing the objects. Divide the set of objects into two subsets: one as the analysis subset and another as the validation subset. At this stage, the values of attributes should be converted to be symbolic, and the lost values should be completed.

(2) Performing rough analysis. At this stage, there are two kinds of strategies: one is to use the existing software such as Rosetta Rough Set Toolkit and Rose, and another is to develop new algorithms as the author of [9] did. At the end of the stage, reducts of the conditional attributes should be obtained.

(3) Evaluating rough rules. For a decision table, the reducts obtained in the former step may generate a lot of decision rules. We can select those rules as our final results according to our preferable criteria such as having higher accuracy, support on the validation data set or similar behavior on both analysis data and validation data.

4 Combining the Guidelines with the Model of Variable Fuzzy Precision Rough Sets

Among the above guidelines, the most important is rough analysis. In the reviewed researches, the rough analysis was performed by either rough set software such as Rose or algorithms developed by their researchers. And, both the rough set software and the above-mentioned algorithms were developed on the bases of either classic rough set theory or its variants such as variable precision rough sets and so on. However, these extensions of classic rough set theory have their shortcomings.

The classic rough set theory only deals with the attributes of symbolic values. So, attributes discretization has to be implemented prior to the rough analysis, which may cause the loss of some information. As for the variants of rough set theory including variable precision rough sets and fuzzy rough sets have a shortcoming in common: sensitivity to noises [12].

As an improvement to them, the fuzzy variable precision rough set model (FVPRS) is proposed in [12]. In this model, the fuzzy lower and upper approximation operators with variable precision $\alpha \in [0,1)$, are respectively defined as follows. $\forall x \in U$

$$\underline{R_{\vartheta_\alpha}} A(x) = \inf_{A(u) \leq \alpha} \vartheta(R(x,u), \alpha) \wedge \inf_{A(u) \leq \alpha} \vartheta(R(x,u), A(u))$$

$$\overline{R_{T_\alpha}} A(x) = \sup_{A(u) \geq N(\alpha)} T(R(x,u), N(\alpha)) \wedge \sup_{A(u) \leq N(\alpha)} T(R(x,u), A(u))$$

$$\underline{R_S}_\alpha A(x) = \inf_{A(u) \leq \alpha} S(N(R(x,u)), \alpha) \wedge \inf_{A(u) \leq \alpha} S(N(R(x,u)), A(u))$$

$$\overline{R_{\sigma_\alpha}} A(x) = \sup_{A(u) \geq N(\alpha)} \sigma(N(R(x,u)), N(\alpha)) \wedge \sup_{A(u) < N(\alpha)} \sigma(N(R(x,u)), A(u))$$

where U is the universe, A is a fuzzy subset on the universe, and ϑ, R,T, N are the residuation implication, the binary fuzzy relation on U, the triangular norm and the negator, respectively. Based on this definition, [12] introduces a concept of attribute reduction and develops an attribute reduction algorithm for a fuzzy decision table with condition fuzzy attributes and decision symbolic attributes as follows:

Algorithm-1: to find all α - reductions of FVPRS:

Step 1: Compute the similarity relation of the set of all condition attributes: SIM (R);

Step 2: Compute $\underline{R_\vartheta}_\alpha ([x]_D)$ for every $x \in U$

Step 3: Compute c_{ij}

$c_{ij} = \{a : T(a(x_i, x_j), \lambda) \leq \alpha\}, \lambda = \underline{R_\vartheta}_\alpha [x_i]_D (x_i)$ for $D(x_i, x_j) = 0$; Otherwise $c_{ij} = \phi$;

Step 4: Compute Core $Core_{D_\alpha}(R) = \{a : c_{ij} = \{a\}\}$; Delete those c_{ij} with nonempty overlap with $Core_{D_\alpha}(R)$;

Step 5: Define $f_\alpha(FD) = \wedge\{\vee(c_{ij})\}$ with nonempty c_{ij} and compute $g_\alpha(FD) = (\wedge R_1) \vee \ldots \vee (\wedge R_l)$ by $f_\alpha(DS)$;

Step 6: Output $Red_{D_\alpha}(R) = Core_{D_\alpha}(R) \cup R_1, \ldots Core \cup R_l$

By use of this algorithm, all α - reducts of a decision can be obtained. Experiments in [12] show this algorithm is less-sensitive to misclassification and perturbation than many others. Combining this algorithm with the guidelines described in section 3, we propose a methodology of applying FVPRS in economic practice as follows.

(1) According the purpose of the analysis, select a proper set of raw data from the history of the enterprise concerned. For example, if we want to predict the coming value of a given economic index, we should select the data occurred recently according to the marginal effect theory of in economics. And then the data set should be divided into analysis subset and validation subset.

(2) Select proper variables as conditional and decision attributes. There are two factors we must take into account here. One is the number of the variables to be chosen as attributes of the decision table. There may being many variables concerned, we must select not much of them as the attributes. Otherwise, it will

take greater amount of time for computer to implement the reduction algorithm. Another factor we must consider is the selection of the decision attribute. For example, when evaluating the performance of insurance companies, the decisive one of the ratios describing an insurance company is not easily recognized [11]. Selection of both the attributes and decision attribute should seek help from the studies on the factors and the knowledge about the insurance sector.

(3) Build a decision table suitable to the algorithm described above. Since the algorithm-1 is developed for a fuzzy decision table with condition fuzzy attributes and decision symbolic attributes, condition attributes must be converted to fuzzy ones by use of existing algorithms and decision attribute should be symbolized. Also, the lost values of attributes should be repaired in advance.

(4) Find the α – reducts of the decision table. The value of the threshold α should be given before the algorithm-1 is operated. However, as pointed out in [12], the larger the threshold is, the smaller the number of elected attributes is; and if the threshold is too large, the accuracy becomes smaller than the accuracy before education. So the value of the threshold α should be given by experiments. The reducts produced in this step an be depicted in the form of decision rules.

(5) Validating the rules obtained in (4). For each rule, hen applying it to the validation set, its support, accuracy, and confidence measures on both analysis and validation data. As [10] suggested, we can choose those rules having similar measures on both analysis and validation sets as the final output.

(6) Interpreting the final output from the economics point of view. As a complementary and empirical research tool in economics, the rough analysis should seek the help f economic theory.

5 Conclusion

Because mathematical model often fails to simulate many complex economic activities, the data mining techniques are widely employed in the economic analysis. As a powerful tool dealing with the vague and incomplete information, rough set theory is used in economic data mining. From several cases of application of rough set in economic analysis, we draw some guidelines to the rough analysis in economics. To avoid the existing methods' drawbacks, by combining with the fuzzy variable precision rough set model, we propose a new methodology of rough analysis in economic practice. In the future work, we will conduct the methodology on concrete data set from the economic sector of Baoding City.

References

1. Dingxiang, M.: Data mining and positive economics. Journal of Shanghai University (social sciences) (2), 120–123 (2006)
2. Pawlak, Z.: Rough Sets: Theoretical Aspects of Reasoning about Data. Kluwer Academic Publishers, Dordrecht (1991)

3. Pawlak, Z.: Rudements of Rough Sets. Information Information Sciences (177), 3–27 (2007)
4. Yongmin, L.: Data mining based on rough sets theory. Journal of Qinghua University (science and technology) (1), 110–113 (1999)
5. Tienyong, Z.: Data classification based on rough set and its application in business bank supervision. Master Dissertation of Chongqing University (2004)
6. Gediga, G., Düntsch, I.: Statistical Techniques for Rough Set Data Analysis, http://www.ggediga.de/nida/papers/stattech.html
7. Jensen, R., Shen, Q.: Fuzzy-rough attributes reduction with application to web categorization. Fuzzy Sets Syst. (141), 469–485 (2004)
8. Dubois, D., Prade, H.: Rough fuzzy sets and fuzzy rough sets. Int. J. Gen. Syst. (17), 191–208 (1990)
9. Pawlak, Z.: Rough sets, decision algorithms and Bayes' theorem. European Journal of Operational Research (136), 181–189 (2002)
10. Herbert, J., Yao, J.: Time-Series Data Analysis with Rough Sets, http://www2.cs.uregina.ca/jtyao/Papers/final_CIEF
11. Segovia, M.J., et al.: Using Rough Sets to predict insolvency of Spanish non-life insurance companies, http://www.invenia.es/oai:dialnet.unirioja.es:ART0000026524
12. Zhao, S., Tsang, E.C.C., Chen, D.: The Model of Fuzzy Variable Precision Rough set. Tran. on Fuzzy set Syst. (to be pulished)

The Study on Adaptive Routing Protocol in Mobile Adhoc Network Based on Rough Set

Pan Shaoming, Cai Qizhong, and Han Junfeng

Department of Electronic Information and Control Engineering,
Guangxi University of Technology, Liuzhou, China

Abstract. This paper introduces the up-to-date research in on-demand routing for ad hoc mobile networks. The on-demand routing protocols arc divided into three categories: flat flooding routing, flat limited routing and hierarchy routing. In flat flooding routing protocols, DSR, ABR, AODV and ZRP arc discussed. In flat limited routing protocols LAR and RDMAR arc discussed. In hierarchy routing protocol CBRP is introduced. For each routing protocol, this paper not only introduces the contents of the routing protocol, but also the merits and drawbacks, and evaluates these protocols based on a given set of parameters such as communication complexity, time complexity, route metric, the range of route discovery, the periodically packets and route maintenance. This paper also evaluates some drawbacks of above routing protocols, for example initial route delay, supporting of unidirectional links and flooding broadcast, and proposes some optimized schema for these drawbacks. These schemas arc cluster based hybrid routing schema to mitigate the flooding broadcast and initial route delay, and unidirectional ad hoc on-demand routing schema to support the ad hoc mobile networks with unidirectional links. Furthermore this paper proposes an optimization schema based on rough set for limited routing protocols, which is that the intermediate nodes update the zone of route discovery by their newer location and/or distance information. Finally authors suggest the research direction in routing for ad hoc mobile wireless networks in the future.

Keywords: Declarative Networking, NetLog, Mobile Ad Hoc Networks, Routing.

1 Introduction

Recently there has been an explosive growth both in the number of mobile computing devices and wireless communication technologies. A mobile wireless ad hoc network is an autonomous collection of nodes that has the ability to dynamically and rapidly form networks without the use of any centralized network infrastructure using wireless communication technologies. Ad hoc routing protocols are at the core of ad hoc networks as they allow remote nodes to communicate without resorting to primitive flooding techniques. It is well established in the literature that a one-size-fits-all approach with regard to the choice of optimum routing protocol does not suffice [1], [2], [6], [7]. Rather, in keeping with the distributed and ad hoc nature of these networking systems, it would be preferable to design a networking system that allows nodes to dynamically configure and utilize the most suitable ad hoc routing protocol [8], [9]. Divergent user scenarios and networking conditions, as exemplified by node mobility, node density and traffic loading, demand networking protocols which are tailored to their requirements.

To this end, a flexible network-layer is proposed which has access to a suite of distributed ad hoc routing protocols at runtime [10], [11].In a distributed system such as a mobile ad hoc network, the routing protocol is a dependent feature of the network-layer, i.e. for two adjacent nodes to communicate they must operate the same protocol. Normally, the static configuration of one protocol across the entire network addresses this issue. The self-stabilizing network-layer auto configuration protocol enables the nodes to collaboratively organize their network-layer routing protocol configuration such that a uniform choice is made across regions of connected nodes.

The network routing protocols are used to organize the network topology through optimization of network metrics, such as load, overhead, energy, link duration, etc.. The declarative approach presents the possibility of the cooperation of different protocols using the same network metrics optimization. The nodes with different protocols can communicate using the same tuples. Each node handles the tuples by executing its declarative protocol.

The paper is organized as follows. In Section II, we briefly introduce the declarative language we used to specify the protocols of MANET. In Section III, we show how to specify a routing protocol in a few rules; and we also show how to update the routing protocol by replacing an module. The Section IV gives the definition with respect to rough set. In this section, we show how to adaptively select a proper protocol using a few rules over decision table. In section V we evaluate the performance of the protocols in section III. Finally, we give the conclusions in section VI.

2 NetLog

NetLog is a declarative language presented for specifying MANET networking. It relies on the deductive languages developed in the 80s in the field of databases, with new constructs ensuring as much as possible a local behavior. The difference is that in NetLog the operator "!" is used to delete data rather than the negation operator in $DataLog^{\neg\neg}$ DataLog when the rule is instantiated, which avoids the double operation semantics of negation operator.

The language NetLog allows local negation well adapted to network processing. Negation is interpreted in classical query language under the closed world assumption. Something which is not known to be True, is considered to be False. Although this assumption makes a lot of sense in the classical centralized setting, it becomes rather tedious in the distributed network environment, where it would require checking the whole network each time the negation is used. For tractability purpose, we consider negation restricted to the local environment, and so locally checkable and valid.

The *forward chaining procedural semantics* is used in our declarative mode. Nodes push data to other nodes, which trigger the forward application of rules, resulting to new data (facts), which can then be pushed to other agents, and so on. The *forward chaining procedural semantics* ensures the local computation.

We prefix a specifier " ↑" to some head relations indicating a broadcast or unicast behavior. The location specifier "@" is required in the unicast relation, and as a matter of notation, it is require to be the first field. The facts derived by the rules whose head is restricted with specifier "↑" will be send to the related neighbors. For example, the derived fact ↑*Hello(A,B)* means that the fact will be broadcast all neighbors while the derive fact ↑*Hello(@A,B)* will be unicast to node A. The derive facts without prefix "↑" will be used to update local database.

3 Declarative Protocols

Network routing protocols are used to optimize the network organization on some etrics. They send messages to other nodes and at the same time they receive neighbors messages to update local tables. For *forward chaining procedural semantics* of the declarative approach, each node executes the rules locally without consider the other nodes. We module the rules and the modules can be divided into two parts. One is operating on the necessary tables, such as link table, route table, etc.. The other is operating on the out-sending tables and auxiliary tables for metric optimization.

The *proactive* protocol is a table-driven protocol for MANET. It periodically broadcasts network information to the network, and after receiving the information each node update local tables. The number of messages in the traffic will increase with the number of nodes. Different proactive protocols are used to decrease the network traffic using different optimizations. DSDV protocol decreases the traffic by limiting the broadcast message to 1-hop message, and OLSR protocol decreases the traffic by reduce the number of neighbors to relay messages.

In the rest of this section, we will use a simple example to show how to use declarative modules to construct a protocol and how to update the protocol just replacing a module.

3.1 Basic Proactive Protocol

Consider a network, the nodes sparsely located in a region and the link between nodes change slowly. The network can be organized using proactive approach without any optimization. The protocol requires each node to broadcast itself to its neighbors periodically for link table update. The process can be expressed by the following rules:

module(LinkTable){
  LT1 : ↑ *Hello(L, n)* : −
      !*timeEvent* ('*hello*') ;
      LinkTable(n, t);
      t > 0.
  LT2 : *LinkTable(n1, T)* : −
      !*Hello(n1, n2)*;
      !*LinkTable(n1, t)*.
}

In the first rule, the relation timeEvent ('*hello*') is a timer event. the relation *LinkTable(n, t)* indicates there is a neighbor n and the link is valid before time *t*. The rule is trigger by the timer event, and for each valid neighbor n it broadcast the fact *Hello(L,n)* to all neighbors. The second rule is used to update local link table when it receives the facts of *Hello(n1,n2)*.

The protocol also requires each node to broadcast itself to the whole network. The nodes need to relay the message when it first receives the message.

module(MsgRelay1){
*MR*1 : ↑ *MsgRoute(@n, L, L, 1, sn1, 1)* : −
    !*timeEvent* ('*MSGB*') ;
    Node(sn);
    LinkTable(n, t);
    *sn*1 := *sn* + 1;
    t > 0.
*MR*2 : ↑ *MsgRoute(@n2, L, src, hops1, sn, 1)* : −
    !*MsgRoute(L, n1, src, hops, sn, 1)*;
    RouteTB(src, nh, hops2, sn2, t2);
    LinkTable(n2, t);
    sn > *sn*2;
    *hops*1 := *hops* + 1;
    t > 0; *n*2! = *n*1.
}

The first rule is used to generate *MsgRoute* tuple for each valid neighbor. The rule *MR2* means that when a node receives a tuple *MsgRoute* which sequence number is bigger than the stored one, the node relays the tuple to each other valid neighbor *n2* except the sending node *n1*.

3.2 Proactive Protocol with Optimization of Message Relay

Consider the increasing of the number of nodes in the region, the node density will also increase correspondingly. The basic proactive protocol will not be fit for the thick network because there is amount of redundant relaying messages. In order to reduce the number of relaying messages, the node requires only a part of neighbors, denoted as relay set, to relay its messages. Compared with the declarative basic proactive protocol, we only need to replace the message relay module. The message relay module with relay set optimization can be expressed by the following rules:

$MR1 : \uparrow MsgRoute(@n, L, L, 1, sn1, 1) : -$
              $!timeEvent('MSGB') ;$
              $Node(sn);$
              $RelaySet(n);$
              $sn1 := sn + 1;$
$MR2 : \uparrow MsgRoute(@n2, L, src, hops1, sn, 1) : -$
              $!MsgRoute(L, n1, src, hops, sn, 1);$
              $RouteTB(src, nh, hops2, sn2, t2);$
              $RelaySet(n2);$
              $sn > sn2;$
              $hops1 := hops + 1;$
              $n2! = n1.$

}

The table *RelaySet* on a node stores the neighbors which is allowed to relay its message. When a node sends out the message, it will inform its relaying neighbors to relay the message to reduce the network traffic.

Consider the continuous increasing of nodes, the fix relay node selection will lead to that a node, such as the node with highest willingness in OLSR, has to relay much messages. By the declarative approach, we can replace the *RelaySet* module to solve the problem.

Let m be the number of the optimization modules on a network metric. Then there will be at most 2^m kinds of composition protocols in theory. For the network organization on the metric, we spend little cost and satisfy the most network demands.

4 Protocol Adaptation Using Rough Set Theory

Consider a universe U of elements, an information table $I = (U, A, d, V, \rho)$, where

- A: the set of attributes of metrics in network
- d: the attribute of protocol identifier
- for any $a_i \in A$, its domain is V_i
- the domain of d is the set of protocol we can compose, denoted as V_d
- $\rho : U \times A \to V$

Let $A = \{a_1, \cdots a_n\}$, and their domains are V_1, ..., V_n respectively. Then we can build the simplified decision table:

U	A			d
	a_1	...	a_n	
$e_1\ e_2\ e_3\ ...$	v_{11} v_{12} v_{13} ...	... ...	$v_{n1}\ vn_2\ v_{n3}$ ...	p_1 p_2 p_3 ...

where $v_{ij} \in V_i, p_i \in V_d$

Let the relation A be defined by the tuple $\{a_1, ..., an, d\}$ The protocol selection module can be expressed as following:

```
module(PSelection){
PS1 : PCount (p, [count]p) : - !PSTrigger (x1, ..., xn);
                              A(x1, ..., xn, p).
PS2 : Decision ([max]cp) : - PCount (p, cp).
PS3 : PSelect (p) : -        ! PCount (p, cp);
                              Decision(cp).
}
```

The *PSTrigger* tuple is used to store the detected information of the network and trigger the protocol selection module. The first rule is used to compute the weight of all protocols. The rules *PS2* and *PS3* are used to choose the protocol with the maximal weight.

Given an explicit trigger tuple *PSTrigger*, there is at most one item $A(x1, ..., xn, p)$ in decision table can be matched, and the result protocol is unique. Given a vague trigger tuple *PSTrigger*, there maybe several protocols satisfy the condition. We count the weight of each candidate protocols using rule *PS1*, and select the protocol with biggest weight suing rules *PS* 2 and *PS* 3.

5 Simulation

5.1 Simulation Environment

The simulation is implemented on the network simulation tool, WSNet. The parameters of the network to be simulated are specified, including MAC protocol, radio range and battery consumption. WSNet, the event-driven simulator for wireless networks, simulates a topology with the first and the second communication layers. In the field with dimensions 360x360, the transmission range of the node is 80. The number of nodes are increased from 10 to 50.

5.2 Simulation Results

For each scenario, the initial positions of the nodes are changed randomly. The following table shows the number of nodes running different protocols with the increase of network size. Let *sparse* be defined that the number of neighbors is smaller than 6, and *thick* that the number of neighbors is bigger than or equals to 6. It is seen that the percent of node with P2 increases with the network size.

Protocol Type	Network Size				
	10	20	30	40	50
P1	10	20	29	27	20
P2	0	0	1	13	30

Fig.1 shows the network traffic increase with the number of nodes. It is seen that The adaptation approach can reduce the number of messages efficiently. Fig.2 shows the average number of queries executed on each node. The protocol adaptation approach can decrease the number of queries execution also when the network size is not too big.

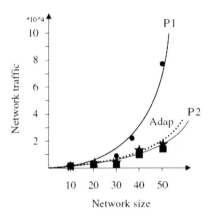

Fig. 1. The network traffic vs. the number of nodes

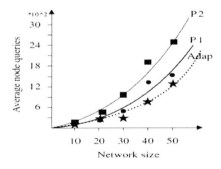

Fig. 2. Per-node query execution vs. the number of nodes

6 Conclusion

The contributions of this paper are (i) the approach of protocol composition; and (ii) the composition routing protocol adaptation using rough set theory. The interests of this system is represented as:

a) Composition Protocol: The routing protocols are composed of modules. A new declarative protocol can be designed by modules composition, and a declarative protocol can be update by replacing some modules.

b) Balance: This method can balance different metrics optimization effectively in a convenient way, such as control traffic and efficiency. It also can satisfy virous demands of the network by adjusting the composition modules.

c) Flexible and Extensible: This method is easy to expand and can fit for the network flexibly. For the explosive growth of MANET, it can effectively update the optimization modules.

d) Adaptation: The rough set theory present the approach to select a protocol from explicit or vague conditions.

Acknowledgement. The paper is supported by Nature Science Foundation of Science Technology Department of GuangXi (NO.0991067, NO.0832066).

References

1. Viennot, L., Jacquet, P., Clausen, T.H.: Analyzing Control Traffic Overhead versus Mobility and Data Traffic Activity in Mobile Ad-Hoc Network Protocols, Hingham, vol. 10, pp. 447–455 (2004)
2. Perkins, D.D., Hughes, H.D., Owen, C.B.: Factors Affecting the Performance of Ad Hoc Networks. In: ICC, vol. 4, pp. 2048–2052 (2002)
3. Loo, B.T., Hellerstein, J.M., Stoica, I., Ramakrishnan, R.: Declarative routing: extensible routing with declarative queries. In: Proceedings of the ACM SIGCOMM 2005 Conference on Applications, Technologies, Architectures, and Protocols for Computer Communications, Philadelphia, Pennsylvania, USA (2005)
4. Loo, B.T., Condie, T., Hellerstein, J.M., Maniatis, P., Roscoe, T., Stoica, I.: Implementing declarative overlays. In: SOSP 2005, Brighton, UK (2005)
5. Loo, B.T., Condie, T., Garofalakis, M.N., Gay, D.E., Hellerstein, J.M., Maniatis, P., Ramakrishnan, R., Roscoe, T., Stoica, I.: Declarative networking: language, execution and optimization. In: Proceedings of the ACM SIGMOD International Conference on Management of Data, Chicago, Illinois, USA, June 27-29 (2006)
6. Broch, J., Maltz, D.A., Johnson, D.B., Hu, Y.-C., Jetcheva, J.: A Performance Comparison of Multi-Hop Wireless Ad Hoc Network Routing Protocols (1998)
7. Das, S.R., Perkins, C.E., Royer, E.M.: Performance Comparison of Two On-demand Routing Protocols for Ad Hoc Networks (2001)
8. Boleng, J., Navidi, W., Camp, T.: Metrics to Enable Adaptive Protocols for Mobile Ad Hoc Networks (2002)
9. Frodigh, M., Parkvall, S., Roobol, C.: Future- Generation Wireless Networks (2001)

10. Forde, T.K., O'Mahony, L.D.D.: Self-Stabilizing Network-Layer Auto-Configuration for Mobile Ad Hoc Network Nodes. In: WiMob, Montreal, Canada, August 22-24 (2005)
11. Forde, T.K., Doyle, L., O'Mahony, D.: Ad Hoc Innovation: Distributed Decision-Making in Ad Hoc Networks. IEEE Communications Magazine (April 2006)
12. Abiteboul, S., Abrams, Z., Haar, S., Milo, T.: Diagnosis of asynchronous discrete event systems: datalog to the rescue! In: Proceedings of the Twenty-fourth ACM SIGACT-SIGMOD- SIGART Symposium on Principles of Database Systems, Baltimore, Maryland, USA (2005)
13. Reiss, F., Hellerstein, J.M.: Declarative Network Monitoring with an Underprovisioned Query Processor. In: ICDE (2006)
14. Alonso, G., Kranakis, E., Sawchuk, C., Wattenhofer, R., Widmayer, P.: Probabilistic Protocols for Node Discovery in Ad Hoc Multi-channel Broadcast Networks. In: Pierre, S., Barbeau, M., Kranakis, E. (eds.) ADHOC-NOW 2003. LNCS, vol. 2865, pp. 104–115. Springer, Heidelberg (2003)

The Evaluation Model of Network Security Based on Fuzzy Rough Sets

Yaolong Qi and Haining An

Computer Center, Hebei University, Baoding 071002, China
dragon@hbu.cn

Abstract. It is very important to know the security status of computer networks accurately. At present, most computer network security evaluation system doesn't analyze the datum thoroughly. Therefore it is difficult to acquire the security status of computer networks at the whole. An algorithm model with fuzzy rough set theory to mine the rules of computer network security evaluation is proposed. A fuzzy rough set knowledge system description of computer network security evaluation is studied. A fuzzy rough set attribute reduction method is given. The decision rules mining method presented in this paper is validated with a simplified network security evaluation data set. The experiment results show that decision rules acquired by the method are in accord with the fact.

Keywords: Network security; Security assessment; Fuzzy sets; Rough sets.

1 Introduction

In the past process of decision-making, people tried to make the decision-making through a certain mode of thinking standardization. However, in the process of participation, how to minimize subjective randomness, thinking, as well as the uncertainty of the fuzzy understanding of the subjective factors such as the adverse impact has always troubled the people. Network security experts Bass put forward the situational awareness of network security (Network Security Situation Awareness, NSSA) research, and established a framework for situational awareness in cyberspace [1]. But there was no specific prototype system to achieve. Stephen G. Batsell et al. developed an integrated network security system of the existing security framework, in order to provide large-scale real-time situational awareness network [2]. But the method had some limited on deployment. Jason Shifflet [3] used the idea of "defense in depth" to integrate network attack detection technologies and had built a framework without modular technology. However, the method can only detect a limited attack and not achieve true assessment for network security.

Thus, many experts and scholars have strong research interest on fuzzy sets and rough sets, and many valuable conclusions have been achieved. As the mentioned above, combining fuzzy sets with rough sets, we propose a model of network security assessment based on fuzzy sets and rough sets in this paper. First of all, using fuzzy sets theory to do fuzzy discretization for information systems, and then connection degree and repeat group of rough sets are applied to compatible and incompatible decision-making information for attribute reduction, in order to mine the decision-making information with a certain credibility.

2 The Related Works

2.1 Fuzzy Sets Theory

Fuzzy sets theory is based on the membership, fuzzy subsets described by the membership function. The fuzzy subset A given on the U domain, any $X \in U$ have identified a, $\mu_A(X)$ is called the membership degree that is X for A. The mapping $X \to \mu_A(X)$ is called as the membership function of the A. The membership function of the A is a continuous characteristic function. The range of the membership function is $\mu_A(X) \in [0,1]$, when $\mu_A(X) = \{0,1\}$, The fuzzy subset A is an ordinary subset. In the fuzzy sets, we can't simply say that whether an element is part of a collection, and can say that what extent the element is part of a collection. If the changing scope of the precise volume x is the [a, b], changing the number x in the [a, b] into the fuzzy number y in the fuzzy domain [E, E] interval, that is:

$$Y = \frac{2E}{b-a}[X - \frac{a+b}{2}] \quad (1)$$

We choose the largest membership degree method as the fuzzy method. This method is selecting the domain element of the largest membership degree as the result of defuzzification. The advantage of this method is simple, but the shortcomings are that the amount of information is rare. This method excludes the influence of the other smaller membership degree elements.

2.2 Rough Sets Theory

Assuming that U is the full-threshold of the discussed individual and R is the fuzzy relationship on the U (or Equivalence relation). It will break down into several mutually exclusive cross subset-E_i (i = 1, 2, ..., n), called the basic set of the R. The "AND" set of the basic sets R that have been included in the X subset on the U is called the lower approximation set X; The intersection of the basic set R and the subset X is not null, called as upon approximation sets. A delineation of the approximation space constitute domain; if R is a equivalence relation on the U. $[X]_R$ describes the R-equivalence classes of the X and U / R describes the collection that the all equivalence classes of the R constitutes.

Setting the R as the equivalence relation family, based $P \subseteq R$ and $P \neq \Phi$, the intersection of all the equivalence relations in the P is called as the not distinguish relations, credited as IND (P), that is:

$$[X]_{INDP} = \bigcap_{R \in P}[X]_R \qquad (2)$$

If <x, y> ∈IND (P), we said that the object x and y is the non-distinction, that is, x and y exist in the same equivalence class of the no distinguish relations IND (P). Based on the classification knowledge formatted by the equivalence relations family P, x and y can't be distinguished. The each equivalence class of the U / IND (P) is called as the basic set P. X ⊆U is the any subset of the U. So we induct two precise collections, respectively, called the upper approximation sets and the lower approximation sets, defined as follows:

The lower approximation sets that the sets X are on the R:

$$R_*(X) = \{X \subseteq U : R(X) \subseteq X\} \qquad (3)$$

That is, if and only if, When R(X) ⊆X ,having the x ∈ R*(X),in other words, R* (X) is on behalf of the largest collections that is composed by the certain elements belonging to X based on the existing knowledge.

The lower approximation sets that the sets X are on the R:

$$R^*(X) = \{XU : R(X) \cap X \neq \Phi\} \qquad (4)$$

That is, if and only if, When R(X)∩ X ≠ Φ, having the x∈R* (X), R* (X) and X represents the non-empty intersection sets that all the equivalent families intersect with X. It can be seen that R*(X) is on behalf of the precision part, but R*(X) is on behalf of the uncertainty part, which led to the concept of the border region:

$$BN_R(X) = R^*(X) - R_*(X) \qquad (5)$$

The border region is the basis that judges the accurate and inaccurate, the clear and non-clear. When BNR (X) = Φ, said that X on R is clear; when the BNR (X) ≠ Φ, said that X is the rough sets of R. Setting that POSR (X) = R* (X) is the positive region of the R on the X; NEGR (X) = U-R*(X) is negative region of the R on the X.

2.3 The Connection Degree in Set Pair Analysis

According to the need of the issue ω, a collection consisted by the A and B starts the analysis on the H, getting a total of N characteristics. The S of the N is the same characteristics on the two sets. The P characteristics of the two sets are relatively independent. In the remaining characteristics F = N-S-P, neither the same nor opposite. So excluding the weight of the characteristics, the ratio – S / N is called as the same degree on the question ω of the set A and B, recorded as a; the ratio of F / N is called as the difference uncertainty on the question ω of the set A and B, recorded as b; the ratio of P / N is called as the confrontation degree on the question ω of the set A and B, recorded as c; That is:

$$u(\omega) = \frac{S}{N} + \frac{F}{N}i + \frac{P}{N}j = a + bi + cj \qquad (6)$$

Comprehensively depicting the contact state of the two sets, i is the difference degree coefficient in the interval [1, 1], the values depending on the circumstances; j is the opposition coefficient, constant value j = -1. Contact mathematics is a mathematical system based on the computing unit-'a + bi + cj'. The rough sets contact degree can be defined as:

$$\mu(X) = a + bi + cj = R(X) + (R^*(X) - R_*(X))i + NEG(X)j$$
$$= POS(X) + BN(X)i + NEG(X)j \qquad (7)$$

The value of the rough sets contact degree, that is, the rough sets numerical contact degree is:

$$\mu(Rough) = \frac{k(POS_R)}{k(U)} + \frac{k(BN_R)}{k(U)} + \frac{k(NEG_R)}{k(U)}$$
$$= \eta + (\eta^* - \eta_*)i + (1 - \eta^*)j \qquad (8)$$

3 The Process of Network Security Assessment Based on Fuzzy Sets and Rough Sets

3.1 Data Preprocessing

In the steps of preprocessing, the original data are handled by fuzzy sets, which are mainly included the Boolean data, numeric data, class attributes and Null. In order to do the further clustering, such four kinds of data should be discretizated.
1) The membership function of Boolean type

Supposed U is the entire data domain. U_i is the ith element in U, i ∈ 1, 2, 3, ..., n. A_j is the jth element in U, j ∈ 1, 2, 3, ..., n. S_{ij} is the attribute value of the ith element, jth attribute. a_{jk} is the kth attribute value in the jth attributes, k ∈ 1,2, 3, ..., t, where t is the class number of one kind of attributes. N (a_{jk}) is the count of a_{jk}, and the dependencies between pairs of attribute values can be expressed by membership function of attribute value, which is:

$$\mu A(S_{ij}) = N(\alpha_{jk})/n, \quad k = 1, 2, \ldots, n \qquad (9)$$

where n is the number of the data.
2) The membership function of numeric type

Supposed l is the class number of attributes, C_l is the lth class, $N(C_l)$ is the number of the attributes in C_l, $C_l^{(i)}$ is the ith attribute value in class l, and the membership function of attribute values is:

$$\mu_A(C_l^{(i)}) = N(C_l)/n, \quad l = 1, 2, 3, \ldots, i = 1, 2, 3, \ldots \qquad (10)$$

3) The membership function of class attributes

$$\mu_A(C_l^{(i)}) = N(C_l)/n, \ l = 1, 2, 3, \ldots, i = 1, 2, 3, \ldots \tag{11}$$

where the variables have the same meanings like the membership function of numeric type described above.

4) The membership function of Null

$$\mu_A(S_{ij}) = \begin{cases} \min(\mu_A(S_{ij})), r_0 \leq l_0 \\ mid(\mu_A(S_{ij})), l_0 < r_0 < h_0 \\ \max(\mu_A(S_{ij})), r_0 \geq h_0 \end{cases} \tag{12}$$

where S_{ij} is the attribute value of the ith element, jth attribute. r_0 is the proportion of the value Null in all data, and h_0 is the corresponding threshold with a high percentage, while l_0 is the corresponding threshold with a low percentage.

3.2 Data Clustering

After data preprocessing, we can design an appropriate fuzzy clustering algorithm for data clustering. Data objects are divided into several categories, and then they can be carried out for various types of mining association rules. By means of this processing, the accuracy of the rules can be improved and the time complexity of the algorithm can also be reduced. The steps of data clustering for the data of network security assessment are as follows:

1) Establish the fuzzy similarity relation R between the elements on the domain U. The order of the R matrix is the | U | and m is the number of attributes. Making use of Euclidean distance formula, calculating 0 elements of the R matrix, r_{ij}.

$$r_{ij} = \begin{cases} 1, i = j \\ \sqrt{\dfrac{1}{m}\sum_{i=1}^{m}(S_{ik} - S_{jk})^2}, i \neq j \end{cases} \tag{13}$$

2) From the R, getting the map G = (V, E) and generating the largest tree - T = (V, TE). We make use of the Prim algorithm to derive the maximum spanning tree T on the Figure G.

3) According to the need of practical problems, an appropriate set, $\lambda \in [0, 1]$, T (e) is the weight value of the edge e, such as T (e) <λ, then, the edge e is deleted. The received branch connectivity is the classification based on λ.

After the cluster analysis, the analysis of the singular category can screen the categories that have fewer elements. Deleting the categories that the ratio of the amount is less than the set threshold, these deleted categories can be analyzed as the singular category.

3.3 Attribute Reduction and Strong Rule Extraction

After the clustering of the initialization data, we can use the attribute reduction and strong rule extraction on the classification data objects can. In the information systems or decision-making system S = (U, A), some attributes of the A is superfluous. We can get the minimum conditions on the attribute set that can't contain the surplus property and can ensure the correct classification. If ind(B) = ind(B-(a)), on the B ⊆ A ,a ∈ A, said that B is reductive. A decision-making table may have several possible reductions at the same time. The properties of the nuclear are an important attribute on the impact of the classification. We can be use distinguish matrix and discernibility function to achieve the attribute reduction.

Reducing the output rule sets and deleting the rules that the frequency probability is less than a given threshold µ0, we make the separate analysis on that rules as singular rules, reducing the complexity of the algorithm. The non-singular rules exported after attribute reduction need an objective degree constraint, outputting strong rules meeting a given support and confidence. The realization of the process:

1) Dividing the system attributes to establish the corresponding conditions attributes and decision attributes.
2) Making use of the distinction matrix and discernibility function to get the reduction set of the attributes [9]. Discernibility matrix as follows:

$$M(B) = \{m(i,j)|_{n \times n} \ 1 \leq i, j \leq n\},$$
$$m(i,j) = \{a \in A \mid a(i) \neq a(j) \ \& \ d(i) \neq d(j)\}, n = |U| \quad (14)$$

Discernibility functions as follows:

$$\Delta = \prod_{(i,j) \in U \times U} \sum m(i,j) \quad (15)$$

where Σ stands for ∨ and Π implies ∧ .

3) Deriving the rule sets using the reduction attributes, we can get the attribute reduction sets and nuclear from the minimal disjunctive paradigm of the discernibility function.
4) Deleting the singular rules that the frequency is less than a given threshold µ0.
5) Reserving the rules that meet the minimum support degree (μ_s) and minimum confidence degree (μ_c) as strong rules.

4 Case Study

In order to test the ability of network security assessment, the experimental environment is set up based on the LAN as: high-performance Gigabit Switch, Firewall, IDS systems and modular multi-service routers. All the devices selected have the function to support multi-source heterogeneous data sources, such as NetFlow, SNMP and log files. In addition, there are no less than 50 PC (Windows XP/1G/160G), Gigabit LAN. At the same time, a variety of network software from the external net are used to attack the local area network, in order to obtain more comprehensive data of network attacks.

4.1 Data Preprocessing of Assessment Samples

By means of attacking the LAN from the external net using network software, we selected the typical network attacks asïthe safety factors. The provisions of the safety factor set F ={D1, D2, D3, D4, D5, D6}, Di stands for the ith attack of the LAN.ïF denotes a group of attacks on LAN,ïdetailed as: F = {V1(Internet worms), V2(Patch case), V3(Portïscan), V4(Trojan), V5(Security vulnerabilities), V6(Bufferïoverflow area)}. The value of each item in F is 0 or 1, where 0ïmeans the attack is not exist while 1 denotes the attack is exist.ïIn accordance with the steps in section A of III, first of all,ïdata preprocessing is carried out. After that, we do theïclustering for the samples that we chosen. 50 samples areïselected into this experiment in a moment and these samplesïare divided into three groups after clustering and one of themïis shown in Table 1. We use H, N, L to justify the degree ofïthe risk in the LAN, which are respectively expressed high risk,ïnormal risk and low risk in LAN.

Table 1. The Decision Table of Network Security Assessment

U	D_1	D_2	D_3	D_4	D_5	D_6	Risk
1	0	0	0	1	1	0	H
2	1	0	0	0	1	1	H
3	0	1	0	0	1	1	N
4	1	0	1	1	1	0	H
5	1	1	1	1	0	0	H
6	1	0	1	0	0	0	H
7	0	0	0	1	0	1	N
8	0	1	1	0	1	0	L
9	1	0	1	0	0	1	H
10	0	1	0	0	0	1	H

4.2 Reduction of Decision Table and Rules Extracting

According to the process given in section C of III, we can do the reduction of decision table generated in the step above. We can also extract the rules of network security assessment from the decision table after attribute reduction, as shown in Table 2. It is seen clearly that the rules of network security assessment in this table and we can pay attention to the degree of risk which is H.

Table 2. The Decision Table after Attribute Reduction

U	Attack attributes				Risk
	D_2	D_3	D_5	D_6	
1	0	0	1	0	H
2	0	0	1	1	H
3	1	0	1	1	N
4	0	1	1	0	H
5	1	1	0	0	L
6	0	1	0	0	H

5 Conclusions

In this paper, a model of network security assessment based on fuzzy sets and rough sets was presented. Firstly, data preprocessing and data clustering were executed, and then the assessment rules were extracted from such data. This method can not only reduce the data size and also reduce the interference of the noise. The classification knowledge of the samples can be obtained and the data objects as rules can also be extracted after clustering. In the process of rules extraction, it is good for improving the efficiency and accuracy of the rules with analysis of singular rule, objective measurement and related analysis of the rules constraints. The experiment results show that the proposed method of network security assessment in this paper was feasible and effective. Furthermore, more studies should be carried on the related technologies, such as: whether to get more samples can obtain a better decision-making rules or not; combining with other soft computing methods (such as fuzzy sets, artificial neural networks, genetic algorithm, etc.), it may achieve a higher IQ Hybrid Intelligent System, and so on.

Acknowledgment. The Project Supported by Science Foundation of Hebei University. ID: 2008Q50.

References

1. Bass, T.: Intrusion Detection System and Multisensor Data Fusion. Communications of the ACM 43(4), 99–105 (2000)
2. Batsell, S.B., Rao, N.S., Shankar, M.: Distributed Intrusion Detection and Attack Containment for Organizational Cyber Security,
   http://www.io-c.ornl.gov/projects/documents/containment.pdf
3. Shifflet, J.: A Technique Independent Fusion Model for Network Intrusion Detection. In: Proceedings of the Midstates Conference on Undergraduate Research in Computer Science and Mathematics, pp. 13–19 (2008)
4. Hu, L.J.: Analysis of Fuzzy Stochastic Systems and Control of Stochastic T-S Fuzzy Systems. Donghua University (2008)
5. Qiao, M.: Research on Knowledge Discovery and Reasoning Methods Based on Rough Set and Database Technology. Tianjin University (2006)

6. Peng, J.S.: Research on the Method of Association rule mining based on Rough Sets. Central South University (2004)
7. Slowinski, R., Vanderpooten, D.: A generalized definition of rough approximation based on similarity. IEEE Trans. on Data and Knowledge Engineering 12(2), 331–336 (2000)
8. Liu, Q.: The interpretation of Rough Sets based on kinds of Inaccurate Theory. Computer Science 26(12), 5–8 (1999)
9. Liu, W.J., Gu, Y.D., Wang, J.Y.: Rough-fuzzy Ideals in Semigroups. Fuzzy Systems and Mathematics 18(3), 21–28 (2004)
10. Chakrabarty, K., Biswas, R., Nanda, S.: Fuzziness in rough sets. Fuzzy Sets and Systems, 247–251 (2000)

A Novel Image Fusion Method Based on Particle Swarm Optimization

Haining An[1], Yaolong Qi[1], and Ziyu Cheng[2]

[1] Computer Center, Hebei University, Baoding 071002, China
[2] Archives, Hebei College of Finance, Baoding 071051, China
yifengniu213@163.com

Abstract. In most of multi object image fusion methods, the parameter configuration of fusion model is usually based on experience. In this paper, a new multi objective optimization method of multi objective image fusion based on APSO (Adaptive Particle Optimization) is presented, which can simplify the model of multi objective image fusion and overcome the limitations of traditional methods. First the proper evaluation indices of multi objective image fusion are given, then the uniform model of multi objective image fusion in DWT (Discrete Wavelet Transform) domain is constructed, in which the model parameters are selected as, the decision variables, and finally APSO is designed to optimize the decision variables. APSO not only uses a mutation operator to avoid earlier convergence, but also uses a crowding operator to improve the distribution of no dominated solutions along the Pareto front, and uses a new adaptive inertia weight to raise the optimization capacities. Experiment results demonstrate that APSO has a higher convergence speed and better search capacities, and that the method of multi objective image fusion based on IMOP- SO achieves the Pareto optimal image fusion.

Keywords: multi-objective image fusion; adaptive particle swarm optimization.

1 Introduction

Image fusion is a valuable process in combining images with various spatial, spectral and temporal resolutions to form new images with more information than that can be derived from each of the source images [1]. Different methods of image fusion have the same objective, that is, to acquire a better fusion effect. Different methods have the given parameters, and different parameters could result in different fusion effects. In general, we establish the parameters based on experience or the parameters adaptively change with the image contents, so it is fairly difficult to gain the optimal fusion effect. If one image is regarded as one information dimension or a feature subspace, image fusion can be regarded as an optimization problem in several information dimensions or the feature space. A better result, even the optimal result, can be acquired through searching the optimal parameters and discarding the given values in the process of image fusion. Therefore, a proper search strategy is very important for the optimization problem. In [2], the optimization of image fusion was primarily explored, and the objective is the mean square error (MSE), but only one objective is too simple to meet the real demands.

In fact, there are various kinds of evaluation indices, and different indices may be compatible or incompatible with one another, so a good evaluation index system of image fusion must balance the advantages of different indices. The traditional solution is to change the multi-objective problem into a single objective problem using weighted linear method. However, the relation of the indices is often nonlinear, and this method needs to know the weights of different indices in advance. So it is highly necessary to introduce multi-objective optimization methods based on Pareto theory to search the optimal parameters in order to realize the optimal image fusion, by which the solutions are more adaptive and competitive because they are not limited by the given weights. At present, multi-objective optimization algorithms include Pareto Archive Evolutionary Strategy (PASE) [3], Strength Pareto Evolutionary Algorithm (SPEA2) [4], No dominated Sorting Genetic Algorithm II (NSGA-II) [5], Non-dominated Sorting Particle Swarm Optimization (NSPSO) [6], Multiple Objective Particle Swarm Optimization (MOPSO) [7], [8], etc. Lots of experiments with the two objective optimization problems show that MOPSO has a better optimization capacity and a higher convergence speed [7]. However, MOPSO uses an adaptive grid [3] to record the searched particles, which need too much calculation time, and may cause failure in allocating memory, even in integer format. Using MOPSO and NSGA-II for reference, we present an adaptive multi-objective particle swarm optimization (AMOPSO) to optimize the fusion model. In contrast to other multi-objective evolutionary algorithms, AMOPSO has a higher convergence speed and better exploratory capabilities. The approach to image fusion based on AMOPSO is more successful. Different from the conventional approaches, the image fusion model is simplified in the approach, and the emphasis is laid on the multi-objective optimization. Through optimizing multiple objectives, the optimal image fusion can be achieved.

2 Image Fusion in Dwt Domain

As shown in Fig. 1, the approach to image fusion based on multi-objective optimization in DWT domain is as follows.

Step 1: Input the registered source images A and B. Find the DWT of each A and B to a specified number of decomposition levels, at each level we will have one approximation sub band and 3×J details, where J is the decomposition level. If the value of J is too high, the pixels in sub images will cause the distortion, otherwise the decomposition can't embody the advantage of multiple scales. In general, J is not greater than 5. When J equals 0, the transform result is the original image and the fusion is performed in spatial domain.

A Novel Image Fusion Method Based on Particle Swarm Optimization

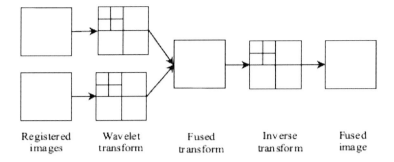

Fig. 1. Illustration of image fusion in DWT domain

Step 2: For the details in DWT domain, salient features in each source image are identified [9], and have an effect on the fused image. The salient feature is defined as a local energy in the neighborhood of a coefficient.

$$S_j(x, y) = \sum\sum W_j(x+m, y+n)^2, j = 1,..., J \quad (1)$$

where $W_j(x, y)$ is the wavelet coefficient at location (x, y), and (m, n) defines a window of coefficients around the current coefficient. The size is typically small, e.g. 3 by 3.

The coefficient with the largest salience is substituted for the fused coefficient while the less salient coefficient is discarded. The selection mode is implemented as

$$W_{FJ}(x, y) = \begin{cases} W_{AJ}(x, y) & S_{Aj}(x, y) \geq S_{Bj}(x, y) \\ W_{Bj}(x, y) & otherwise \end{cases} \quad (2)$$

where $W_{Fj}(x, y)$ are the final fused coefficient in DWT domain, W_{Aj} and W_{Bj} are the current coefficients of A and B at level j.

Step 3: For approximations in DWT domain, use weighted factors to calculate the approximation of the fused image of F. Let C_F, C_A, and C_B be the approximations of F, A, and B respectively, two different fusion rules will be adopted. One rule called "uniform weight method (UWM)" is given by

$$CF(x, y) = w_1 \cdot C_A(x, y) + w_2 \cdot C_B(x, y) \quad (3)$$

where the weighted factors of w_1 and w_2 are the values in the range [0, 1]. w_1 and w_2 are the decision variables.

The other rule called "adaptive weight method (AWM)" is given by

$$CF(x, y) = w_1(x, y) \cdot C_A(x, y) + w_2(x, y) \cdot C_B(x, y) \quad (4)$$

where $w_1(x, y)$ and $w_2(x, y)$ are the decision variables.

Using multi-objective optimization methods, we can find the optimal decision variables of image fusion in DWT domain, and realize the optimal image fusion.

Step 4: The new sets of coefficients are used to find the inverse transform to get the fused image F.

3 Apso Algorithm

J. Kennedy and R. C. Eberhart brought forward particle swarm optimization (PSO) inspired by the choreography of a bird flock in 1995 [14]. Unlike conventional evolutionary algorithms, PSO possesses the following characteristics: 1) Each individual (or particle) is given a random speed and flows in the decision space; 2) each individual has its own memory; 3) the evolutionary of each individual is composed of the cooperation and competition among these particles. Since the PSO was proposed, it has been of great concern and become a new research field. PSO has shown a high convergence speed in single objective optimization [15], and it is also particularly suitable for multi-objective optimization [7], [8], [16]. In order to improve the performances of the algorithm, we present a proposal, called "adaptive multi objective particle swarm optimization" [17], in which not only an adaptive mutation operator is used to avoid earlier convergence, but also a crowding distance operator is used to improve the distribution of no dominated solutions along the Pareto front and maintain the population diversity [5], and an adaptive concave exponent inertia weight is used to raise the searching capacity. The flow of APSO is as follows.

3.1 Algorithm Initialization

Initialize the population and algorithm parameters.

1) Initialize the position of each particle: pop[i], where i=1,...,NP, NP is the particle number.
2) Initialize the speed of each particle: vel[i]=0
3) Initialize the record of each particle: pbests[i]=pop[i]
4) Evaluate each of the particles in the POP: fun[i, j], where j=1,...,NF, and NF is the objective number.
5) Store the positions that represent no dominated particles in the repository of the REP according to the Pareto optimality.

3.2 Program Execution

Before the maximum number of cycles is reached, do

1) Update the speed of each particle using (5).

$$vel[i] = W \cdot vel[i] + c_1 \cdot rand_1 \cdot (pbests[i] - pop[i]) \\ + c_2 \cdot rand_2 \cdot (rep[h] - pop[i]) \tag{5}$$

where W is the inertia weight [18]; c1 and c2 are the learning factors [19], rand1 and rand2 are random values in the range [0, 1]; pbests[i] is the best position that the particle i has had; h is the index of the maximum crowding distance in the repository that implies the particle locates in the sparse region, as aims to maintain the population diversity; pop[i] is the current position of the particle i.

2) Update the new positions of the particles adding the speed produced from the previous step

$$pop[i] = pop[i] + vel[i] \quad (6)$$

3) Maintain the particles within the search space in case they go beyond their boundaries. When a decision variable goes beyond its boundaries, the decision variable takes the value of its corresponding boundary, and its velocity is multiplied by (-1).
4) Adaptively mutate each of the particles in the POP at a probability of P_m.
5) Evaluate each of the particles in the POP.
6) Update the contents in the REP, and insert all the current no dominated positions into the repository.
7) Update the records, when the current position of the particle is better than the position contained in its memory, the particle's position is updated.

$$pbests[i] = pop[i] \quad (7)$$

8) Increase the loop counter of g.

4 Results

The performance of the proposed image fusion approach was tested and compared with that of different fusion schemes. The image "plane" was selected as the standard image R. Through image processing, we got two source images A and B. As shown in Fig. 2(a) and (b), the left region of A is blurred, while the right region of B is blurred. We used APSO to search the Pareto optimal weights of the fusion model and compared the results with those of MOPSO based method and simple wavelet method (SWM) [11]. The parameters of APSO are as follow: the particle number of NP is 100; the objective number of NF is 6; the inertia weight of W_{max} is 1.2, and W_{min} is 0.2; the learning factor of c_1 is 1, and c_2 is 1; the maximum cycle number of Gmax is 100; the allowed maximum capacity of MEM is 100; the mutation probability of Pm is 0.05. The parameters of MOPSO are the same, while the inertia weight of W is 0.4, the grid number of N_{div} is 20, for a greater number may cause the failure of program execution, e.g. 30. The sum of the weights at each position of two source images is limited to 1. All approaches are run for a maximum of 100 evaluations. Since the solutions to the optimization of image fusion are no dominated by one another, we give preference to the six indices so as to select the Pareto optimal solutions to compare, e.g. one order of preference is SSIM, MI, Entropy, PSNR, Gradient, IS. When the standard image doesn't exist, the MI will become the principal objective. The fused images from the Pareto optimal solutions are shown in Fig. 2(c), (d), (e), and (f). It can be seen that the fused image of AWM at decomposition level 5 is the best. Table I shows

Fig. 2. Input and fused images

Table 1. Evaluation Indices of the Fused Images From Different Schemes

Schemes	Level	Entropy	Gradient	MI	IS	PSNR	SSIM	Time(s)
ImageA	0	6.4272	6.7631	29.7198	1.9998	25.8193	0.955731	-
ImageB	0	6.4195	6.9881	29.7043	1.9996	26.5760	0.963261	-
UWM	0	6.6826	6.3220	28.1494	1.9995	29.1992	0.979466	199.72
AWM	0	6.6929	6.7477	28.2180	1.9995	28.8614	0.977869	1920.14
AWM	1	6.6433	8.2222	28.4209	1.9992	29.7771	0.982455	850.75
AWM	2	6.6591	8.9455	28.1787	1.9995	31.9960	0.990060	397.94
AWM	3	6.6685	9.1581	28.1816	1.9995	34.7226	0.995364	303.17
AWM	4	6.6718	9.1853	28.2376	1.9993	35.6570	0.996478	286.50
UWM	5	6.6675	9.1465	28.1496	1.9994	34.6555	0.995207	587.70
AWM I	5	6.6832	9.1876	28.2640	1.9992	35.6598	0.996492	347.50
AWM II	5	6.6845	9.1857	28.2656	1.9992	35.6342	0.996497	341.73
AWM III	5	6.6837	9.1865	28.2782	1.9992	35.6588	0.996510	285.17
AWM	5	6.6849	9.1868	28.2804	1.9993	35.6666	0.996525	284.11
MOPSO	5	6.6921	9.1883	28.2799	1.9993	35.5331	0.996437	1375.30
SWM	5	6.6839	9.1831	28.2434	1.9994	35.4461	0.996240	1.26

the evaluation indices of the fused images from different schemes, where AWM I is the AWM with a linear inertia weight (see [18]), AWM II is the AWM without the mutation operator, AWM III is the AWM with the crowding distance of NSGA-II (see [5]), MOPSO denotes the AWM based on MOPSO (see [7]), SWM uses the schemes in [11].

From Table I, we can see that when the decomposition level equals zero in DWT domain, which is in spatial domain, the indices of AWM (Fig. 2(d)) is inferior to those of UWM (Fig. 2(c)). The reason is that the decision variables of AWM in spatial domain are too many and AWM can't reach the Pareto optimal front in a limited time, e.g. the number of iteration is 100. The run time of AWM must increase with the number of decision variables, so AWM can only be regarded as an ideal method of image fusion in spatial domain. The advantage of spatial fusion is easy to realize, however the simple splice of pixels smoothes the image and is not convenient for the later processing, such as comprehension. In DWT domain, the indices of AWM at level 5 (Fig. 2(f)) are superior to those of AWM at other levels. The higher the decomposition level is, the better the fused image is, for a higher level decreases the decision variables and improves the adaptability. Moreover, the indices of AWM are superior to those of UWM because the weights of AWM are adaptive in different regions. The indices of SWM are inferior to our results except IS and time. In fact, IS can't be used as an important objective, e.g. IS reaches the maximum in A and B before fusing. The less time of SWM is due to the simplicity of algorithm. The indices of AWM I, II, III and MOPSO are inferior to those of AWM at level 5, which indicates the adaptive concave exponent weight is superior to the linear weight; the mutation operator can avoid earlier convergence; the new crowding distance can increase the running speed and achieve better results, MOPSO needs too much memory because the grid [3] is worse for too many objectives, e.g. 6. Therefore, the approach to image fusion that uses APSO to search the adaptive fusion weights at level 5 in DWT domain is the optimal. This approach can save up the optic features of the images in contrast to the spatial approach, overcome the limitations of given parameters, and obtain the optimal fusion performances.

5 Conclusions

The approach using APSO to optimize the model parameters of image fusion is feasible and relatively effective, and can get the Pareto optimal fusion result. Multi-objective optimization for the fusion parameters can avoid the limitations of too heavy dependence on the experience and simplify the algorithm design for image fusion. Once the valid evaluation indices are established, the method of multi objective optimization can be used to deal with these objectives that could conflict with one another and eliminate the influence of preference effectively. The proposed APSO is an effective algorithm to solve the multi objective problem, which can get to the Pareto front of optimization problems quickly and attain the optimal solutions.

One aspect that we would like to explore in the future is the analysis for the evaluation indices system using PCA to acquire a meaningful measurement. This would improve the performance of image fusion. We are also considering the possibility of improving the crowding distance to become more effective. Finally it is desirable to study the applications of the optimization algorithm in a color image with other fusion methods.

Acknowledgment. The Project Supported by Science Foundation of Hebei University. ID: 2008Q50.

References

1. Pohl, C., Van Genderen, J.L.: Mulitisensor image fusion in remote sensing: concepts, methods and applications. Int. J. Remote Sensing 19(5), 823–854 (1998)
2. Qin, Z., Bao, F.M., Li, A.G., et al.: Digital image fusion. Xi'an Jiaotong University Press (2004)
3. Knowles, J.D., Corne, D.W.: Approximating the nondominated front using the Pareto archived evolution strategy. Evol. Comput. 8(2), 149–172 (2000)
4. Zitzler, E., Laumanns, M., Thiele, L.: SPEA2: improving the strength Pareto evolutionary algorithm. In: EUROGEN (2001)
5. Deb, K., Pratap, A., Agarwal, S., Meyarivan, T.: A fast and elitist multiobjective genetic algorithm: NSGA-II. IEEE Trans. Evol. Comput. 6(2), 182–197 (2002)
6. Li, X.: A non-dominated sorting particle swarm optimizer for multiobjective optimization. In: Cantú-Paz, E., Foster, J.A., Deb, K., Davis, L., Roy, R., O'Reilly, U.-M., Beyer, H.-G., Kendall, G., Wilson, S.W., Harman, M., Wegener, J., Dasgupta, D., Potter, M.A., Schultz, A., Dowsland, K.A., Jonoska, N., Miller, J., Standish, R.K. (eds.) GECCO 2003. LNCS, vol. 2723, pp. 37–48. Springer, Heidelberg (2003)
7. Coello, C.A., Pulido, G.T., Lechuga, M.S.: Handling multiple objectives with particle swarm optimization. IEEE Trans. Evol. Comput. 8(3), 256–279 (2004)
8. Sierra, M.R., Coello, C.A.: Improving PSO-based multi-objective optimization using crowding, mutation and e-dominance. In: Coello Coello, C.A., Hernández Aguirre, A., Zitzler, E. (eds.) EMO 2005. LNCS, vol. 3410, pp. 505–519. Springer, Heidelberg (2005)
9. Huang, X.S., Chen, Z.: A wavelet-based image fusion algorithm. In: Proc. IEEE TENCON 2002, October 2002, pp. 602–605 (2002)
10. Qu, G.h., Zhang, D.l., Yan, P.f.: Information measure for performance of image fusion. IEE Electron. Lett. 38(7), 313–315 (2002)
11. Ramesh, C., Ranjith, T.: Fusion performance measures and a lifting wavelet transform based algorithm for image fusion. In: Proc. ISIF 2002, July 2002, vol. 1, pp. 317–320 (2002)
12. Wang, Z., Bovik, A.C., Sheikh, H.R., Simoncelli, E.P.: Image quality assessment: from error visibility to structural similarity. IEEE Trans. Image Processing 13(4), 600–612 (2004)
13. Wang, Z.J., Ziou, D., Armenakis, C., et al.: Comparative analysis of image fusion methods. IEEE Trans. GeoRS 43(6), 1391–1402 (2005)

14. Kennedy, J., Eberhart, R.C.: Particle swarm optimization. In: Proc. IEEE Int. Conf. Neural Networks, December 1995, vol. 4, pp. 1942–1948 (1995)
15. Kennedy, J., Eberhart, R.C.: Swarm intelligence. Morgan Kaufmann, San Mateo (2001)
16. Ray, T., Liew, K.M.: A swarm metaphor for multiobjective design optimization. Eng. Opt. 34(2), 141–153 (2002)
17. Niu, Y.F., Shen, L.C.: Multi-Objective Deformable Template for Forward Looking Object Tracking. Comp. Issue, Dynam Cont. Dis. Imp. Sys., Ser. B (unpublished)
18. Shi, Y., Eberhart, R.C.: A modified particle swarm optimizer. In: Proc. IEEE Int. Conf. Evol. Comput., May 1998, pp. 69–73 (1998)
19. Eberhart, R.C., Shi, Y.: Particle swarm optimization: development, applications and resources. In: Proc. IEEE Conf. Evol. Comput., pp. 81–86 (2001)

Relational Database Semantic Access Based on Ontology

Shufeng Zhou

College of Mathematics Science, Liaocheng University
Liaocheng 252059, P.R. China
zhousfpk@163.com

Abstract. Ontology plays more and more roles in creating semantic data in the Semantic Web. Now, because large amounts of data in legacy system are stored and managed in relational database, so there are many existing approaches used to transform relational databases into to ontologies, including ontologies definition and instances. But when vast amount of instance data are stored in OWL files, it is very difficult to manage and query. So, an ontology-based approach for accessing relational database is proposed in this paper, which acquires OWL ontology based on relational database schema, and the instances data still store in database. The ontology is viewed as mediated schema providing shared vocabularies for Semantic Web to be queried using SPARQL.

Keywords: Ontology; Relational Database; SPARQL; RDF; SQL; OWL.

1 Introduction

The Semantic Web [3] thought up by Tim Berners-Lee is a major research initiative of the World Wide Web Consortium (W3C) to create a metadata-rich Web of resources that can describe themselves not only by how they should by displayed or syntactically, but also by the meaning of the metadata.

Ontology [22], a formal explicit specification of a shared conceptualization [23], provides a way to make the data readable and understandable by machines, whereby it improves system interoperation and knowledge share, and has been applied in many fields, such as Semantic Web and information retrieval, etc. But the construction of ontology in special domain is very time-consuming and error-prone. So more and more experts and researchers begin to give attention to ontology research.

The Web Ontology Language (OWL) [19] is a knowledge representation language to create ontologies, playing an important role in an increasing number and range of applications. The OWL language is intended to provide a language that can be used to describe the classes and relations between them that are inherent in Web documents and applications. The OWL builds on Resource Description Framework (RDF) [15](a language of representing Web resources in a triple format of subject-predicate-object) and RDF Schema (RDFS) [6], and facilitates much machine interpretability by adding more common vocabularies for describing properties, classes, cardinality and so on.

In most current enterprise environments, large amounts of data in legacy systems are stored and managed in relational database (RDB); because relational database management system technology is very well fit to manage large amounts of data. RDB may be used as a kind of key resource for ontology construction at present. So how to transfer the data in RDB to ontology is the key issue of ontology research. Specially, the RDB2RDF WG [11] is trying to standardize a language for mapping RDB data, schemas into RDF, OWL.

Now, there are existing tools used to transform relational database into ontologies, including ontologies definitions and instances. But when vast amount of instance data are stored in OWL files, it is very difficult to manage and query data. So, an ontology-based approach for accessing of relational database is proposed in this paper, which firstly acquires OWL ontology based on relational database schema. But the instance data still store in relational database; and the ontology is viewed as mediated schema [7] for the explicit description of database, providing shared vocabularies for the specification of the semantics. So the Semantic Web application can query the mediated ontology using SPARQL [20] which is a query language for RDF. Then the query is translated into equivalent SQL executed against original relational database and the results set are then expressed in RDF format.

The remainder of the paper is organized as follows: Section 2 introduces some basic definitions about concepts and Section 3 gives rules for ontology construction. In Section 4 and 5, the paper depicts the design and implementation and introduces query rewriting respectively. Section 6 analyzes related works and draws a conclusion in the end.

2 Basic Definitions

Definition 1. A relational schema is defined as a tuple $R(U, D, dom, I)$. Here R is the name of relation (or table) in database; U is a set of attributes of relation in database; D is the domain of discourse of attributes; dom represents function that maps U to D; and I represents a finite set of integrity constraint.

A relation is defined as a set of tuples with the same attributes. A tuple usually describes an object and information about that object. A relation is also called a table, which is organized into rows and columns. A domain describes the set of possible values for a given attribute. Mathematically, attaching a domain to an attribute means that all values for this attribute must be an element of the specified set.

In relational schema, primary key is a candidate key to uniquely identify each row in a table. Primary key usually consists of a single column or set of columns. Foreign key is a referential constraint between two tables, which identifies a column or a set of columns in one table that refers to a column or set of columns in another table. Thus, a row in the referencing table cannot contain values that don't exist in the referenced table.

Constraints further restrict the data stored in relations. Furthermore, constraints can apply to single attribute, to a tuple or to an entire relation.

Given a relational database schema used below is shown in Tab.1. There are three columns: Relations, Primary Key and Foreign Key.

Table 1. Excerpt of Relational Database Schema

Relations	Primary Key	Foreign Key
Customer(cid string, name string, city string)	cid	city referring to City.cid
Product(pid string, name string, destription string)	pid	No
Order(cid string, pid string, number int)	cid, pid	cid referring to Customer.cid; pid referring to Product.pid
Supplier(sid string, name string, description string)	sid	No
City(cid string, name string)	cid	No

In this database, the customer relation consists of information about customer. The cid column denotes the identifier of customer; the name denotes the name of customer; and the city denotes the location of customer.

Similarly, the product relation contains information about product; the order relation contains customer's ordering information, etc.

Next, ontology is a formal, explicit specification of a shared conceptualization of a domain of interest [23]. Ontology is a description of concepts and relationships between them.

Definition 2. An ontology is a 4-tuple $O(C, P^C, R, H)$. Here O is name of ontology in special domain; C is a finite set of concepts (usually be called classes); P^C is a finite set of properties of concepts; R represents relationships between concepts; H is hierarchy in classes or properties.

This paper adopts the OWL as the ontology description language. The basic elements of an OWL ontology are classes properties and relationships between these instances. The most basic concepts in a domain should correspond to classes that defines individuals belong to together; because they share some properties. For instance, Cat and Dog are both members of the class Animal.

Properties in OWL ontology provide us a way of asserting general facts about the members of classes and specific facts about individuals. There are two different types of properties in OWL: datatype properties and object properties. Datatype properties are relations between instances of classes and RDF literals and XML Schema [12] datatypes; object properties are relations between instances of two classes in a particular direction. Furthermore, property characteristics and cardinality describe the mechanisms used to further specify properties.

3 Rules for Ontology Construction

In this paper, the OWL ontology is constructed from relational database according to the rules below. The rules are organized in three groups.

3.1 Rules for Classes

The rules for classes are shown below.

Rule 1. An ontology class can be created unless all its two columns are foreign keys referring to two other relations.

Rule 2. An ontology class can be created based on a relation, if there is a single-attribute key in this relation. And the name of relation plus the name of primary key is used as the name of the class.

According to Rule 2, four ontology classes *CustomerCid*, *ProductPid*, *SupplierSid* and *CityCid* can be created based on the relation *Customer*, *Product*, *Supplier* and *City* in database as following:

```
<owl:Class rdf:ID= "CustomerCid"/>
<owl:Class rdf:ID= "ProductPid"/>
<owl:Class rdf:ID= "SupplierSid"/>
<owl:Class rdf:ID= "CityCid"/>
```

3.2 Rules for Properties

In OWL, there are two kinds of properties: datatype properties and object properties.

Rule 3. For a relation in database, supposed that the relation has a single-attribute primary key, then object property can be created. The domain is the class corresponding to the relation, the range the class created applying Rule 2 respectively.

According to Rule 3, for example, object properties *CustomerPk* and *ProductPk* can be created based on the relations *Customer* and *Product* in database.

```
<owl:ObjectProperty rdf:ID= "CustomerPk">
   <rdf:domain rdf:resource= "#Customer"/>
   <rdf:range rdf:resource= "#CustomerCid"/>
</owl:ObjectProperty>
<owl:ObjectProperty rdf:ID= "ProductPk">
   <rdf:domain rdf:resource= "#Product"/>
   <rdf:range rdf:resource= "#ProductPid"/>
</owl:ObjectProperty>
```

Rule 4. For relations in database, supposed that one relation has a foreign key referring to other relation, then two object properties can be created.

According to Rule 4, for example, two object properties should be created between relations customer and city.

```
<owl:ObjectProperty rdf:ID= "locatedIn">
   <rdf:domain rdf:resource= "#CustomerCid"/>
   <rdf:range rdf:resource= "#CityCid"/>
</owl:ObjectProperty>
<owl:ObjectProperty rdf:ID= "hasCustomer">
   <rdf:domain rdf:resource= "#CityCid"/>
   <rdf:range rdf:resource= "#CustomerCid"/>
</owl:ObjectProperty>
```

Rule 5. For a relation in database, supposed that each column belongs neither to primary key nor foreign key, then a datatype property can be created. The domain is the class corresponding to the relation, range the XML datatype corresponding to the type of column in database respectively.

According to Rule 5, for example, the name in customer, the name and description in product should be datatype properties of corresponding ontology classes.

```
<owl:DatatypeProperty rdf:ID= "CustomerName">
   <rdf:domain rdf:resource= "#CustomerCid"/>
   <rdf:range rdf:resource= "&xsd;string"/>
</owl:DatatypeProperty>
<owl:DatatypeProperty rdf:ID= "ProductName">
   <rdf:domain rdf:resource= "#ProductPid"/>
   <rdf:range rdf:resource= "&xsd;string"/>
</owl:DatatypeProperty>
<owl:DatatypeProperty rdf:ID= "ProductDescription">
   <rdf:domain rdf:resource= "#ProductPid"/>
   <rdf:range rdf:resource= "&xsd;string"/>
</owl:DatatypeProperty>
```

3.3 Rules for Constraints

The rules for property constraints are shown below.

Rule 6. For a relation, a minimum cardinality of property is set 1, if column is declared as NOT NULL.

Rule 7. For a relation, a maximum cardinality of property is set 1, if column is declared as UNIQUE.

Rule 8. For a relation, cardinality of property is set 1, if column is primary key or foreign key.

4 Design and Implementation

The approach proposed in this paper firstly constructs OWL ontology as mediated schema to provide shared vocabularies for Semantic Web applications. Then

Semantic Web applications could query the ontology using SPARQL and acquires RDF instances. The approach is implemented based on Jena 2.6.0 [16] in JSDK 1.6 [25] development platform. Jean is an open source and grown out of work with the Hewlett Packard Labs Semantic Web Programme. Jena is a Java framework for building Semantic Web applications. We use MySQL [24] database as the original database.

First, based on the rules introduced above, this system analyses the database structure and extracts metadata from relational database using reverse engineering technology [1], [8], then applies mapping rules to create mediated ontology. At the same time, the mapping information between relational database and mediated ontology is saved in an XML document.

Second, this system gets the SPARQL-like syntax query using terms of mediated ontology and then analyzes the query and rewrites them into SQL query if the query is valid. After that this system executes the query against backend relational database to acquire result sets.

Final, formatting algorithm transforms the result sets into RDF instances according to correspondences between mediated ontology and relational database.

5 Query Rewriting

In [13], [14] the problem of query rewriting considers how to reformulate a query expressed in SPARQL over mediated schema into an equivalent SQL query targeting the underlying relational database. The mediated schema provides uniform query interface for user access or Semantic Web applications. In addition, the SPARQL language is based on matching graph pattern. Graph pattern contains a set of triples (subject, predicate and object). So based on the mediated schema and mappings defined between relational database and ontology, the query rewriting can be divided into three steps below. (For the sake of space, we only give the main steps.)

Grouping Triples: This step begins with looking at the body of the query and then groups the triples by subject name in triples.

Replacing Predicate: This step replaces all predicates in triples with corresponding relation columns in database because the every group describes one entity. So we can acquire instance data from one or more relation.

Constructing Query: For each separate group, a sub-query clause is created, which consists of three parts: select, from and where clause in query.

After the three steps, an equivalent SQL is created. The SQL is executed against RDB and acquire result sets. Then result sets are expressed in RDF form according mediated ontology generated using rules above.

6 Related Work

In the last few years, many approaches have been proposed to construct RDF or OWL ontology from relational database. Especially, literature [18] lists approaches for addressing this problem of ontology construction from relational database and the problem of query rewriting.

In [9], Cullot describes DB2OWL using the tables to concepts and columns to predicates approach; the process depends on particular database table cases. And the mappings are stored in a R2O [2] document. In [21], Martin presents approach to converts selected data from relational database to a RDF/OWL ontology document based on a defined template. In [17], Li proposes an automatic ontology learning approach to acquire OWL ontology from relational database automatically be using a group of rules. In [10], this paper presents a representation format Relational.OWL for schema and data information based on the OWL. So we are able to represent and to transfer every schema and data component of a database without having to define a data and schema exchange format. D2R MAP [4], D2RQ [5] provides conversion approaches from relational databases to RDF.

There are approaches investigating the construction of ontology from relational database and the query transformation. However, the approach proposed in this paper is different from those existing solutions. Our approach acquires ontology by using rules directly and generates mapping information saved in XML document, which can be used in query transformation. Moreover, the instance data still store in relational database, and the ontology is viewed as mediated schema providing shared vocabularies for Semantic Web.

7 Conclusion

Ontology provides a way to make the data readable and understandable by machines. Study on ontology construction is becoming increasingly widespread in computer science community. In this paper, we present an approach of construction of ontology from relational database and view it as the mediated schema providing shared vocabularies. At the same time, the ontology could be used as local ontology and integrated into a global ontology in future.

Acknowledgments. This work was supported by: Natural Science Foundation of Liaocheng University.

References

1. Astrova, I.: Reverse Engineering of Relational Databases to Ontologies. In: Bussler, C.J., Davies, J., Fensel, D., Studer, R. (eds.) ESWS 2004. LNCS, vol. 3053, pp. 327–341. Springer, Heidelberg (2004)
2. Barrasa, J., Corcho, O., Gómez-Pérez, A.: R2O, an Extensible and Semantically based Database-to-Ontology Mapping Language. In: Bussler, C.J., Tannen, V., Fundulaki, I. (eds.) SWDB 2004. LNCS, vol. 3372, pp. 1–17. Springer, Heidelberg (2004)
3. Berners-Lee, T., Hendler, J., Lassila, O.: The semantic web. J. Scientific american 284(5), 28–37 (2001)

4. Bizer, C.: D2R MAP-A database to RDF mapping language. In: The 12th International World Wide Web Conference, WWW 2003 (2003)
5. Bizer, C., Seaborne, A.: D2RQ: Treating Non-RDF Databases as Virtual RDF Graphs. In: McIlraith, S.A., Plexousakis, D., van Harmelen, F. (eds.) ISWC 2004. LNCS, vol. 3298. Springer, Heidelberg (2004)
6. Brickley, D., Guha, R.V.: RDF Vocabulary Description Language 1.0: RDF Schema. W3C Recommendation (2004), http://www.w3.org/TR/2004/REC-rdf-schema-20040210/ (Retrieved February 10, 2004)
7. Calì, A., Calvanese, D., De Giacomo, G., et al.: Accessing Data Integration Systems through Conceptual Schemas. In: Kunii, H.S., Jajodia, S., Sølvberg, A. (eds.) ER 2001. LNCS, vol. 2224, pp. 270–284. Springer, Heidelberg (2001)
8. Chiang, R.H.L., Barron, T.M., Storey, V.C.: A Framework for the Design and Evaluation of Reverse Engineering Methods for Relational Databases. J. Data & Knowledge Engineering 21(1), 57–77 (1996)
9. Cullot, N., Ghawi, R., Yétongnon, K.: DB2OWL: A Tool for Automatic Database-to-Ontology Mapping. In: Proceedings of 15th Italian Symposium on Advanced Database System, Torre Canne, Italy, pp. 491–494 (2007)
10. De Laborda, C.P., Conrad, S.: Relational.OWL: A Data and Schema Representation Format based on OWL. In: Proceedings of the 2nd Asia-Pacific conference on Conceptual modelling, APCCM 2005, Newcastle, New South Wales, Australia, vol. 43, pp. 89–96. Australian Computer Society, Inc. (2005)
11. Ezzat, A., Hausenblas, M., Halpin, H.: W3C RDB2RDF Working Group. World Wide Web Consortium (W3C), http://www.w3.org/2001/sw/rdb2rdf/
12. Fallside, D.C., Walmsley, P.: XML Schema: Primer Second Edition. World Wide Web Consortium (W3C)
13. Halevy, A.Y.: Answering Queries using Views: A survey. J. Journal of Very Large Database 10(4), 270–294 (2001)
14. Harris, S., Shadbolt, N.: SPARQL Query Processing with Conventional Relational Database Systems. In: Dean, M., Guo, Y., Jun, W., Kaschek, R., Krishnaswamy, S., Pan, Z., Sheng, Q.Z. (eds.) WISE 2005 Workshops. LNCS, vol. 3807, pp. 235–244. Springer, Heidelberg (2005)
15. Hayes, P., Mcbride, B.: RDF Semantics. World Wide Web Consortium (W3C), http://www.w3.org/TR/2004/REC-rdf-mt-20040210/
16. Hewlett-Packard Development Company, Jena Semantic Web Framework. Hewlett-Packard Development Company, http://jena.sourceforge.net/
17. Li, M., Du, X.Y., Wang, S.: Learning ontology from relational database. In: Proceedings of 2005 International Conference on Machine Learning and Cybernetics (ICML 2005), pp. 3410–3415 (2005)
18. Noy, N.F., Mcguinness, D.L.: Ontology development 101: A guide to creating your first ontology. SMI tecnical report SMI-2001-0880, Citeseer (2001)
19. Patel-Schneider, P.F., Hayes, P., Horrocks, I.: OWL Web Ontology Language Semantics and Abstract Syntax World Wide Web Consortium (W3C), http://www.w3.org/TR/owl-absyn/
20. Prud'hommeaux, E., Seaborne, A.: SPARQL Query Language for RDF. W3C Recommendation (2008), http://www.w3.org/TR/rdf-sparql-query/
21. Šeleng, M., Laclavík, M., Balogh, Z., et al.: RDB2Onto: Approach for creating semantic metadata from relational database data. In: The ninth international conference on Informatics (2007), http://www.w3.org/TR/rdf-sparql-query/

22. Smith, M.K., Welty, C., Mcguinness, D.L.: OWL Web Ontology Language Guide. World Wide Web Consortium (W3C), http://www.w3.org/TR/owl-guide/
23. Studer, R., Benjamins, V.R., Fensel, D.: Knowledge engineering: Principles and methods. J. Data & Knowledge Engineering 25(1-2), 161–197 (1998)
24. Sun Microsystems, MySQL Guide to MySQL Documentation. SUN, http://dev.mysql.com/doc/mysqldoc-guide/en/index.html
25. Sun Microsystems, The Source for Java Developers. SUN, http://java,sun.com/

Dynamic Analysis of Transmission Shaft of Width Equipment for Square Billet

Xianzhang Feng and Hui Zhao

School of Mechatronics Engineering,
Zhengzhou Institute of Aeronautical Industry Management,
Zhengzhou, Henan, China, 450015
Phdfxz@163.com

Abstract. With the rapid development of science and technology, the quality of machinery products have been greatly improved, the market competition becomes increasingly fierce, the product quality of life become an important indicator. The modernization of production facilities not only improve the level of production, but also greatly reduce the labor intensity. Machinery products in the practical application will often appear a variety of issues, which is not a simple theoretical analysis. Practical experience is a very effective way to be solved the problem, but in the rapid development of technology today, it is relatively backward, computers and the software can quickly identify emerging problems. In the field of engineering, the finite element analysis method can be used in products which can increase strength and avoid defects. It can be a simple model that gives the reason of the problem, the research results can show that the method solve the reality engineering problem, and it can reduce the product time of testing and improve product life.

Keywords: Material; Matrix analysis; Universal Joint; Transmission shaft; Displacement cloud.

1 Introduction

In the mechanical field, the axis is often used as a part, to find the reasons for failure appear shaft [1], the research can improve the life of the shaft and reduce the capital investment. This is indeed a very effective way to gain experience by practicing. However, it can get a more effective method to study the shaft parts failure, by use of computer technology and related software. General axis is unable to complete such a function, under the conditions that the shaft part not only needs to pass the shaft torque, but also there is a certain angle between the axes. The universal couplings can achieve this functionality, in the transmission process, due to the complexity of the environment will lead to damage for the universal couplings. Therefore, it can quickly find out the reasons of destruction of universal coupling by use of finite element theory, the research results can effectively improve the efficiency of the production line, and supply reference value for the actual production. [2-10]

For the shaft part, the primary role is supporting transmission parts, transmission movement and power. In their work process, there is composite action with variety of stress in the shaft part. Therefore, from the perspective of the analysis of material selection, the material has high mechanical properties [11], localized by the friction of the parts require a certain degree of hardness in order to enhance its anti-wear ability. [12-18]

2 Theoretical Foundation

The main structure of single universal coupling is as shown in Figure 1.

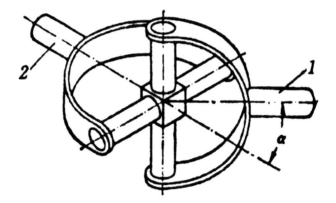

Fig. 1. The structure of universal coupling

Based on the Figure 1, it can be established according to the following coordinate system:

$OX_1Y_1Z_1$ and $OX_2Y_2Z_2$ are expressed relative to the fixed rack connected to a fixed coordinate system, $ox_1y_1z_1$ and $ox_2y_2z_2$ are moving coordinate system for the driving shaft 1 and driven shaft 2. When the driving shaft turn the angle θ_1, the driven shaft 2 also turned the corresponding angle θ_2, according to the coordinate transformation matrix, in the location coordinate of $OX_1Y_1Z_1$ and $OX_2Y_2Z_2$, The points (x_1,y_1,z_1) and (x_2,y_2,z_2) can follows that:

$$\begin{cases} \begin{vmatrix} X_1 \\ Y_1 \\ Z_1 \end{vmatrix} = \begin{vmatrix} \cos\theta_1 & -\sin\theta_1 & 0 \\ \sin\theta_1 & \cos\theta_1 & 0 \\ 0 & 0 & 1 \end{vmatrix} \begin{vmatrix} x_1 \\ y_1 \\ z_1 \end{vmatrix} \\ \\ \begin{vmatrix} X_2 \\ Y_2 \\ Z_2 \end{vmatrix} = \begin{vmatrix} \cos\theta_2 & -\sin\theta_2 & 0 \\ \sin\theta_2 & \cos\theta_2 & 0 \\ 0 & 0 & 1 \end{vmatrix} \begin{vmatrix} x_2 \\ y_2 \\ z_2 \end{vmatrix} \end{cases} \quad (1)$$

Dynamic Analysis of Transmission Shaft of Width Equipment for Square Billet 549

Taking $OX_1Y_1Z_1$ as uniform coordinate reference system, by the matrix transformation, the point (x_2, y_2, z_2) can map the coordinate system $OX_1Y_1Z_1$.

$$\begin{vmatrix} X_1 \\ Y_1 \\ Z_1 \end{vmatrix} = \begin{vmatrix} 1 & 0 & 0 \\ 0 & \cos\alpha & \sin\alpha \\ 0 & -\sin\alpha & \cos\alpha \end{vmatrix} \begin{vmatrix} X_2 \\ Y_2 \\ Z_2 \end{vmatrix} = \begin{vmatrix} 1 & 0 & 0 \\ 0 & \cos\alpha & \sin\alpha \\ 0 & -\sin\alpha & \cos\alpha \end{vmatrix} \begin{vmatrix} \cos\theta_2 & -\sin\theta_2 & 0 \\ \sin\theta_2 & \cos\theta_2 & 0 \\ 0 & 0 & 1 \end{vmatrix} \begin{vmatrix} x_2 \\ y_2 \\ z_2 \end{vmatrix} \quad (2)$$

As the universal coupling is a special spherical rod bodies, which cross the center axis angle and the angle between the center of the cross are at right angles. Taking the point $(x_1, y_1, z_1)=(0,1,0), (x_2, y_2, z_2)=(1,0,0)$, these two points were substituted into the formula (1) and (2).

According to the direction of the provisions of positive matrix, the α shall be taken as negative in the formula (2), the equation is described as following:

$$\begin{cases} \begin{vmatrix} X_1 \\ Y_1 \\ Z_1 \end{vmatrix} = \begin{vmatrix} -\sin\theta_1 \\ \cos\theta_1 \\ 0 \end{vmatrix} \\ \\ \begin{vmatrix} X_1 \\ Y_1 \\ Z_1 \end{vmatrix} = \begin{vmatrix} \cos\theta_2 \\ \sin\theta_2\cos\alpha \\ \sin\theta_2\sin\alpha \end{vmatrix} \end{cases} \quad (3)$$

According to the structural characteristics of universal coupling, type and style of expression vector should always remain vertical, so it will have the following equation.

$$\sin\theta_1 \cos\theta_2 - \sin\theta_2 \cos\theta_1 \cos\alpha = 0 \quad (4)$$

The formula (4) can also be written as.

$$tg\theta_1 = \cos\alpha \, tg\theta_2 \quad (5)$$

It can derivative with the time for the equation (5), considering $\dfrac{d\theta_1}{dt} = \omega_1$, $\dfrac{d\theta_2}{dt} = \omega_2$, at the same time eliminate θ_2, the results are as follows

$$\frac{\omega_2}{\omega_1} = \frac{\cos\alpha}{1-\sin^2\alpha\cos^2\theta_1} \quad (6)$$

The formula (7) is expression of transmission ratio for a single universal coupling. The universal coupling model of two-way is as shown in Figure 2.

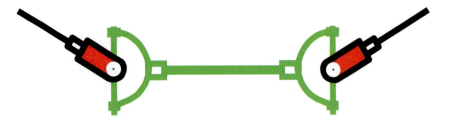

Fig. 2. Intermediate shaft at both ends of forks in the same plane surface

The figure 2 is the fixed coordinate system to rack-phase fixed connection for the pairs of universal couplings $oX_3Y_3Z_3$. $ox_2y_2z_2$ and $ox_3y_3z_3$ are the fixed coordinate system between the second shaft and third 3.When the second shaft turn the angle θ_2, the third shaft also turned the corresponding angle θ_3, using the same method, in the coordinates of $OX_2Y_2Z_2$ and $OX_3Y_3Z_3$, the location of the points (x_2,y_2,z_2) and (x_3,y_3,z_3) can be written:

$$\begin{cases} \begin{vmatrix} X_2 \\ Y_2 \\ Z_2 \end{vmatrix} = \begin{vmatrix} \cos\theta_2 & -\sin\theta_2 & 0 \\ \sin\theta_2 & \cos\theta_2 & 0 \\ 0 & 0 & 1 \end{vmatrix} \begin{vmatrix} x_2 \\ y_2 \\ z_2 \end{vmatrix} \\ \\ \begin{vmatrix} X_3 \\ Y_3 \\ Z_3 \end{vmatrix} = \begin{vmatrix} \cos\theta_3 & -\sin\theta_3 & 0 \\ \sin\theta_3 & \cos\theta_3 & 0 \\ 0 & 0 & 1 \end{vmatrix} \begin{vmatrix} x_3 \\ y_3 \\ z_3 \end{vmatrix} \end{cases} \quad (7)$$

Taking the coordinate system of $OX_2Y_2Z_2$ as a unified reference coordinate system, in this coordinate system of $OX_2Y_2Z_2$, the point of (x_3,y_3,z_3) can be expressed as:

$$\begin{vmatrix} X_2 \\ Y_2 \\ Z_2 \end{vmatrix} =$$

(8)

$$\begin{vmatrix} 1 & 0 & 0 \\ 0 & \cos\alpha & \sin\alpha \\ 0 & -\sin\alpha & \cos\alpha \end{vmatrix} \begin{vmatrix} x_3 \\ y_3 \\ z_3 \end{vmatrix} = \begin{vmatrix} 1 & 0 & 0 \\ 0 & \cos\alpha & \sin\alpha \\ 0 & -\sin\alpha & \cos\alpha \end{vmatrix} \begin{vmatrix} \cos\theta_3 & -\sin\theta_3 & 0 \\ \sin\theta_3 & \cos\theta_3 & 0 \\ 0 & 0 & 1 \end{vmatrix} \begin{vmatrix} x_3 \\ y_3 \\ z_3 \end{vmatrix}$$

Taking the point $(x_2, y_2, z_2) = (1,0,0)$, $(x_3, y_3, z_3) = (0,1,0)$, substitution the points into the formula (7) the results can be expressed as:

$$tg\theta_3 = \cos\alpha tg\theta_2 \tag{9}$$

Because of the relationship : $\theta_3 = \theta_1$

Thence

$$\frac{\omega_3}{\omega_1} = 1 \tag{10}$$

As intermediate shaft at both ends of the vertical fork physiognomy, as shown in Figure 3.

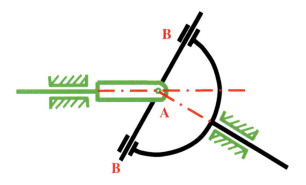

Fig. 3. The surface at both ends of the middle fork

Taking the point $(x_2, y_2, z_2) = (0,1,0), (x_3, y_3, z_3) = (1,0,0)$. Substitution the points into the formula (8), the results can be expressed as:

$$tg\theta_2 = \cos\alpha tg\theta_3 \tag{11}$$

Analysis of Formula 1:

$$tg\theta_1 = \cos^2\alpha tg\theta_3 \tag{12}$$

Their relationship between ω_3 and ω_1 can be expressed as:

$$\frac{\omega_3}{\omega_1} = \frac{\cos^2\alpha}{1 - \cos^2\theta_1 \sin^2\alpha(1 + \cos^2\alpha)} \tag{13}$$

The formula (13) is expression of transmission ratio for the universal coupling of two-way.

3 Example

For the universal coupling, the design parameters is $n = 40 rpm$, $W = 13.718 \, Kg \bullet m^2$ (Inertia), $M = 1.7 \times 10^3 \, Kg$, $T = 34300 N \bullet m$, $T_{max} = 68600 N \bullet m$, and $\alpha_{max} = 9.6°$ for empty load.

By using the Finite element theory, building the calculated force model with the tetrahedral solid element for the rotating shaft. The grid density is bigger in the relative movement of the region, the model has 57212 nodes and 292856 units. The material of the shaft is Q235, Elastic modulus E=206Gpa, Poisson's ratio μ=0.30, Material density ρ=7850 Kg/m³. The established finite element model is shown in Figure 4.

Fig. 4. The finite element model for Universal coupling

It can see from Figure 4, according to the working state to add constraints conditions, taking into account the torque load to bear crosshead, the underside of each node in the region are defined as fixed displacement constraints.

Based on the results of finite element analysis, in the x direction, the shaft head displacement cloud as shown in Figure 5.

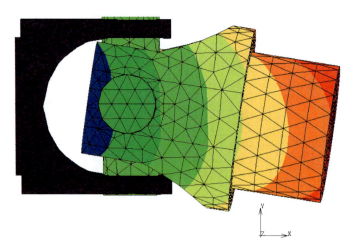

Fig. 5. The displacement cloud of shaft head in the x direction

Analysis of Figure 5, after applied the certain load, there is the greatest peak of displacement in the head shift in the y direction, its peak mainly located in the crosshead of the contact area for the universal coupling. According to the theory of fatigue, it easily leads to fatigue failure theory in the direction.

Based on the finite element model of the bi-directional universal coupling, the cloud of equivalent stress in fork top edge is as shown in Figure 6.

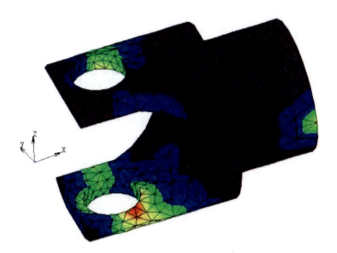

Fig. 6. The cloud of equivalent stress in fork top edge

Analysis of Figure 6, it we can see that the theoretical surface contact area from the surface into a point or line. The primarily stress is also focus on those areas. There is phenomenon of non-uniform stress distribution in shaft head, and produced the stress concentration, thus the Results can affect the life and application efficiency efficiency for the universal coupling in process of working.

4 Conclusion

Whether the results meet requirements in the process of analyzing for the shaft parts. At first, it can consider the components appeared on the maximum stress value is less than the material allowable stress, moreover, the parts of the deformation amount is less than the allowable deformation. Based on the necessity conditions, the distribution of stress has a tendency to uniformly in the components.

Acknowledgments. This work is partially supported by the Foundation of Science and Technique of Henan Province (092102210272). Foundation of Aviation Science (2008165500), Projects of International Cooperation Science and Technology of Ministry of Science and Technology (2006DFA72470), Foundation of Science and Technique of the Education Department of Henan Province (2006460018).

References

1. Whitehead, J.C.: Four Wheel Steering Maneuverability and High Speed Stabiliza-Tion. SAE Technical Paper No. 980642 (1998)
2. Yunchao, W.: Research on Dynamic Steering Control Strategy of Multi-steering Vehicle. Technic Forum 8, 46–50 (2009)
3. Kawasaki, H., Major, D.J.: Estimation of Friction Levels between Tire and road. SAE Paper 2002-01-1198
4. Yong, L., Xicheng, Z.: On-line Configuration the Logic of Regulating-valve Open Sequence Reducing the Vibration of Turbine Axle. China Instrumentation 8, 74–75 (2009)
5. Song, J.G., Yoon, Y.S.: Feedback Control of Four-Wheel Steering Using Time Delay Control. International Journal of Vehicle Design 19(3), 282–298 (1998)
6. Yang, L., Du, X., Yang, Y., Wang, L.: Operation Performance Analysis of Axial Flow Fan Cluster in Direct Air-cooled System. In: Proceedings of the CSEE, July 15, 2009, vol. 29(20), pp. 1–5 (2009)
7. Wenping, Z., Shengli, T.: Coupled flow-field calculation of a direct air-cooled condenser's component unit. Journal of Power Engineering 27(5), 766–770 (2007)
8. Hac, A., Simpson, M.D.: Estimation of Vehicle Side Slip Angle and Yaw Rate. SAE Paper 2000-01-0696
9. Lijun, Y., Xiaoze, D., Yongping, Y., et al.: Designing and verifying calculations and performance analysis for direct air-cooled steam condensers in power plant. Modern Electric Power 23(6), 50–53 (2006)
10. Wilber, K.R., Zammit, K.: Development of procurement guidelines for air-cooled condensers. In: Advanced Cooling Strategies/ TechnologyConference, Sacramento CA, USA (2005)
11. Duvenhage, K., Vermeulen, J.A., Meyer, C.J., et al.: Flow distortions at the fan inlet of forced-draught air-cooled heat exchangers. Applied Thermal Engineering 16(8-9), 741–752 (1996)
12. Duvenhage, K., Kroger, D.G.: The influence of wind on the performance of forced draught air-cooled heat exchangers. Journal of Wind Engineering and Industrial Aerodynamics 62(2-3), 259–277 (1996)
13. Maulbetsch, J.S., DiFilippo, M.N., Zammit, K.D., et al.: Spray cooling-an approach to performance enhancement of air-cooled condensers. In: Proceedings of EPRI Cooling Tower Technology Conference, Charleston, SC, USA (2003)
14. Bredell, J.R., Kroger, D.G., Thiart, G.D.: Numerical investigation of fan performance in a forced draft air-cooled steam condenser. Applied Thermal Engineering 26(8-9), 846–852 (2006)
15. Meyer, C.J., Kroger, D.G.: Air-cooled heat exchanger inlet flow losses. Applied Thermal Engineering 21(7), 771–786 (2001)
16. Meyer, C.J., Kroger, D.G.: Numerical investigation of the effect of fan performance on forced draught air-cooled heat exchanger plenum chamber aerodynamic behavior. Applied Thermal Engineering 24(2-3), 359–371 (2004)
17. Meyer, C.J., Kroger, D.G.: Plenum chamber flow losses in forced draught air-cooled heat exchangers. Applied Thermal Engineering 18(9-10), 875–893 (1998)
18. Xiaolin, W.: The status of dry cooling technology of thermal power plants in China. International Electric Power for China 9(1), 15–18 (2005)

Mobile Software Testing Based on Simulation Keyboard

Hua Ji

Department of Computer Science,
Shandong Normal University,
Jinan, Shandong, China
Jihua_sdnu@hotmail.com

Abstract. Confronted with the increase in system complexity, the code quantity of mobile software has doubled every two years. Simultaneously, the system also requires that the application should be concise, efficient, stable and reliable, so that the time of mobile software development gets longer and longer in the whole system, and the quality of software plays a decisive role in the final product. In the market, prevail available testing tools are generally classified into pure software tools and pure hardware tools. Pure software testing tools are generally used in the stub function and preprocessed task, and the pure hardware testing tools generally used for the hardware design and testing task including debugging phase of driver. This paper presents a software testing method based on simulation keyboard for the mobile terminal with the combination of hardware and software.

Keywords*:* Mobile software; software testing; test tool; simulation keyboard; ICE.

1 Introduction

Mobile terminal system is one of specific computer system, based on computer technology, in which software and hardware can be tailored to meet the function of application, and it requires strictly the reliability, cost, size, power consumption. Mobile software is special in some particular devices, which usually have very limited hardware resources, and are very sensitive to the cost.

With the development of mobile software, it's an urgent need for a kind of tool to guarantee system performance and reliability, for real-time online testing and analysis in the single-board phase, integration phase, the system phase and other phases of software development. Prevail testing tools available in the market is general classified into pure software testing tools and pure hardware testing tools.

Pure software testing tools adopt inserting the testing code into the software code, checking the change of the data in the shared memory of the target system, and processing data via the target machine on the network or serial ports to the host platform. All these need to occupy the resource of the objectives. Through the above process, the user can get to know the present running status of the program. While using pure software testing tools for testing on the target system, the user target system is not true in a running environment, and the captured data is not precise enough. Therefore, pure software tools can not be used for the target system in the function and tasks of running time analysis indicators. Pure software testing tools are generally used in the stub function and preprocessed task.

Pure hardware tools are used in hardware design and testing task. In instance, the logic analyzing monitor the instruction of system, and capture these signals, then determine the status of currently running process. At present, the In Circuit Emulator (ICE) is used only in embedded design and debug phase of driver, to track and extract the source code, also to monitor the activities of the bus.

This paper proposes a kind of software testing method based on the simulation keyboard with the combination of hardware and software, which can be used in the majority testing of mobile software. At present, mobile software is still rely on manual testing mode, and the defects of manual operation include massive randomness, weak traceability, easily leaking testing points. If the simulation keyboard is used in simulation testing, the defects of manual testing can be overcome.

2 The Basic Architecture

2.1 The Testing Architecture

In order to overcome the low testing efficiency caused by the pure software testing and the compatibility problems caused by pure hardware testing, this paper proposes mobile software testing based on the simulation keyboard with the combination of hardware and software. Its organization chart is shown in Figure 1:

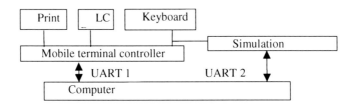

Fig. 1. The architecture of mobile software testing

The testing for mobile software based on the simulation keyboard mainly include mobile terminal controller (such as controller board, keyboard, LCD, Printer etc..), simulation keyboard, computer and serial lines. Computer connect to simulation keyboard via the serial port, to send the testing script to the simulation keyboard, which translate the testing script into keyboard commands, and send the commands to controller via the serial cable to simulate the manual work operate the keyboard.

Mobile Software Testing Based on Simulation Keyboard 557

The controller send the execution data to the computer via the serial port, then the computer judge in accordance with the testing data to form the automatic testing.

2.2 The Design of Simulation Keyboard

Keyboard connects to the controller with line scanning signal H_1, H_2 ... H_m, and column scanning signal V_1, V_2 ... V_n. The simulation keyboard also receive those signals, then the simulation keyboard take place of the keyboard. When to be pressed a key, the manual simulation keyboard begin to work, and the organization chart is shown in figure 2.

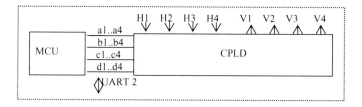

Fig. 2. The architecture of simulation keyboard

Based on the signal a1 .. a4, b1 .. b4, c1 .. c4, d1 .. d4 by MCU,CPLD creates V1 .. V4 signal, and the logic realization is shown in figure 3.

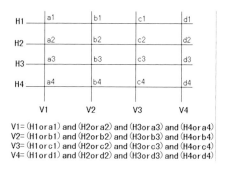

Fig. 3. The logic of simulation keyboard

For example, to input keys a1 and c3, simulation keyboard must set a1 and c3 to be 0, and thus the signal V_1 turns to low level will causes the signal H_1 turns to low level, the signal V_3 turns to low level will causes the signal H_3 turns to low level, the signal sequence chart is shown in figure 4. The controller scans such line and column signals, and read the key a1 and c3.

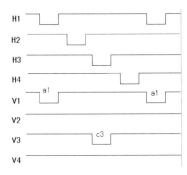

Fig. 4. The simulation of keyboard

2.3 The Insertion Point of the Mobile Controller

Compared with the PC software, the largest difference of mobile software lies in the fact: mobile software can not control the expressions of the progressive implementation in real-time, because it does not have debug mode, so the mobile software can only be compiled in computer environment, then download to the mobile terminal and run, then man-made "insertion point" should be marked in mobile application software. The "insertion point" software will run, and monitor the content of key data areas, then send the data to the computer. Functions to achieve are in the following:

1) Analysis of key data areas: critical data, such as changes in the content area, then distributed to the computer in real-time.

2) Trace: the parameters of the upper function, operation status and other information are sent to the computer.

2.4 The Testing Software of Computer

In the client computer, requires to join the following modules: receive and analysis testing data modules, testing cases editor module, testing cases management and sending module, and the structure is shown in Figure 5.

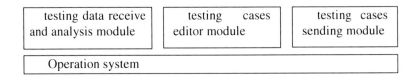

Fig. 5. The architecture of mobile software testing

Testing data receive and analysis modules receive the testing data sent from the mobile software, such as the critical data of the data area changes, the parameters of the upper function, and analysis of these data and stores the records into the database.

Testing cases editor module edits the testing cases quickly and easily, and the formation of engineering documents, to facilitate the preparation of the testing staff.

Testing cases sending module sends to the testing cases into the simulation keyboard automatically, so that the simulation keyboard can translated into the keyboard simulation timing sequence, to imitate the manual keyboard.

3 Implementation Steps

The concrete steps of testing based on simulation keyboard are as follows:

Step1: According to the requirements of debugging, it determine the transport protocol between the mobile software and computer communication, as well as instructions for each procedure of the command words, parameters, return values. The mobile software should be added a debug subroutine to achieve the above-mentioned functions.

Step2: If the testing case inputted into the computer, the computer will send it to the simulation keyboard via the serial port

Step3: Simulation keyboard will translate the testing case into the keyboard scanning signal, instead of manual keyboard operation.

Step4: Pseudo-breakpoint routine in the mobile terminal software will receive the testing cases via the keyboard, and run the top routine, then send the debugging information the computer.

Step5: the computer accepts debug information, which will be stored into the database and displayed on the screen in real-time.

Step6: If the debug information is abnormal, computer will stop sending testing cases to the simulation keyboard.

As an example for mobile system presented below, and it's keyboard is shown in table 1:

Table 1. The function of key

key	function	key	function
A1	0	C1	8
A2	1	C2	9
A3	2	C3	Enter
A4	3	C4	sum
B1	4	D1	Category one
B2	5	D2	Category two
B3	6	D3	Category three
B4	7	D4	Landing

The steps of realizing these testing methods are shown as follows:

1) D4 (press the landing key)
2) A2,A3,A4,A5,C3 (press the key "1234")
3) C3 (press Enter key)
4) D1 (press category one)
5) A4 (press the key "3")
6) C3 (press Enter key)
7) A2,A1,A1(press the key "100")
8) C3 (press Enter key)
9) D2 (press category two)
10) A3 (press the key"2")
11) C3(press Enter key)
12) A4,A3,A2(press the key "321")
13) C3(press Enter key)
14) C4 (press sum key)
15) C3(press Enter key)

4 The Effect of the Mobile Software Testing Based on Simulation Keyboard

The mobile software testing methods based on simulation keyboard can improve efficiency substantially, and change the original manual testing to automated testing, to achieve reusable testing cases. Simultaneously, manual testing errors caused by misoperation can be solved.

Although well-designed testing cases still leads to many loopholes in testing. For example, the adopting of manual testing usually only test 30% of the code, and the test time can not protect the project plan. As a result of the use of simulation testing of the keyboard can speed up the progress of the test, and expand the scope of the testing and code coverage, This method has been applied into many complex mobile softwares, and achieved good results.

Acknowledgments. This research is funded by the special project of high-technology independent innovation engineering, Shandong Province, China (No.2007ZZ17).

References

1. Kamal, R.: Embedded Softwares: Architecture, Programming. Tsinghua University Press, Beijing (2005)
2. Pasetti, A.: Software Frameworks and Embedded Control Systems. Hunan Literature and Art Publishing House, Hunan (2002)
3. Chen, D.: Real-time tracking method with embedded software debugging: Electronic Engineering Times (June 2002)

4. Zhongming, Z., Tianhui, L.: The design and implementation of the embedded-test for DCS based on CodeTest: Application of electronic technique (April 2006)
5. Limei, L., Shibiao: Research and realize of embedded software testing system: Foreign Electronic Measurement Technology (January 2009)
6. Wenjun, Q., Xiaodong, W.: Research of Coverage Test Tool on Embedded Software: Computer Measurement & Control (September 2007)

VLSI Prototype for Mpeg-4 Part 2 Using AIC

Kausalya Gopal, Kanimozhi Ilambarathi, and Riaz Ahmed Liyakath

Final Year, Dept of ECE, Sri Venkateswara College of Engineering,
Chennai, Tamil Nadu, India
kausalya16@gmail.com, kanibarathi89@gmail.com,
riaznumerouno@gmail.com

Abstract. This paper presents a VLSI prototype for the 2*2 Hadamard transform that is applied to the DC coefficients of each image component as described in the MPEG-4 part 2 Advanced Image Coding (AIC).

A VLSI prototype for the quantization process that is accompanied with the transform operation is given as well. It mainly deals about reducing the loss of clarity of the image when it is compressed. This is achieved by using hadamard transform. This transform is computed using add operations only. Thus the computational requirements of the design are reduced. Using the appropriate transforms and coding technique we can actually reduce the clarity loss of an image at each stage of either transforms or coding techniques. This is because at each stage we compare the actual and the predicted value. Thus the residual value is just taken which is loss free or with minimal loss.

Thus VLSI implementation of degraded images using AIC improves the features of the image to a very large extent. This sort of implementation using AIC can be used in variety of applications such as HD and digital photo applications.

Keywords: Hadamard Transform, Advanced Image Coding (AIC), VLSI prototype.

1 The Invention

We use a VLSI prototype for the 2*2 hadamard transform. This in turn is applied to the DC coefficients of the AIC. The AIC uses Discrete Cosine Transform for predicting the coefficients and performing the binary arithmetic coding. We use this coding for enhancing the RGB values in the AIC.

2 Working

The VLSI prototype for hadamard transform is done. This is applied to the corresponding DC coefficients of each image component of the AIC. AIC basically uses Discrete Cosine Transform which calculates the DCT coefficients after the colour conversion and prediction process. The coefficients thus calculated are coded using Binary arithmetic coding and formed as a bit stream.

This Bit Stream is encoded and then given to the decoder. On the decoder side, Binary arithmetic decoding is done and Inverse DCT is applied and corresponding predicted values are found. These values are compared an the residual value is converted back into RGB enhanced components. The whole AIC process is simulated using ModelSim ver5.4. It is found that only 2*2 hadamard transform takes the minimal time on comparison with other transforms. Thus this forms the working of the VLSI prototype Implementation.

AIC Block Diagram:

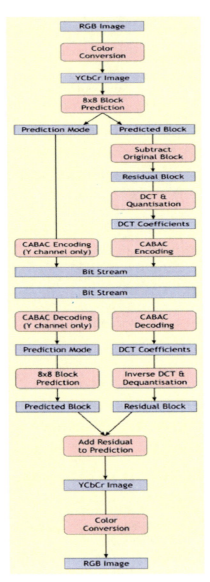

Both encoder and decoder share the same functional elements:

Color Conversion

A source RGB image is transformed to the YCbCr color space that has less entropy in the chominance channels (Cb and Cr) and thus makes these channels better compressible. This only applies to color images. Grayscale images have only 1 channel.

Block Prediction

Every channel (Y, Cb and Cr for color images, or Y for grayscale images) is split into blocks of 8x8 pixels. The contents of each block is predicted from previously encoded and decoded blocks in the channel. There are 9 ways to predict the current block. The prediction mode that minimizes the differences between the original and predicted block is chosen. To save bits on coding the prediction mode, the blocks in the Cb and Cr channels use the same prediction modes as the corresponding blocks in the Y channel. The values in the predicted block are subtracted from the ones in the original block to form a residual block. When a good prediction mode is chosen, the values in this residual block are smaller than the original pixel values and thus better compressible.

Discrete Cosine Transform (DCT) and Quantisation

To further reduce the entropy of the residual block, this block is transformed using the Discrete Cosine Transform. The resulting DCT coefficients are usually smaller than the original values. By quantising these coefficients with a certain value based on a chosen quality level, the coefficients are reduced even further. As a result, many of the coefficients in the block will be zero, and most others will be close to zero.

3 Context Adaptive Binary Arithmetic Coding (CABAC)

The prediction mode and quantised DCT coefficients are finally encoded to the stream using a binary arithmetic coding scheme that adapts to the context in which these values are coded. This means that commonly used prediction modes and DCT coefficients are coded using fewer bits than rare prediction modes and coefficients.

4 Appendix

4.1 Advanced Image Coding

Advanced Image Coding (AIC) is an experimental still image compression system that combines algorithms from the H.264 and JPEG standards. More specifically, it combines intra frame block prediction from H.264 with a JPEG-style discrete cosine transform, followed by context adaptive binary arithmetic coding as used in H.264. The result is a compression scheme that performs much better than JPEG and close to JPEG-2000.

AIC in a nut shell

- For photographic images, AIC performs much better than JPEG and close to JPEG-2000. For typical bit rates, AIC sometimes even outperforms JPEG-2000.
- For graphic images, the gap between JPEG-2000 and AIC grows, in the favor of JPEG-2000.
- For small images, the gap between JPEG-2000 and AIC also grows, however this time in the favor of AIC. For image sizes up to 100x100 (or 10,000 pixels), AIC performs much better for typical bit rates. This would make an AIC-like codec ideal for images on web pages.
- AIC is somewhat slower than JPEG, but faster than JPEG-2000, even without speed optimizations in the software.

5 Hadamard Transform

The Hadamard transform H_m is a $2^m \times 2^m$ matrix, the Hadamard matrix (scaled by a normalization factor), that transforms 2^m real numbers x_n into 2^m real numbers X_k. The Hadamard transform can be defined in two ways: recursively, or by using the binary (base-2) representation of the indices n and k.

Recursively, we define the 1×1 Hadamard transform H_0 by the identity $H_0 = 1$, and then define H_m for m > 0 by:

$$H_m = \frac{1}{\sqrt{2}} \begin{pmatrix} H_{m-1} & H_{m-1} \\ H_{m-1} & -H_{m-1} \end{pmatrix},$$

where the $1/\sqrt{2}$ is a normalization that is sometimes omitted. Thus, other than this normalization factor, the Hadamard matrices are made up entirely of 1 and −1.

The Hadamard transform is also used in many signal processing and data compression algorithms, such as HD Photo and MPEG-4 AVC. In video compression applications, it is usually used in the form of the sum of absolute transformed differences.

6 Context-Adaptive Binary Arithmetic Coding

Context-adaptive binary arithmetic coding *(CABAC)* is a form of entropy coding used in H.264/MPEG-4 AVC video encoding. As such it is an inherently lossless compression technique. It is notable for providing considerably better compression than most other encoding algorithms used in video encoding and is considered one of the primary advantages of the H.264/AVC encoding scheme. CABAC is only supported in Main and higher profiles and requires a considerable amount of processing to decode compared to other similar algorithms. It is also difficult to parallelize and vectorize. As a result, Context-adaptive variable-length coding (CAVLC), a lower efficiency entropy encoding scheme, is sometimes used instead to increase performance on slower playback devices.

7 Discrete Cosine Transform (DCT)

A discrete cosine transform (DCT) expresses a sequence of finitely many data points in terms of a sum of cosine functions oscillating at different frequencies. DCTs are important to numerous applications in science and engineering, from lossy compression of audio and images(where small high-frequency components can be discarded), to spectral methods for the numerical solution of partial differential equations. The use of cosine rather than sine functions is critical in these applications: for compression, it turns out that cosine functions are much more efficient (as explained below, fewer are needed to approximate a typical signal), whereas for differential equations the cosines express a particular choice of boundary conditions.

The DCT is used in JPEG image compression, MJPEG, MPEG, DV, and Theora video compression.

8 Conclusion

Thus using AIC and corresponding VLSI prototype MPEG-4 part2 can be used to improve enhancement features to a very great extent. This sort of implementation using AIC can be used in variety of applications such as HD and digital photo applications.

References

[1] http://www.bilsen.com/index.htm?http://www.bilsen.com/aic/
[2] http://en.wikipedia.org/wiki/Discrete_cosine_transform
[3] http://en.wikipedia.org/wiki/CABAC
[4] A VLSI Prototype with Application to MPEG-4 Part 10 2004. In: IEEE International Conference on Multimedia and Expo. (ICME) PG, pp. 1523–1526 (2004)

Author Index

An, Haining 517, 527

bing, Yan 459

Cai, Songmei 9
Cao, Zaihui 411
Che, Yi 389
Chen, Chao 333
Chen, Jun-jie 35, 209
Chen, Shuai 19
Chen, Shuang 419
Cheng, Ye 397
Cheng, Ziyu 527
Chinnappen-Rimer, Suvendi 183
Cui, Delong 309

Dai, Yiqi 333
Deyi, Li 371
Dong, Yu-you 419

Fei, Gao 177
Feng, Shuo 289
Feng, Xianzhang 547
Feng, Zishuo 143
Fu, Jianfeng 327

Gao, Fei 289
Ge, Hongyi 389
Gong, Songjie 381
Gopal, Kausalya 563
Gu, Junhua 235
Gu, Junzhong 9
Gu, Yungao 317
Guo, Lejiang 87
Guo, Zhitao 235

Hai-lian, Gui 93
Han, Yuexiao 317
Hancke, Gerhard P. 183
Heng, Li 177
Hongxing, Wang 475
Hu, Hailin 245, 253
Hu, Zhongyan 411
Hu, Zong 281
Huang, Dongming 281
Huang, Wei 19

Ilambarathi, Kanimozhi 563

Ji, Hua 555
Ji-yin, Sun 101
Jiang, Yuying 389
Jun, Cai 359
Junfeng, Han 507
Junying, Zhang 483

Keji, Mao 61

Li, Bingfeng 161
Li, Cui-hua 119
Li, Heng 289
Li, Jian 317
Li, Pengfei 343
Li, Wenguo 131
Li, Xiao 43
Li, Xiaoxia 153
Li, Yanheng 137
Li, Yonghua 77
Li, Zheng 295
Liao, Shuren 333
Libiao, Jin 69

Liu, Jianguo   245, 253
Liu, Min   53
Liu, Xiu-ju   425
Liu, Yanwu   201
Liu, Zhikai   235
Liyakath, Riaz Ahmed   563
Lu, Lan   43
Lu, Zhao   9
Luo, Hanyang   351

Naiqian, Zhang   69
Nan, Huang   435

Qi, Yaolong   517, 527
Qian, Wenxue   171
Qiang, Yan   35, 209
Qing-xue, Huang   93
Qizhong, Cai   507
Quan, Sun   27

Ru, Xing   371
Ruixia, Yang   259

Shan, Chen   101
Shao, Kang   397
Shaoming, Pan   507
Shengmin, Luo   499
Shi, Chenghua   317
Shi, Qianzhu   365
Shun-Zheng, Yu   359
Shuo, Feng   177
Shuqing, Zhang   371
Sun, Aifen   193, 265
Sun, Hongchun   227
Sun, Jianhua   411
Sun, Xiuyan   467

Tang, Qian   87
Tao, Hui   161
Tao, Xue-li   217

Wan, Huojin   245, 253
Wang, Bengwen   87
Wang, Binli   303
Wang, Jia-zhuo   119
Wang, Lina   303
Wang, Nianpeng   137
Wang, Qifeng   451
Wang, Xinying   111
Wang, Yan   295
Wei, Hongjun   153
Wei, Jinyu   193, 265

Wei, Xiaoxiao   343
Wei-bo, Zhang   491
Weifeng, M.A.   61
Wu, Ming-hui   1
Wu, Yingjie   77

Xia, Jiang   259
Xiao, Ming   309
Xinjiang, Luo   259
Xie, Liyang   171
Xu, Jinqiu   275
Xue, Yanming   289

Yan, Jun   153
Yan, Lei   111
Yancang, Li   405
Yang, Hongyong   43
Yang, Jinxian   161
Yang, Xiangchi   343
Yanming, Xue   177
Yanni, Han   371
Yi, Ping   327
Yin, Xiaowei   171
Yin, Zhongbin   303
Ying, Jing   1
Yingxia, Jian   435, 443
Yuan, Jinli   235

Zhang, Bofeng   327
Zhang, Changming   295
Zhang, Dongyang   111
Zhang, Hong   131, 137
Zhang, Qiuyan   333
Zhang, Zhongzhen   201
Zhao, Hui   547
Zhao, Juan-juan   35, 209
Zhao, Shukuan   143
Zhen, Tong   389
Zheng, Gengsheng   19
Zheng, Yan-bin   217
Zhengping, Zhao   259
Zhenguo, Hou   405
Zhiguo, Zhang   259
Zhihong, Feng   259
Zhou, Haikun   131
Zhou, Shufeng   537
Zhou, Wen   327
Zhu, Fan-wei   1
Zhu, Ying   327
Zhuang, Jing   265
Zuo, Jinglong   309